Innovation in Chemical Plant Design

Innovation in Chemical Plant Design

Editors

Roberta Campardelli
Paolo Trucillo

MDPI

Basel • Beijing • Wuhan • Barcelona • Belgrade • Novi Sad • Cluj • Manchester

Editors
Roberta Campardelli
University of Genoa
Genoa
Italy

Paolo Trucillo
University of Naples
Federico II
Napoli
Italy

Editorial Office
MDPI
St. Alban-Anlage 66
4052 Basel, Switzerland

This is a reprint of articles from the Special Issue published online in the open access journal *Processes* (ISSN 2227-9717) (available at: https://www.mdpi.com/journal/processes/special_issues/Chemical_Plant_Design).

For citation purposes, cite each article independently as indicated on the article page online and as indicated below:

Lastname, A.A.; Lastname, B.B. Article Title. *Journal Name* **Year**, *Volume Number*, Page Range.

ISBN 978-3-0365-9722-5 (Hbk)
ISBN 978-3-0365-9723-2 (PDF)
doi.org/10.3390/books978-3-0365-9723-2

Cover image courtesy of Paolo Trucillo

Contents

About the Editors

Roberta Campardelli

Roberta Campardelli has been Associate Professor in the scientific field of Chemical Plants (ING-IND/25) at the Department of Civil, Chemical and Environmental Engineering (DICCA) of the Polytechnic School at the University of Genoa since 2022. She is the professor of the course of Unit Operations, Bachelor's Degree in Chemical Engineering, of the course Chemical Plant Design, Master's Degree in Chemical Engineering and of the course of Innovation in Unit Operations for the PhD course of the DICCA. She is a member of the Research Center of Biologically Inspired Engineering in Vascular Medicine and Longevity (BELONG) at the University of Genoa, a joint center with the school of Medicine where innovative and functionalized biomaterials are investigated to develop new biodegradable vascular prosthesis and hydrogel-based wound dressing. She is co-founder of an academic spin-off, BEST SRL, for innovation in the agricultural field obtained by novel biostimulants. Roberta Campardelli's research activity has been continually focused on Chemical Engineering, Food Engineering and Biomaterials particularly on process and product innovation through the design, development and implementation of new plant solutions. In particular, her research activity is focused on the nanotechnology field used for the controlled and targeted release of drugs using polymeric nanoparticles or nanoemulsions and the development of innovative food packaging materials derived from natural sources.

Paolo Trucillo

Paolo Trucillo (1988) is a chemical engineer; he received his PhD in industrial engineering at the University of Salerno, Italy. He is assistant professor (ING-IND/22) in Materials Science and Engineering at the University of Naples Federico II, Italy. He delivers lectures on "Innovative Materials for Design" of the master's degree "Design for the Built Environment" at the Department of Architecture of the same University. His research is continuously focused on innovation in plant design, merging the knowledge achieved on processes with material science. In particular, he has worked on the following topics: development of a supercritical fluid-assisted process for the production of liposomes for pharmaceutical, nutraceutical, and textile applications; extraction of precious metals from spent catalysts and electronic waste, as well as the recovery of dyes from wastewater in the textile industry; pyrolysis and co-pyrolysis of biomass aimed at producing bio-oils for fuels; production of polymeric foams for drug delivery and aerospace applications; sustainable indicators for materials and processes.

Preface

In the vast landscape of chemical engineering, the impact of innovation in chemical plant design is propelling remarkable advancements in the realms of efficiency, safety, and environmental sustainability. The genesis of this progress is rooted in a complex web of dynamic factors, each wielding a pivotal influence. Safety standards serve as the bedrock, ensuring that the industry remains a paragon of responsible practices that prioritize the well-being of workers and the broader community. Environmental regulations play a complementary role, stressing the importance of reducing ecological footprints in chemical manufacturing, thereby weaving environmental stewardship into the fabric of the industry. Meanwhile, market dynamics keep the industry responsive to the ever-evolving needs of a fast-paced world, driving the quest for greater efficiency and competitiveness. Technological marvels serve as catalysts, ushering in a new era of possibilities in chemical plant design.

These innovations have cast their net wide, manifesting across an array of applications that touch upon the very essence of human well-being and industrial progress. The pharmaceutical and biomedical realm has borne witness to groundbreaking endeavors, none more promising than the supercritical fluid-assisted production of liposomes loaded with antibodies. This innovation is poised to yield a profound transformation in the efficacy of drug delivery systems, promising not just improved therapeutic outcomes but also sustainability as it aligns with environmentally responsible practices. In the textile industry, strides in natural dyeing and the infusion of antibacterial properties in cotton fabrics results in a sustainable and hygienic future. Meanwhile, the synergistic relationship between spray-drying and extraction processes reverberates through numerous sectors, enhancing product development, improving quality, and promoting sustainability. The confluence of remote sensing and microelectronics is illuminating our understanding of land surface temperature, with far-reaching applications spanning agriculture and climate change studies. The adoption of Convolutional Neural Networks in material science is revolutionizing material identification, accelerating progress in areas as diverse as energy storage and lubrication. In parallel, the study of techno-economics in the realm of gas-to-liquid processes for natural gases is not only vital but emblematic of the industry's relentless commitment to energy and environmental sustainability.

The evolution of the field of chemical engineering is not merely a matter of choice; it is an imperative, etched in the very core of this discipline. It is a quest to craft chemical processes that are not only safer but also more sustainable and economically viable, reducing their environmental impact and contributing to the prosperity and sustainability of our world. The future of chemical plant design is a beacon of promise, heralding a new era of industrial chemistry that harmonizes with our environment, bolsters our society, and ensures a more sustainable and prosperous future for all. This continuous pursuit of innovation is the lifeblood that ensures that we remain at the cutting edge of technology and best practices, ever adaptable to new challenges and opportunities that lie ahead

Roberta Campardelli and Paolo Trucillo
Editors

processes

MDPI

Editorial

Editorial Overview of the Special Issue "Innovation in Chemical Plant Design"

Paolo Trucillo [1,*] and Roberta Campardelli [2]

1 Department of Chemical, Materials and Industrial Production Engineering, University of Naples Federico II, Piazzale Vincenzo Tecchio, 80, 80125 Naples, Italy
2 Department of Civil, Chemical and Environmental Engineering, University of Genoa, via Opera Pia, 15, 16145 Genoa, Italy; roberta.campardelli@unige.it
* Correspondence: paolo.trucillo@unina.it

Innovation in chemical plant design stands at the forefront of transformative advancements in the dynamic field of chemical engineering [1]. The intricate process of designing and constructing chemical plants holds pivotal significance, as it fundamentally governs the trifecta of efficiency, safety, and environmental sustainability within the realm of chemical manufacturing operations. Over the course of years, this field has borne witness to a striking metamorphosis, catalyzed by an intricate interplay of factors including the evolution of stringent safety standards, increasingly demanding environmental regulations, shifting market dynamics, and the advent of groundbreaking technological marvels [2–4].

From the pioneering embrace of process intensification and modular design principles to the seamless integration of cutting-edge automation, data analytics, and sustainability-centric solutions, the landscape of chemical plant design has undergone profound evolution. This evolution is an ardent response to the multifaceted challenges and golden opportunities in the contemporary era. The papers presented in this Special Issue provide a vivid and dynamic panorama of innovative approaches, breakthrough technologies, and groundbreaking methodologies that are actively shaping the future of chemical plant design. Collectively, they pave the way for the realization of chemical processes that are not only more efficient and competitive but also significantly more sustainable, in alignment with the pressing needs of our modern world.

The topics involved in this Special Issue concern different fields of applications, such as the production of drug delivery systems, textile technologies, spray drying, extraction, active packaging, land surface temperature monitoring, microelectronics industry, neural networks, sustainability indicators, economics, and drug release modeling. A brief description of these topics is provided in the following paragraphs.

The field of pharmaceutical and biomedical research continues to push the boundaries of innovation and precision to develop novel drug delivery systems that can revolutionize medical treatments [5]. Among these groundbreaking endeavors, the production of liposomes loaded with antibodies using a supercritical fluid-assisted process is a pivotal pursuit [6]. This intricate undertaking involves a meticulous optimization process, wherein researchers embark on a journey to masterfully adjust various critical variables. These variables include temperature, pressure, flow rates, and the composition of supercritical fluids, all of which play important roles in shaping the outcome of this cutting-edge technique. Through rigorous and systematic experimentation coupled with comprehensive data analysis, researchers are committed to achieving a delicate equilibrium that yields liposomal formulations of the utmost quality. The promise of this endeavor extends far beyond the laboratory, holding the potential to enhance therapeutic outcomes and deepening our comprehension of the transformative power of supercritical fluid technology in the realm of drug delivery applications. The production of liposomes loaded with antibodies via a supercritical fluid-assisted process is a crucial endeavor in pharmaceutical and biomedical research. This complex optimization process involves meticulously fine-tuning various

Citation: Trucillo, P.; Campardelli, R. Editorial Overview of the Special Issue "Innovation in Chemical Plant Design". *Processes* **2023**, *11*, 3023. https://doi.org/10.3390/pr11103023

Received: 5 October 2023
Accepted: 18 October 2023
Published: 20 October 2023

variables, such as temperature, pressure, flow rates, and the composition of supercritical fluids. Using systematic experimentation and data analysis, researchers strive to strike a delicate balance that yields the highest quality liposomal formulations, promising improved therapeutic outcomes and a deeper understanding of the potential of supercritical fluid technology in drug delivery applications.

Supercritical fluid technologies can be used to produce liposomes loaded with various compounds. Supercritical fluids, often carbon dioxide in a supercritical state, offer a unique set of properties that make them ideal for liposome production [7]. In their supercritical state, these fluids exhibit both gas-like and liquid-like properties, enabling precise control over critical parameters such as temperature and pressure. This level of control is paramount for crafting liposomal formulations with exceptional precision and consistency. The beauty of supercritical fluid technology lies in its ability to serve as a clean and green solvent, eliminating the need for toxic organic solvents typically used in traditional liposome production methods. This not only enhances the safety of the process but also aligns with sustainability goals by reducing the environmental impact. Furthermore, the versatility of supercritical fluids allows researchers to encapsulate a wide array of compounds within the liposomes. The controlled expansion of supercritical solutions into aqueous phases or other suitable media facilitates liposome formation. This controlled encapsulation process ensures high encapsulation efficiency, uniform liposome size distribution, and enhanced stability of the loaded compounds.

The utilization of supercritical fluid technologies to produce liposomes loaded with diverse compounds represents a pivotal convergence of science and technology, offering a promising pathway for the development of advanced, efficient, and environmentally friendly drug delivery systems and other innovative products across multiple industries. This approach not only enhances the precision and sustainability of liposome production but also expands the horizons of what can be achieved in the realms of pharmaceuticals, biotechnology, and beyond. In particular, the supercritical-assisted production of lutein-loaded liposomes and the modeling of drug release represent an innovative approach in drug delivery systems [8]. Moreover, modeling the drug release kinetics is essential to predict and control the release profile over time. This combination of supercritical technology and drug release modeling not only enhances the precision and efficiency of drug delivery but also opens avenues for the development of targeted and patient-specific therapies, promising significant advancements in the pharmaceutical field. However, supercritical-assisted processes need to be evaluated in terms of sustainability. The environmental and sustainability analysis of supercritical carbon dioxide ($scCO_2$)-assisted processes for pharmaceutical applications is a crucial aspect of modern drug manufacturing. The analysis encompasses various factors, including energy consumption, waste generation, and overall process efficiency. By evaluating these parameters, it was possible to quantify the environmental benefits, such as reduced greenhouse gas emissions and waste disposal costs, associated with $scCO_2$-based processes.

The incorporation of antibacterial properties via enhanced natural dyeing processes not only adds value to textiles but also extends their potential applications to innovative therapeutics, where infection control and patient comfort are of paramount importance. The enhanced natural dyeing and antibacterial properties of cotton using physical and chemical pretreatments represent significant advancements in textile technology. This innovation not only offers sustainable alternatives to synthetic dyes but also addresses hygiene concerns, making cotton fabrics ideal for various applications, including healthcare and fashion. These combined advancements underscore the potential of interdisciplinary research at the intersection of textiles, chemistry, and microbiology, opening doors to more environmentally friendly and versatile cotton-based products in the market.

Spray-drying and extraction processes are also objects of this issue. Spray-drying and extraction processes are closely intertwined in the field of chemical engineering, as spray-drying can be employed to extract valuable compounds from various substances, making them integral steps in the development of many products and materials. Understanding the mechanism of particle agglomeration in spray drying, particularly in the context of both single- and multi-nozzle atomization techniques, is a critical topic explored in various research studies. Insights gained from such studies contribute significantly to optimizing spray-drying processes for a wide range of applications, including the production of powders, pharmaceuticals, and food products, ultimately leading to improved product quality and process efficiency. The high-performance extraction process of anthocyanins represents a promising approach in the field of natural product extraction and bioactive compound recovery. This methodology not only ensures the preservation of the integrity and bioactivity of the extracted compounds but also aligns with the principles of green chemistry and sustainable extraction practices.

This Special Issue began with topics regarding purely therapeutic approaches and continued with the definition and description of innovative ways of food preservation. The optimization of poly(ε-caprolactone) loaded with alpha-tocopherol polymeric films as an active packaging component via a combination of green processes represents a significant advancement in sustainable packaging technology. The resulting active packaging material offers effective protection against the oxidative degradation of packaged goods while minimizing the environmental footprint associated with conventional packaging processes, emphasizing the importance of green chemistry and sustainable practices in the packaging industry.

Other topics are related to the advancement of the integration of remote sensing and microelectronics industries. Earth observations can mitigate the limitations of each sensor individually and produce a more reliable representation of the surface temperature across different landscapes. This evaluation not only enhances our ability to monitor and understand the dynamics of land surface temperature but also contributes to various applications, including agriculture, urban planning, and climate change studies, and provides a more robust and comprehensive Earth observation dataset. In the microelectronics industry, joints serve as critical connections within electronic devices, and their integrity directly influences the overall performance and durability of the components. The evaluation of the interfacial microstructure involves examining the composition, morphology, and distribution of the phases at the solder-to-substrate interface. This assessment is pivotal for understanding the bonding mechanisms and potential issues like intermetallic compound formation, preventing premature failure in electronic devices, and maintaining their functionality throughout their intended lifespan.

The classification of graphite based on an improved Convolutional Neural Network (CNN) represents a cutting-edge approach in material science and image analysis. The use of CNNs, a type of deep learning algorithm, allows researchers to train models capable of recognizing subtle differences in graphite morphology, layering, and other important features. This innovative approach not only expedites material identification but also contributes to advancements in fields such as energy storage, lubrication, and composite materials by facilitating precise material selection and quality control.

The techno-economics of gas-to-liquid (GTL) processes for the associated natural gas, biogas, and landfill gas are pivotal in the realm of energy and environmental sustainability. These GTL technologies provide a viable solution to harness and convert these gas resources into valuable liquid fuels, such as diesel or synthetic crude oil. The economic analysis involves assessing the capital investment, operating costs, and revenue potential associated with each gas feedstock, considering factors like gas composition, availability, and market demand.

In conclusion, innovation in chemical plant design is a dynamic and essential aspect of the chemical engineering landscape. It is clear from the various innovative approaches and technologies discussed that the industry is continuously evolving to meet the challenges of safety, sustainability, and efficiency. The adoption of modular design, process intensification, digitalization, and sustainable practices all contribute to a more resilient and competitive chemical manufacturing sector. These innovations not only improve the economic viability of chemical processes but also reduce environmental impacts and enhance safety standards.

However, it is important to emphasize that innovation should not be viewed as a one-time effort but rather as an ongoing commitment to staying at the cutting edge of technology and best practices. The future of chemical plant designs holds tremendous promise. By embracing innovative approaches and technologies, the chemical engineering community is poised to create safer, more sustainable, and economically viable chemical processes that meet the ever-evolving needs of society while minimizing their environmental footprint. This journey of innovation is not only exciting but also imperative in the pursuit of a more sustainable and prosperous future for the chemical industry and our planet.

In conclusion, this insightful analysis underscores the dynamic and indispensable nature of innovation in chemical plant design within the broader landscape of chemical engineering. It is evident from the diverse array of innovative approaches and cutting-edge technologies discussed that the industry is in a perpetual state of evolution, driven by the imperatives of safety, sustainability, and efficiency. The strategic embrace of modular design principles, the implementation of process intensification strategies, the integration of digitalization, and the promotion of sustainable practices collectively contribute to forging a chemical manufacturing sector that is not only robust but also highly competitive. One notable aspect of these innovations is their potential to yield significant economic benefits concurrently with their capacity to mitigate environmental impacts and elevate safety standards. This dual focus on profitability and responsibility reflects a conscientious commitment to balancing progress with ethical stewardship. However, it is paramount to emphasize that innovation should not be perceived as a one-time endeavor but rather as an enduring dedication to remaining at the forefront of technological advancements and best practices. The ever-evolving nature of science and technology necessitates a perpetual quest for improvement to keep chemical engineers vigilant and adaptable in the face of new challenges and opportunities.

The future of chemical plant design is promising, holding the potential to reshape the landscape of industrial chemistry [9,10]. By embracing innovative approaches and cutting-edge technologies, the chemical engineering community stands poised to usher in a new era of safer, more sustainable, and economically viable chemical processes. This transformation is vital not only for meeting the evolving demands of society but also for minimizing the environmental footprint of the industry, thereby contributing to a more sustainable and prosperous future for both the chemical sector and our planet. The journey of innovation is not only exciting but also imperative as we strive for harmonious coexistence with our environment and seek solutions that benefit both industry and society.

List of Contributions

1. Ennaceur, S.; Bouaziz, A.; Gargoubi, S.; Mnif, W.; Dridi, D. Enhanced Natural Dyeing and Antibacterial Properties of Cotton by Physical and Chemical Pretreatments. *Processes* **2022**, *10*, 2263.
2. Trucillo, P.; Martino, M.; Reverchon, E. Supercritical assisted production of lutein-loaded liposomes and modelling of drug release. *Processes* **2021**, *9*, 1162.
3. Zbicinski, I.; Ciesielski, K.; Ge, B. Mechanism of particle agglomeration for single and multi-nozzle atomization in spray drying: a review. *Processes* **2022**, *10*, 727.
4. Vannuchi, N.; Braga, A.R.C.; De Rosso, V.V. High-performance extraction process of anthocyanins from jussara (*Euterpe edulis*) using deep eutectic solvents. *Processes* **2022**, *10*, 615.
5. Drago, E.; Campardelli, R.; Marco, I.D.; Perego, P. Optimization of PCL polymeric films as potential matrices for the loading of alpha-tocopherol by a combination of innovative green processes. *Processes* **2021**, *9*, 2244.
6. Li, S.; Wang, J.; Li, D.; Ran, Z.; Yang, B. Evaluation of Landsat 8-like land surface temperature by fusing Landsat 8 and MODIS land surface temperature product. *Processes* **2021**, *9*, 2262.

7. Chen, C.; Wang, C.; Sun, H.; Yin, H.; Gao, X.; Xue, H., Ni, B; Bian, K.; Gu, Q. Interfacial microstructure and mechanical reliability of Sn-58Bi/ENEPIG solder joints. *Processes* **2022**, *10*, 295.
8. Liu, G.; Xu, X.; Yu, X.; Wang, F. Graphite classification based on improved convolution neural network. *Processes* **2021**, *9*, 1995.
9. Trucillo, P.; Campardelli, R.; De Marco, I. Environmental and Sustainability Analysis of a Supercritical Carbon Dioxide-Assisted Process for Pharmaceutical Applications. *Processes* **2021**, *9*, 1788.
10. Galli, F.; Lai, J.J.; De Tommaso, J.; Pauletto, G.; Patience, G.S. Gas to liquids techno-economics of associated natural gas, biogas, and landfill gas. *Processes* **2021**, *1*, 1568.

Funding: This research received no external funding.

Acknowledgments: The co-guest editors thank the authors for providing excellent papers and sharing their knowledge and experience.

Conflicts of Interest: The authors declare no conflict of interest.

References

1. Keil, F.J. Process intensification. *Rev. Chem. Eng.* **2018**, *34*, 135–200. [CrossRef]
2. Pasman, H.; Sripaul, E.; Khan, F.; Fabiano, B. Energy transition technology comes with new process safety challenges and risks. *Process Saf. Environ. Prot.* **2023**, *177*, 765–794. [CrossRef]
3. Sofri, S.; Reddy Prasad, D.M.; Bin Azri, M.H.; Timbang, A. Analysis of chemical engineering curriculum to improve process safety competency. *Int. J. Occup. Saf. Ergon.* **2023**, *29*, 642–650. [CrossRef] [PubMed]
4. Qian, Y.; Vaddiraju, S.; Khan, F. Safety education 4.0–A critical review and a response to the process industry 4.0 need in chemical engineering curriculum. *Saf. Sci.* **2023**, *161*, 106069. [CrossRef]
5. Bamrungsap, S.; Zhao, Z.; Chen, T.; Wang, L.; Li, C.; Fu, T.; Tan, W. Nanotechnology in therapeutics: A focus on nanoparticles as a drug delivery system. *Nanomedicine* **2012**, *7*, 1253–1271. [CrossRef] [PubMed]
6. Ferrari, P.F.; Trucillo, P.; De Negri Atanasio, G.; Bufalini, C.; Campardelli, R.; Perego, P.; Palombo, D.; Reverchon, E. Operating Parameters Optimization for the Production of Liposomes Loaded with Antibodies Using a Supercritical Fluid-Assisted Process. *Processes* **2023**, *11*, 663. [CrossRef]
7. Ramsey, E.; Qiubai, S.; Zhang, Z.; Zhang, C.; Wei, G.O.U. Mini-Review: Green sustainable processes using supercritical fluid carbon dioxide. *J. Environ. Sci.* **2009**, *21*, 720–726. [CrossRef] [PubMed]
8. Maja, L.; Željko, K.; Mateja, P. Sustainable technologies for liposome preparation. *J. Supercrit. Fluids* **2020**, *165*, 104984. [CrossRef]
9. Porter, M.E.; Heppelmann, J.E. How smart, connected products are transforming competition. *Harv. Bus. Rev.* **2014**, *92*, 64–88.
10. Smith, R. *Chemical Process: Design and Integration*; John Wiley & Sons: Hoboken, NJ, USA, 2005.

processes

MDPI

Article

Enhanced Natural Dyeing and Antibacterial Properties of Cotton by Physical and Chemical Pretreatments

Soukaina Ennaceur [1,†], Aicha Bouaziz [2,3,†], Sondes Gargoubi [4], Wissem Mnif [5,6,*] and Dorra Dridi [7]

1 Department of Public Health, College of Health Sciences, Saudi Electronic University, Jeddah 23442, Saudi Arabia
2 Bio-Resources, Integrative Biology & Valorization (BIOLIVAL, LR14ES06), Higher Institute of Biotechnology of Monastir, University of Monastir, Monastir 5000, Tunisia
3 Higher School of Health Sciences and Techniques of Sousse, University of Sousse, Sousse 4054, Tunisia
4 Textile Engineering Laboratory—LGTex, University of Monastir, Monastir 5000, Tunisia
5 Department of Chemistry, Faculty of Sciences and Arts in Balgarn, University of Bisha, P.O. Box 199, Bisha 61922, Saudi Arabia
6 BVBGR-LR11ES31, Higher Institute of Biotechnology of Sidi Thabet, Biotechpole SidiThabet, University of Manouba, Ariana 2020, Tunisia
7 Unit of Analysis and Process Applied to the Environment (UR17ES32) Issat Mahdia, University of Monastir, Monastir 5000, Tunisia
* Correspondence: wmoneef@ub.edu.sa
† These authors contributed equally to this work.

Abstract: In this study, pomegranate peel extract was used to dye cotton fabrics. Generally, natural dyes have low affinity for cellulosic fibers and therefore need treatment for attachment. Air atmospheric plasma was applied to lower the concentrations of mordants needed and enhance dyeability. The outcomes showed that the residual concentrations of metallic ions decreased when applying the plasma treatment. ATR-FTIR and SEM analysis were performed to evaluate the effect of plasma modification on the surface chemistry and topography of the cellulosic fibers. The traces of metals released from finished fabric were quantified using ICP OES spectrometry. Color fastnesses (wash, rubbing and light) were evaluated to assess the mordanting step effect and the durability of the color. All tested fastnesses were good to very good. The concentrations of metal ions in finished fabrics were 32.4 for Fe, 0.28 for Al, 0.29 for Cr and 1.21 for Cu. Plasma treatment can be considered an efficient strategy to reduce the concentration of toxic metallic mordants.

Keywords: air-atmospheric plasma; mordant; natural dye; cotton; antibacterial activity

Citation: Ennaceur, S.; Bouaziz, A.; Gargoubi, S.; Mnif, W.; Dridi, D. Enhanced Natural Dyeing and Antibacterial Properties of Cotton by Physical and Chemical Pretreatments. *Processes* **2022**, *10*, 2263. https://doi.org/10.3390/pr10112263

Academic Editors: Roberta Campardelli and Paolo Trucillo

Received: 30 April 2022
Accepted: 26 October 2022
Published: 2 November 2022

1. Introduction

The textile finishing industry is moving towards cleaner production strategies since it is under increasing social pressure to lower the use of hazardous chemicals [1], especially substances classified as carcinogenic, mutagenic and allergenic [2]. Recent studies have developed new nontoxic, sustainable, ecofriendly, and cost-effective processes [3]. New restrictions related to health hazards and environmental impacts have resulted in the replacement of hazardous substances by safer alternatives [4,5]. Recently, the use of natural dyes has been an emerging trend for textile finishing. These dyes are derived from plants, animals and minerals [6,7] and they offer the advantage of being safe, ecofriendly and sustainable [8,9]. However, it is well known that they have a very low attachment to cotton, the most used natural fiber in the world [10,11].

Textile fibers and especially cellulosic ones do not present sufficient affinity for numerous natural dyes, which are nonsubstantive [12]. Cellulosic fibers, when immersed in the dyeing bath, generate ionized hydroxyl groups characterized by a negative charge, whereas most natural dyes are anionic when immersed in water. The negative charge on the cotton fabric is responsible of the repellency between fibers and dye [13].

Cotton, a carbohydrate polymer that is made up of cellulose, is the most used fiber in day-to-day life. It is a soft polymer. Cotton fibers can hold more water than their weight and they are considered as absorbent and breathable materials.

The application of natural dyes for cotton dyeing is always associated with the application of a pretreatment to improve dye uptake and fastness [14]. The fixing substances, called mordants, contain heavy metals such as iron, copper, chromium, aluminum, cobalt or nickel [15]. After dyeing, the residual amounts of such chemicals in wastewater cause serious health and environmental problems [16]. Mordants are able to form an insoluble complex with the dye used to intensify and fix the color [17]. Colorimetric parameters such as color strength (K/S), lightness (L), blueness-greenness (b) and redness-yellowness (a) significantly vary relying on the used mordant and the mordanting method [18].

The data generated from a literature survey demonstrated significant enhancement of color strength during the dyeing step when using mordants [19].

Meena et al. reported that the color yield of cotton fabrics, dyed with pomegranate peel increased with rising mordant concentration [20]. Ibrahim et al. showed that mordants are important to fix the natural dye extracted from the plant Bisham (*Commiphora gileadensis*) [21]. Moniruzzaman et al. investigated the influence of mordant concentration on the dyeing efficiency of cotton fabric using tea extract and found that a darker shade was obtained when using a mordant concentration of 3% [22]. Mulec et al. carried out cotton dyeing with turmeric extract. The results revealed the effect of mordants in enhancing the color yield of dyed fabrics [23]. Other researchers reported that natural dyeing of cotton fabrics using metal mordants leads to a significant amelioration in color depth [15].

Generally, metallic mordants are considered as hazardous chemicals. Metal ions are known to induce organ damage. They are considered as human carcinogens according to the U.S. Environmental Protection Agency, and the International Agency for Research on Cancer [24,25]. However, the degree of danger depends on the concentration of the mordant. This is the parameter that defines the hazards it presents [15]. The amount of mordant left in wastewater or released from finished fabrics has to be considered according to the limit values of ecological standards [26].

Some strategies have been proposed to reduce the toxic effects of metal mordants while keeping their efficiency, such as using auxiliaries or pretreating textile materials by physical or chemical treatments such as UV, ultrasound and microwave [15].

Plasma is an ecofriendly initiating system to achieve cotton surface modification without changing its inner properties. It is a good alternative to the use of chemicals for the activation of textile surfaces. Plasma treatment is a technique with low environmental pollution, low maintenance costs, and is a quick and simple process with an energy and water-saving effect with high productivity [27,28]. Compared to conventional wet methods, air plasma treatment reduces the use of chemicals, energy and water. The elimination of chemicals and water is cost-effective [29]. Materials treated with plasma show changes in surface properties resulting in enhanced attachment of metals on the surface. The attachment of metal ions on the surface is an important issue. In fact, when the textile is in contact with liquids or with the skin, the metal ions leach from the surface causing undesirable effects and leading to serious consequences, especially when the leached concentrations are high [30].

During this study, we aimed to investigate an eco-friendly physical treatment using air atmospheric plasma to enhance uptake of metal ions while keeping good coloring and antibacterial properties using pomegranate peel extract, which is known as a good natural dye for textile dyeing [31–33]. By enhancing the metal ions uptake using a plasma physical treatment, their concentrations in wastewater are reduced. Indeed, the concentrations of metal ions discharged from the finished fabric must be in the safe zones to avoid human health risks. In addition, metals are known to exhibit antibacterial activities. They were applied in medicine before antibiotics were discovered. The rapid increase in resistance of bacteria, fungi and viruses has brought into focus the lack of new antimicrobial agents.

To avoid the transmission of infectious diseases, there has been a revival of interest in the application of metals as antimicrobial agents.

To our knowledge, little research has been done on air-atmospheric plasma and mordants preparation for dyeing.

2. Materials and Methods

2.1. Materials

Pomegranate peels were collected and impurities were removed by washing with water. After washing, the peels were dried at room temperature (25 ± 2 °C) for 72 h and powdered using a grinder. An aqueous extract of powdered pomegranate peel was used as a natural dye for finishing cotton fabrics. Commercially available woven cotton fabrics were used. Alum ($KAl(SO4)_2\ 12H_2O$), potassium dichromate ($K_2Cr_2O_7$), Iron sulfate hydrate ($FeSO_4\ 7H_2O$) and copper sulphate hydrate ($CuSO_4\ 5H_2O$) were kindly provided by Chimitex plus company. Analytical grade chemicals were employed for the entire experiments.

2.2. Plasma Treatment

A plasma machine Coating Star (Ahlbrandt System, Lauterbach, Hessen, Germany) was used to treat cotton fabrics before mordanting and dyeing steps. Plasma discharge was created at atmospheric pressure by a mechanism composed of three ceramic electrodes. The parameters of the plasma machine were adjusted as following: ambient air as working gas, a frequency of 30 kHz, electrical power of 1 kW, inter-electrode distance of 1.5 mm and an electrode length of 0.5 m. The power was adjusted to 1000 W and the treatment speed to 2 m·min^{-1}. Each side of the fabric (30 cm × 30 cm) was treated twice.

2.3. Cotton Surface Characterization

To examine the effects of plasma treatment, the wettability of cotton samples was evaluated using a Drop Shape Analyzer-DSA25. The contact angle of the immediate contact with ultrapure water (2 μL) was recorded. The presented values are an average of three measurements.

To monitor the difference between cotton samples before and after plasma treatment, attenuated total reflectance Fourier-transform infrared (ATR-FTIR) spectra were recorded using a commercial ATR-FTIR attachment (Spectrum Two™ FTIR, Perkin Elmer, Waltham, Massachusetts, United States). Spectra were recorded between 4000 cm^{-1} and 400 cm^{-1}, resolution = 4 cm^{-1}, 400 scans. The peak area at 3271 cm^{-1} was calculated for the untreated and plasma treated samples using the software OriginPro 2019.

Morphological changes of different cotton samples were evaluated using a pressure-controlled scanning electron microscope (JEOL JSM5400). Cotton fabric samples (5 mm × 10 mm) were mounted on the specimen stabs and then coated with a thin film of gold using a sputtering method.

2.4. Dyeing Process

Cotton fabric dyeing was carried out using an Ahiba Datacolor machine using a bath ratio of 40:1. After adding the fabrics, the dyeing baths were heated to 100 °C and held for 45 min. The beakers of the machine were used for controlling fabrics and the dyeing bath throughout the dyeing process. Subsequently, the cotton samples were washed with water to remove residual dye and dried at room temperature.

During this study, a simultaneous mordanting method was applied (dyeing and mordanting carried out in the same bath). Three-percent mordant was used at 40 °C for 20 min before adding the natural dye (Figure 1).

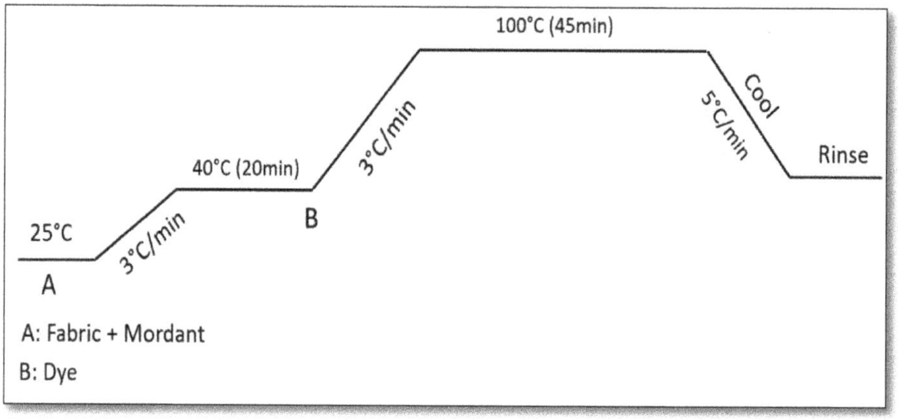

Figure 1. Simultaneous mordanting dyeing process.

2.5. Concentrations of Metal Ions in Dyed Fabrics

Before the experiments, prepared samples were dried for 24 h at 65 °C. Metal ions were extracted from the fabrics using an artificial sweat solution prepared according to the ISO 3160/2 standard as follows: 17.5 g NH_4Cl, 20 g NaCl, 5 g CH_3COOH, and 15 g lactic acid were dissolved in 1 L of ultrapure water. The pH was regulated at 4.7. A mass of 1 g of each sample was mixed with 30 mL of artificial sweat solution for 24 h. Then, the solutions were filtered and analyzed by ICP OE (Perkin Elmer, Norwalk, CT, USA) [34]. The operating conditions were adjusted as follow: RF power 1450 W, plasma gas flow rate 15 L min^{-1}, auxiliary gas flow rate 0.2 L min^{-1}, sample flow rate 1.5 mL min^{-1}, Read delay (s) 50 s, replicates 3 and plasma gas: Argon.

2.6. Assessment of Natural Dye Attachment

Natural dye attachment was evaluated based on colorimetric data using a laboratory machine (Data Color 650®, Datacolor, Suzhou, China) under the following parameters: illuminant D65 and 10° standard observer. Color parameters (L, a, b, and K/S) were evaluated. The parameters L, a, b and K/S refer to lightness-darkness, red-green share yellow-blue share and the color yield, respectively.

2.7. Color Fastness

For assessment of color fastness levels of dyed fabrics, standard methods were used. Color fastness (washing, rubbing and light) were carried out by using scales of color change according to ISO 105-C06 (washing), ISO 105-X12 (dry and wet rubbing) and ISO 105-B02 (light).

2.8. Antibacterial Activity

The antibacterial activity of dyed samples was evaluated according to the ASTM E2149 standard. *Staphylococcus aureus* (*ATCC 6538*) and *Escherichia coli* (*ATCC 8739*) bacteria strains were used during experiments. A mass of 1 g of the dyed samples was placed in a contact with bacteria (3×10^5 UFC/mL). Treated and control samples were shaken with the bacterial suspension for 1 h at 200 rpm. Then, 0.1 mL of prepared suspension was distributed over a petri dish. All petri dishes were incubated for 24 h at 37 °C. Finally, the formed bacteria colonies were counted. Results are presented as viable colonogenic cell numbers in CFU/mL.

3. Results and Discussion

3.1. Surface Characterization

ATR-FTIR spectra of the samples before and after plasma treatment are presented in Figure 2. Qualitative analysis showed the characteristic peaks of cellulose [35]. The broad bands at 3271 cm^{-1} corresponds to hydroxyl groups (O-H) in cellulose [36]. The peak at 2896 cm^{-1} is characteristic of stretching vibration of C-H in cellulose molecules [37].

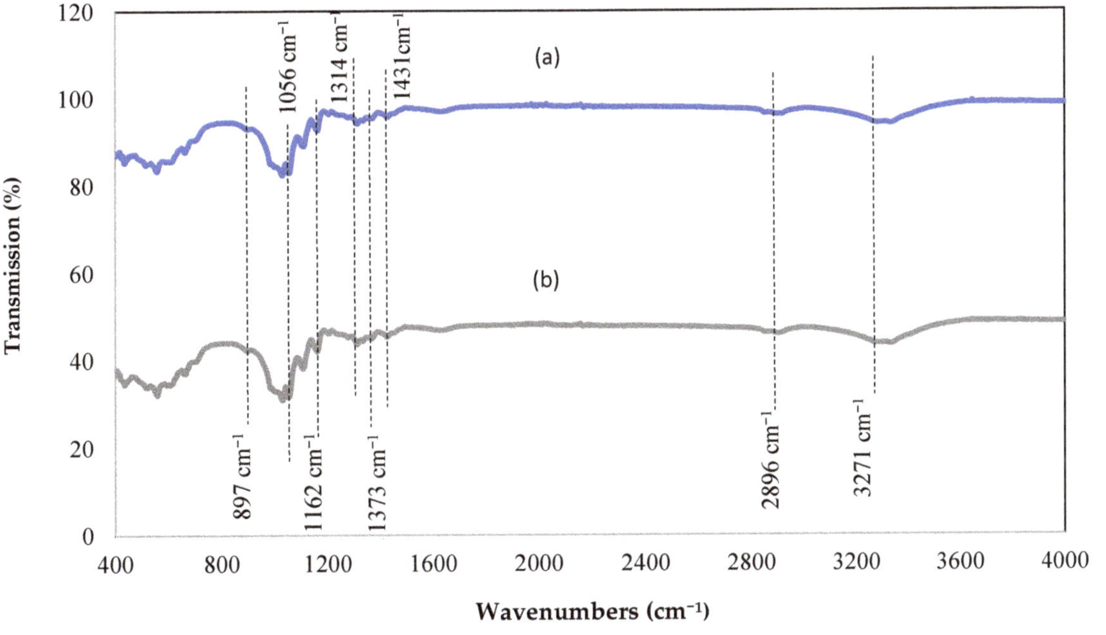

Figure 2. ATR-FTIR spectra of (**a**) untreated cotton and (**b**) plasma treated cotton.

Characteristic peaks at 1431 and 1373 cm^{-1} are attributed to bending vibrations of methyl and ethyl groups $\nu(CH_3)$ and $\nu(CH_2)$ [38]. The peaks at 1314, 1162–1056 and 897 cm^{-1} are assigned to C–O stretching of ether, C–O–C stretching, and glycosidic C-H deformation, respectively [39,40]. The spectrum of cotton fabrics before plasma treatment did not show any significant difference when compared to the untreated fabric. This outcome is due to the high penetration depth (around 1 μm) of ATR-FTIR analysis. This depth cannot detect a fine surface modification produced by plasma treatment. However, there was a difference between both spectra in terms of peak intensity at 3271 cm^{-1}.

Table 1 show that plasma surface activation generates additional C–OH groups. This finding confirms the results of previous investigations [41].

Table 1. ATR-FTIR analysis: peak area.

Sample	3271 cm^{-1}
Untreated	1.75
Plasma treated	2.13

In general, under air atmospheric plasma, CH/C-C groups decreased while the oxygen groups increased. Thus, it is evident from ATR-FTIR analysis that the plasma treatment increased the oxygen functional groups, which made the cotton surface more active.

The wettability of the cellulosic fibers measured as water contact angle indicated the hydrophilic behavior of the untreated cotton and cotton treated with plasma. Initial contact angle values were considered to evaluate the wettability of different cotton fabrics. Figure 3 shows that plasma-treated cotton fabric had a lower contact angle compared to that of untreated fabric. This result proves that plasma treatment enhances the hydrophilicity of the treated fabric. This result is in agreement with the ATR-FTIR results. The plasma treated cotton fabric had higher percent oxygen. The surface was richer in functional groups associated with a hydrophilic property (C-O and C-OH).

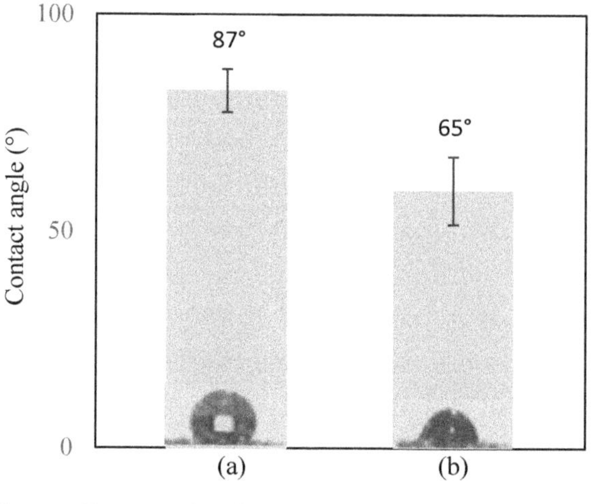

Figure 3. Contact angle values of (**a**) untreated cotton and (**b**) plasma treated cotton.

The normal spiral structure of cotton fibers was evidenced by SEM images (Figure 4). The surface of the untreated sample appeared relatively smooth (Figure 4a). However, the plasma treated surface was rougher than that of the untreated sample (Figure 4b). No significant differences were visible except for the more distinct macrofibrilar structure of cellulosic fibers.

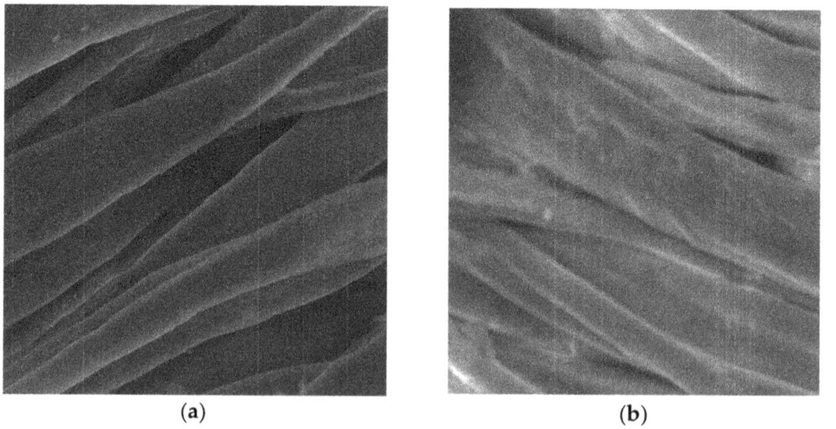

(**a**) (**b**)

Figure 4. SEM images of (**a**) untreated cotton and (**b**) plasma-treated cotton.

3.2. Trace Metal Ions in Finished Samples

During this study, the mordants used were aluminum sulfate, iron sulfate, potassium dichromate and copper sulfate. Metals extraction was conducted using a standard for fabrics used in direct contact with the human skin. The results are presented in Table 2. Results were compared with previous studies. The contents of trace element of Al, Cr and Cu range from 0.11–1.58 µg/mL, 0.01–0.32 µg/mL, and 0.05–1.95 µg/mL, respectively [26]. These concentrations are below the concentration limits set by ecological standards such as Öko Tex [26,34,42]. There are no any limit values for the element Fe, since the use of this metal is not prohibited [43].

Table 2. Metals concentrations expressed as µg of metal/mL.

Mordanted Sample	Element	Concentration	
		Plasma treated	Untreated
Iron sulfate	Fe	32.4 ± 2.4	69.3 ± 0.4
Aluminum sulfate	Al	0.28 ± 0,04	1.11 ± 4.2
Potassium dichromate	Cr	0.29 ± 0.11	0.36 ± 1.4
Copper sulfate	Cu	1.21 ± 0.53	2.14 ± 25

± Standard deviation.

3.3. Colorimetric Parameters

Table 3 shows the influence of cotton treatment on dyed samples. A difference in colorimetric parameters was detected. This was due to the effect of different mordants. The uses of copper sulfate and iron sulfate affected the shades of the dyed fabrics. Many flavonoids can coordinate various metal ions to form metal complexes due to the presence of oxo and hydroxyl groups [44]. The formed complex acts as a binding agent and thus produces deeper color on the fabric [45]. It is clear from the values presented in Table 2 that metal mordants affected the colorimetric data (L, a and b).

Table 3. Colorimetric data of cotton samples.

Sample	K/S	L	a	b
Without mordant	1.37	78.24	2.78	18.03
Iron sulfate	0.85	64.94	0.43	0.99
Aluminium sulfate	0.78	83.71	0.43	19.68
Potassium dichromate	0.24	87.76	−0.27	11
Copper sulfate	1.93	71.69	2.94	20.88

The different obtained colors were expressed in CIELAB (Figure 5); L expresses lightness while a and b parameters define the red/green and yellow/blue value, respectively. A value movement in the direction +a indicates a tendency toward reddish hue. For the b axis, a value movement in the direction + b represents a shift toward yellow. Values moving in the directions -a and -b reflect a tendency towards greenish and bluish shades, respectively. The L scale (degree of lightness) is also presented in Figure 5. Lightness varies from 0 (black) to 100 (white).

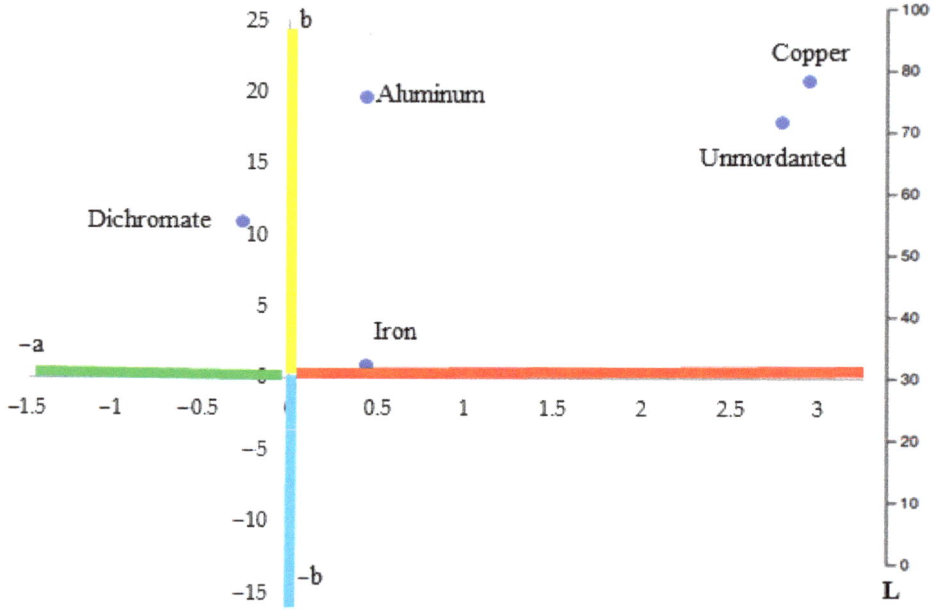

Figure 5. The L, a and b values expressed in CIELAB (CIE, International Commission on Illumination).

3.4. Color Fastness

The results of washing fastness are shown in Table 4. The untreated sample showed a poor color fastness rating. For all mordanted samples, the color change rating was 4–5. The treated cotton fabrics exhibited very good washing fastness.

Table 4. Color fastness.

Sample	Washing	Rubbing (Dry)	Rubbing (Wet)	Light
Without mordant	3	4	4	2
Iron sulfate	4–5	4–5	4–5	2–3
Aluminium Sulfate	4–5	4–5	4–5	2–3
Potassium dichromate	4–5	4–5	4–5	2–3
Copper sulfate	4–5	4–5	4–5	2–3

The rubbing fastnesses (dry and wet) of all mordanted fabrics were very good. Untreated dyed samples exhibited good fastness for dry and wet rubbing properties. Concerning the color fastness to light, the unmordanted sample showed a rating of 2. This result indicates a poor light fastness property. Mordanted samples displayed light fastness ratings of 2–3, which means light fastnesses were slightly improved when using mordants.

The mordant-dye complex has excellent cross-linking performance onto the cotton surface. So, an almost equivalent color fastness was observed for natural dyed cotton as with synthetic dye. However, the natural dye had an inferior light fastness property compared to synthetic dyes [46].

3.5. Antibacterial Activity

Antibacterial activity results after 1 h of contact are shown in Table 5. The counting test results of mordanted samples showed better activity against bacteria, causing bacterial reduction. It is known that metals such as aluminum, iron and copper are effective agents in reducing the growth of a wide range of microorganisms.

Table 5. Antibacterial activity results.

Sample	CFU/mL	
	(*E. coli*)	**(*S. aureus*)**
Without mordant	126.67 ± 8.62	142.11 ± 3.42
Iron sulfate	11.14 ± 5.05	35.14 ± 2.61
Aluminium sulfate	29.2 ± 5.75	38.2 ± 4.12
Potassium dichromate	31.33 ± 1.17	51.21 ± 3.21
Copper sulfate	21 ± 2.79	42 ± 1.64

± Standard deviation.

The mechanism of bacterial inhibition consists of the entry of metal ions into the metabolic system of bacteria with consequent formation of secondary metabolites. These secondary metabolites contain metal ions and are toxic to the organism [47]

Heavy metals cause a change in bacterial cell enzyme activities, growth inhibition and inhibition of replication. These mechanisms lead to cell lysis [48].

4. Conclusions

In this work, an aqueous extract of pomegranate peel was used for dyeing cotton fabrics. To improve the attachment of mordants metal ions, plasma treatment of cotton samples was used as a safe and ecofriendly pretreatment. Contact angle, ATR-FTIR, and SEM analysis showed the effects of plasma treatment on the surface morphology and chemistry of cotton fibers. Colorimetric data were evaluated. The results prove that the attachment of pomegranate peel dye was enhanced using metal mordants. Mordanting with aluminum sulfate, potassium dichromate, ferrous sulfate and copper sulfate enhanced the color yield and fastness properties of dyed samples.

The air atmospheric plasma treatment showed an effect on dying ability of cotton with natural dye and mordant. This was due to an etching effect and the introduction of functional groups on the surface of cotton fibers that enhanced mordant absorption into the plasma treated fibers. Plasma pretreatment is a promising method to lower the impact of toxic metal ions usually released after natural dyeing processes.

Author Contributions: Conceptualization, A.B. and S.G.; investigation, S.E. and S.G.; data curation, A.B. and D.D.; writing—original draft preparation, S.E., A.B. and S.G.; writing—review and editing, W.M. and S.G.; visualization, S.G. and D.D.; supervision, W.M.; project administration, S.E.; funding acquisition, S.E. All authors have read and agreed to the published version of the manuscript.

Funding: This research was supported by the Deanship of Scientific Research in the Saudi Electronic University.

Institutional Review Board Statement: Not applicable.

Informed Consent Statement: Not applicable.

Data Availability Statement: The data presented in this study are available on request from Dr. Sondes Gargoubi.

Acknowledgments: The authors extend their appreciation to the Deanship of Scientific Research in the Saudi Electronic University.

Conflicts of Interest: The authors declare no conflict of interest.

References

1. Gulzar, T.; Farooq, T.; Kiran, S.; Ahmad, I.; Hameed, A. *Green Chemistry in the Wet Processing of Textiles, in the Impact and Prospects of Green Chemistry for Textile Technology*; Elsevier: Amsterdam, The Netherlands, 2020; pp. 1–20.
2. Kishor, R.; Purchase, D.; Saratale, G.D.; Saratale, R.G.; Ferreira, L.F.R.; Bilal, M.; Chandra, R.; Bharagava, R.N. Ecotoxicological and health concerns of persistent coloring pollutants of textile industry wastewater and treatment approaches for environmental safety. *J. Environ. Chem. Eng.* **2021**, *9*, 105012. [CrossRef]
3. Islam, S.U.; Mohammad, F. *Emerging Green Technologies and Environment Friendly Products for Sustainable Textiles*; Springer: Singapore, 2014; pp. 63–82.
4. Sheriff, I.; Debela, S.A.; Kabia, O.A.; Ntoutoume, C.E.; Turay, M.J. The phase out of and restrictions on per-and polyfluoroalkyl substances: Time for a rethink. *Chemosphere* **2020**, *251*, 126313. [CrossRef] [PubMed]
5. Saxena, S.; Raja, A.S.M.; Arputharaj, A. *Challenges in Sustainable Wet Processing of Textiles*; Springer: Singapore, 2016; pp. 43–79.
6. Adeel, S.; Rehman, F.-U.; Rafi, S.; Zia, K.M.; Zuber, M. *Environmentally Friendly Plant-Based Natural Dyes: Extraction Methodology and Applications*; Springer: Dordrecht, Germany, 2019; pp. 383–415.
7. Eid, B.M.; Ibrahim, N.A. Recent developments in sustainable finishing of cellulosic textiles employing biotechnology. *J. Clean. Prod.* **2020**, *284*, 124701. [CrossRef]
8. Arora, J.; Agarwal, P.; Gupta, G. Rainbow of Natural Dyes on Textiles Using Plants Extracts: Sustainable and Eco-Friendly Processes. *Green Sustain. Chem.* **2017**, *07*, 35–47. [CrossRef]
9. Adeyemo, S.M.; Akinloye, A.J.; Adekanmi, G.B. The Use of Plant Dyes for Microbial Staining and Identification: An Eco-friendly and Non-Toxic Alternative Method. *J. Adv. Biol. Biotechnol.* **2018**, *16*, 1–10. [CrossRef]
10. Haji, A. Dyeing of Cotton Fabric with Natural Dyes Improved by Mordants and Plasma Treatment. *Prog. Color Color. Coat.* **2019**, *12*, 191–201.
11. Khan, M.A.; Wahid, A.; Ahmad, M.; Tahir, M.T.; Ahmed, M.; Ahmad, S.; Hasanuzzaman, M. *World Cotton Production and Consumption: An Overview*; Springer: Singapore, 2020; pp. 1–7.
12. Omotosho, O.O.; Ameuru, U.S. Synthesis and Dyeing Properties of Acid Dyes Derived from 1-amino-2-naphthol-4-sulphonic Acid. *World J. Appl. Chem.* **2019**, *4*, 63. [CrossRef]
13. Teklemedhin, T.B.; Gopalakrishnan, L.H. Environmental Friendly Dyeing of Silk Fabric with Natural Dye Extracted from Cassia singueana Plant. *J. Text. Sci. Eng.* **2018**, *1*, 1–6. [CrossRef]
14. Kan, C.; Lo, C.K.Y.; Man, W.S. Environmentally friendly aspects in coloration. *Color Technol.* **2016**, *132*, 4–8. [CrossRef]
15. İşmal, Ö.E.; Yıldırım, L. *Metal Mordants and Biomordants, in the Impact and Prospects of Green Chemistry for Textile Technology*; Elsevier: Amsterdam, The Netherlands, 2019; pp. 57–82.
16. Repon, M.R.; Islam, M.T.; Al Mamun, M.A. Ecological risk assessment and health safety speculation during color fastness properties enhancement of natural dyed cotton through metallic mordants. *Fashion Textiles* **2017**, *4*, 1–17. [CrossRef]
17. Saxena, S.; Raja, A.S.M. Natural dyes: Sources, chemistry, application and sustainability issues. In *Roadmap to Sustainable Textiles and Clothing*; Springer: Singapore, 2014; pp. 37–80.
18. Rani, K.; Aliya, R.; Solangi, B.A.; Pervez, M.K.; Akhtar, N.; Ahmed, F. Antimicrobial textile dyeing by applying natural colorants of brown seaweed (*cystoseira indica*). *Pakistan J. Weed Sci. Res.* **2020**, *26*, 403–414. [CrossRef]
19. Kasiri, M.B.; Safapour, S. Natural dyes and antimicrobials for green treatment of textiles. *Environ. Chem. Lett.* **2013**, *12*, 1–13. [CrossRef]
20. Meena, V.; Sheikh, J. Multifunctional modification of knitted cotton fabric using pomegranate peel waste. *Cell. Chem. Technol.* **2018**, *52*, 883–889.
21. Ibrahim, F.; El-Amoudy, E.S.; Shady, K. Thermal Analysis and Characterization of Some Cellulosic Fabrics Dyed by a New Natural Dye and Mordanted with Different Mordants. *Int. J. Chem.* **2011**, *3*, 40. [CrossRef]
22. Moniruzzaman, M.; Mondal, M.S.; Hossain, M.N. The influence of mordant and mordanting techniques on ecofriendly dyeing of cotton fabric by extracted used tea. *J. Eng. Sci.* **2018**, *9*, 111–117.
23. Mulec, I.; Gorjanc, M. The Influence of Mordanting on the Dyeability of Cotton Dyed with Turmeric Extract. *TEKSTILEC* **2015**, *58*, 199–208. [CrossRef]
24. Tchounwou, P.B.; Yedjou, C.G.; Patlolla, A.K.; Sutton, D.J. Heavy metal toxicity and the environment. *Mol. Clin. Environ. Toxicol.* **2012**, *101*, 133–164.
25. Baysal, A.; Ozbek, N.; Akman, S. Determination of trace metals in waste water and their removal processes. *Waste Water-Treat. Technol. Recent Anal. Dev.* **2013**, *1*, 145–171.
26. Iva Rezić, I.S. ICP-OES determination of metals present in textile materials. *Microchem. J.* **2007**, *85*, 46–51. [CrossRef]
27. Park, Y.; Koo, K. The eco-friendly surface modification of textiles for deep digital textile printing by in-line atmospheric non-thermal plasma treatment. *Fibers Polym.* **2014**, *15*, 1701–1707. [CrossRef]
28. Yaman, N.; Özdoğan, E.; Seventekin, N.; Ayhan, H. Plasma treatment of polypropylene fabric for improved dyeability with soluble textile dyestuff. *Appl. Surf. Sci.* **2008**, *255*, 6764–6770. [CrossRef]
29. Zille, A. *Plasma Technology in Fashion and Textiles, in Sustainable Technologies for Fashion and Textiles*; Elsevier: Amsterdam, The Netherlands, 2020; pp. 117–142.
30. Woskowicz, E.; Łożyńska, M.; Kowalik-Klimczak, A.; Kacprzyńska-Gołacka, J.; Osuch-Slomka, E.; Piasek, A.; Gradon, L. Plasma deposition of antimicrobial coatings based on silver and copper on polypropylene. *Polimery* **2020**, *65*, 33–43. [CrossRef]

31. Davulcu, A.; Benli, H.; Şen, Y.; Bahtiyari, M.I. Dyeing of cotton with thyme and pomegranate peel. *Cellulose* **2014**, *21*, 4671–4680. [CrossRef]
32. Ajmal, M.; Adeel, S.; Azeem, M.; Zuber, M.; Akhtar, N.; Iqbal, N. Modulation of pomegranate peel colourant characteristics for textile dyeing using high energy radiations. *Ind. Crop. Prod.* **2014**, *58*, 188–193. [CrossRef]
33. Basak, S.; Ali, S.W. Wastage pomegranate rind extracts (PRE): A one step green solution for bioactive and naturally dyed cotton substrate with special emphasis on its fire protection efficacy. *Cellulose* **2019**, *26*, 3601–3623. [CrossRef]
34. Sungur, Ş.; Gülmez, F. Determination of Metal Contents of Various Fibers Used in Textile Industry by MP-AES. *J. Spectrosc.* **2015**, *2015*, 1–5. [CrossRef]
35. Liu, P.; Chen, Q.; Wu, S.-S.; Shen, J.; Lin, S.-C. Surface modification of cellulose membranes with zwitterionic polymers for resistance to protein adsorption and platelet adhesion. *J. Membr. Sci.* **2010**, *350*, 387–394. [CrossRef]
36. Li, P.; Wang, Y.; Hou, Q.; Liu, H.; Lei, H.; Jian, B.; Li, X. Preparation of cellulose nanofibrils from okara by high pressure homogenization method using deep eutectic solvents. *Cellulose* **2020**, *27*, 2511–2520. [CrossRef]
37. Kim, M.H.; Jun, Y.J.; Elzatahry, A.; Alothman, Z.A.; Vinu, A.; Bin Choy, Y.; Choy, J.-H. Hydrophobic Guest Mediated Micellization and Demicellization of Rationally Designed Amphiphilic Poly(organophosphazene) for Efficient Drug Delivery. *Sci. Adv. Mater.* **2016**, *8*, 1553–1562. [CrossRef]
38. El-Gaoudy, H.; Kourkoumelis, N.; Varella, E.; Kovala-Demertzi, D. The effect of thermal aging and color pigments on the Egyptian linen properties evaluated by physicochemical methods. *Appl. Phys. A* **2011**, *105*, 497–507. [CrossRef]
39. Liang, C.Y.; Marchessault, R.H. Infrared spectra of crystalline polysaccharides. II. Native celluloses in the region from 640 to 1700 cm.1. *J. Polym. Sci.* **1959**, *39*, 269–278. [CrossRef]
40. Pappas, C.; Tarantilis, P.; Daliani, I.; Mavromoustakos, T.; Polissiou, M. Comparison of classical and ultrasound-assisted isolation procedures of cellulose from kenaf (*Hibiscus cannabinus L.*) and eucalyptus (*Eucalyptus rodustrus Sm.*). *Ultrason. Sonochemistry* **2002**, *9*, 19–23. [CrossRef]
41. Kan, C.-W.; Man, W.-S. Surface Characterisation of Atmospheric Pressure Plasma Treated Cotton Fabric—Effect of Operation Parameters. *Polymers* **2018**, *10*, 250. [CrossRef]
42. Tuzen, M.; Onal, A.; Soylak, M. Determination of trace heavy metals in some textile products produced in Turkey. *Bull. Chem. Soc. Ethiop.* **2008**, *22*, 379–384. [CrossRef]
43. Standard, G.O.T., Global Organic Textile Standard. Recuperado el, 27 June 2008. Available online: http://www.sopurefashion.com/images/certificated/What-is-GOTS-english.pdf (accessed on 17 October 2022).
44. Kasprzak, M.M.; Erxleben, A.; Ochocki, J. Properties and applications of flavonoid metal complexes. *RSC Adv.* **2015**, *5*, 45853–45877. [CrossRef]
45. Uddin, M.G. Effects of Different Mordants on Silk Fabric Dyed with Onion Outer Skin Extracts. *J. Text.* **2014**, *2014*, 1–8. [CrossRef]
46. Repon, R.; Repon, R.; Tauhidul Islam, M.; Al Mamun, A.; Abdur Rashid, M. Comparative study on natural and reactive dye for cotton coloration. *J. Appl. Res. Technol.* **2018**, *16*, 160–169. [CrossRef]
47. Chudobova, D.; Dostalova, S.; Ruttkay-Nedecky, B.; Guran, R.; Rodrigo, M.A.M.; Tmejova, K.; Krizkova, S.; Zitka, O.; Adam, V.; Kizek, R. The effect of metal ions on Staphylococcus aureus revealed by biochemical and mass spectrometric analyses. *Microbiol. Res.* **2014**, *170*, 147–156. [CrossRef]
48. Silver, S.; Ji, G. Newer systems for bacterial resistances to toxic heavy metals. *Environ. Health Perspect.* **1994**, *102*, 107–113.

processes

MDPI

Article

High-Performance Extraction Process of Anthocyanins from Jussara (*Euterpe edulis*) Using Deep Eutectic Solvents

Nicholas Vannuchi [1], Anna Rafaela Cavalcate Braga [2] and Veridiana Vera De Rosso [3,*]

[1] Programa de Pós-Graduação Interdisciplinar em Ciências da Saúde, Universidade Federal de São Paulo (UNIFESP), Rua Silva Jardim 136 CEP, Santos 11015-020, São Paulo, Brazil; nickvannuchi@gmail.com
[2] Department of Chemical Engineering, Campus Diadema, Universidade Federal de São Paulo (UNIFESP), Rua São Nicolau 201 CEP, Diadema 09972-270, São Paulo, Brazil; ana.braga@unifesp.br
[3] Nutrition and Food Service Research Center, Universidade Federal de São Paulo (UNIFESP), Rua Silva Jardim 136 CEP, Santos 11015-020, São Paulo, Brazil
* Correspondence: veridiana.rosso@unifesp.br; Tel.: +55-11-99658-3459

Abstract: New strategies for obtaining target bioactive compounds and natural pigments with the use of "green solvents" are consistently being developed, and deep eutectic solvents are (DES) a great alternative. This work established the significant variables and models for anthocyanin extraction, using DES and experimental design, of *Euterpe edulis* Mart. (jussara) fruit pulp, an endangered palm tree from the Brazilian Atlantic Forest. From a screening of seven initially tested DES, choline chloride/xylitol-based solvents had the best results with up to 42% increase in the total anthocyanin yield compared to methanolic extraction. Antioxidant assays also revealed a maximum antioxidant capacity of 198.93 mmol Trolox/100 g dry weight basis. The DES extract showed slower degradation to heat at 60° and 90 °C (2.5 times) and indoor constant light source (1.9 times) than methanolic extracts. The optimal extract also revealed slight inhibition of *S. enterica* and *S. aureus* growth in the agar plate.

Keywords: deep eutectic solvent; sustainable chemistry; antioxidant activity; *Euterpe edulis*; experimental design

Citation: Vannuchi, N.; Braga, A.R.C.; De Rosso, V.V. High-Performance Extraction Process of Anthocyanins from Jussara (*Euterpe edulis*) Using Deep Eutectic Solvents. *Processes* **2022**, *10*, 615. https://doi.org/10.3390/pr10030615

Academic Editors: Roberta Campardelli and Paolo Trucillo

Received: 28 January 2022
Accepted: 16 March 2022
Published: 21 March 2022

Publisher's Note: MDPI stays neutral with regard to jurisdictional claims in published maps and institutional affiliations.

1. Introduction

Volatile Organic Solvents (VOS) are estimated to account for 60% of all industrial emissions and 30% of all volatile organic compounds emitted worldwide [1,2]. Although commonly used to extract organic compounds, such solvents cause environmental and social damage due to their limitations on reuse, toxicity, flammability, volatility, and presence of residues in the final product [3–5]. These risks have led to restrictive legislation such as Directive 2010/75/EU proposed by the European Parliament, which aims to limit the emission of certain VOS by industries.

Seeking to work around this problem, new technologies aiming at safer and more sustainable alternatives for VOS use have emerged, among them the so-called deep eutectic solvents (DES). These liquids have more desirable solvent characteristics such as negligible volatility, non-flammability, chemical, thermal, and electrochemical stability, and the ability to be synthesized from non-petroleum derived organic and/or inorganic ionic species [6]. Considering their unique properties and attractive price, DES have become a promising topic for academic research and industry for both separation processes and extraction of plant bioactive compounds [7,8].

Anthocyanins are a class of natural pigments with underexplored use in the food and cosmetics industry, which commonly uses artificial colorants. DES have been successfully tested for anthocyanin extraction, contributing to the "clean label" of the product, a strategy used by industry to attract increasingly demanding consumers seeking environmentally responsible consumption and healthy eating choices [7,9]. Potential health risks involving artificial colorants, including allergic reactions and carcinogenic evidence, particularly of blue dyes [10,11], among other risks, have led to the banning of various artificial colorants in various countries [11,12].

Based on the above, DES are ideal candidates for developing new techniques for the extraction and preparation of poorly explored fruits, like the jussara (*Euterpe edulis* Mart.). This palm tree from the *Arecaceae* family has a large range of distribution, residing mostly along the Brazilian coastal Atlantic Forest, with occurrences from northeast Brazil to eastern Paraguay and northern Argentina. The tree is an important food source for many birds and mammals, who work as seed dispersers [13]. Despite the large area that it has occupied, jussara is at risk of extinction because of deforestation, defaunation, and clandestine heart of palm extraction (which kills the plant), and in some cases, it has even been locally eliminated [14]. Alternative sustainable use of the palm include the preparation of frozen pulp from the fruit, employed by small communities that live near the forest and consumed locally in sorbets, juices, and children's snacks [15,16]. Sustainable use of the plant by collecting its berries is an alternative of great economic and environmental potential, which is being pursued by these communities, mostly in southern states of Brazil, where they use the discarded jussara seeds to aid reforestation [17,18].

The pulp has been studied recently, with most studies evaluating the composition and different methods of extracting the phenolic and anthocyanin content. Large concentrations of anthocyanins and other antioxidant compounds, high lipid content and good quantities of minerals such as potassium, zinc, copper, and cobalt were identified, increasing the interest for the fruit [19,20]. Some studies regarding extraction methods to obtain anthocyanins and phenolic were conducted [21–23], but mostly using methanol and ethanol, with the best results coming from ultrasound assisted extraction [16].

This study proposes to develop processes for obtaining anthocyanins from jussara pulp, using DES and water as co-solvents for the first time, providing an alternative and exploration technique for the fruit through green extraction methods, and optimizing the process using experimental design. The resulting extract of concentrated anthocyanins presents a new technology for fruit usage, as a natural colorant, further encouraging their exploration over the heart of palm extraction.

2. Material and Methods

2.1. Fruit Gathering

The *Euterpe edulis* fruits were collected in São Paulo, Brazil (Ubatuba city: 23°19′01.47″ S, 44°52′37″ W) in January 2014 by Instituto de Permacultura e Ecovilas da Mata-atlântica (IPEMA). All fruits were sanitized with running water. The fruit was depulped in specific equipment used to separate jussara pulp from seed and peel (the same used by açaí), aided by warm water (45 °C maximum, to protect the seed and the anthocyanins) resulting in a thick purple juice. This juice is then immediately frozen at −40 °C, lyophilized for 48 h, and then stored at −40 °C to preserve the anthocyanin content. The moisture content was performed according to the methods of AOAC [24].

2.2. Eutectic Solvent Synthesis

The eutectic solvents were synthesized by a heating method [25]. Compounds used in the study are described in Table 1, based on the literature [25–28]. They are Chlorine Chloride (Ch), L-proline (Lp), Levulinic acid (Leu), Butene-1,4-diol (but), and Glycerol (Gly), Xylitol (Xyl). The DES mixtures were stirred at 80 °C for 1 h until a clear liquid was obtained with minimum addition of water. They were subsequently frozen at −40 °C and then lyophilized to ensure removal of water. A total of seven mixtures of DES were prepared using methods and molarities already described in the literature as follows: [Ch-Leu 1:2], [Ch-But 1:2], [Ch-Gly 1:2], [Ch-Xyl 5:2], [Lp-Leu 1;2], [Lp-But 2:5], and [Lp-Gly 2:5].

Table 1. Properties of the compounds used to create deep eutectic solvents.

Compound	Chemical Structure	Molar Mass (g/mol)	Density (g/mL) at 20 °C	Chemical Characteristic
1-5 Chorine chloride	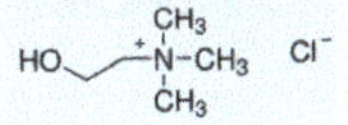	139.62	1.10 (aq)	Hydrogen bond acceptor
L-proline	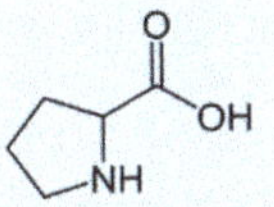	115.13	1.35	Hydrogen bond acceptor
Levulinic acid	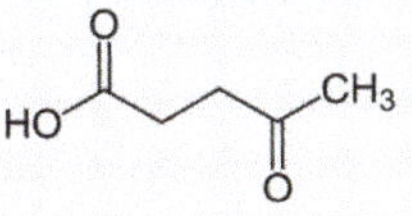	116.116	1.14	Hydrogen bond donor
Butane-1,4-diol	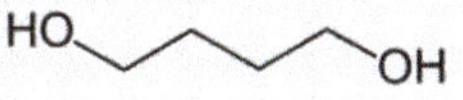	90.121	1.02	Hydrogen bond donor
Glycerol	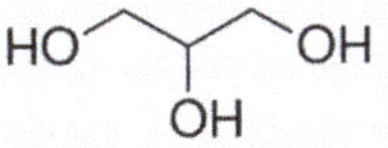	92.0776	1.26	Hydrogen bond donor
Xylitol	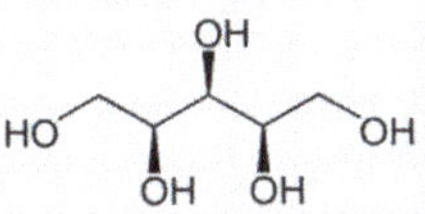	152.12	0.77	Hydrogen bond donor

2.3. Experimental Design: Response Surface Methodology

We evaluated the seven synthetized DES potentials for extracting anthocyanins by ultrasound-assisted extraction (Eco-sonics Ultronique, Campinas, Brazil), with fixed potency of 400 W for all experiments. An initial screening to choose 2 of the seven initial DES was conducted comparing anthocyanin content in the DES extracts and control (MeOH) with 30% co-solvent (acidified water), 1:15 sample:solvent ratio, and 3 extraction repetitions, during 5 min in the ultrasonic probe. Two of the seven initial DES were chosen (DES1 and DES2), based on anthocyanin yield, then two experimental designs were conducted: first, a fractional factorial experimental design 2^{5-1} (FFED) to evaluate possible non-significant variables, and second, a central composite rotatable design (CCRD) with remaining significant variables. The FFED 2^{5-1} with three central points and three levels for each independent variable was adopted, giving a total of 19 trials (Table 2 and S1A). The independent variables evaluated for each extraction were the following: (i) eutectic solvent composition; (ii) solid–liquid ratio ($R_{(S/L)}$), meaning the mass (g) of fruit pulp per mass of solution (g), (iii) number of extraction repetitions, (iv) time of extraction, and (v) co-solvent content (%) were chosen. The co-solvent (water) was acidified with HCl 3% to ensure the protonated state of anthocyanins [29]. Methanol with the same co-solvent proportion and parameters was used as a control. After the extractions, the extracts were filtered with filter paper (Unifil qualitative 80 g/m^2) and under a vacuum, then concentrated using a rotary evaporator (<37 °C), removing the water content of the co-solvent for quantification. The results were evaluated using ANOVA followed by Dunnett's test considering a confidence level of 95%, comparing the DES results to the control (MeOH extractions) in terms of total anthocyanin level (TAL). The response was expressed in $\mu g_{anthocyanins}/mL$.

An estimate of the main effect was obtained by evaluating the difference in the total anthocyanin yield caused by a change from the low (−1) to the high (+1) levels of the corresponding variable. The fractional factorial experimental design's focus was to discover the significant variables and their main interactions with the total yield of extracted anthocyanins. Therefore, after analyzing the data, the variables that had a significative effect on the total yield of the process were fixed and the others removed (including the removal of DES2 from further trials) for an additional assay CCRD 2^2 plus axial and central composites, with three replicates at the central point, giving a total of 11 additional trials. Table 3 shows the real values (coded values in Table S1B) and levels used at CCRD 2^2. After analyzing the CCRD results, the best conditions for the anthocyanin extraction with the best DES was determined, and the model was validated in triplicate to be used in further trials.

Table 2. Independent variable values of fractional factorial experimental design (2^{5-1}) with three central points for DES1 and DES2 were used as independent variables as well as $R_{(S/L)}$, Repetitions, time, and $R_{(CoSolvt\%)}$.

Assay	DES	$R_{(S/L)}$	Extraction Repetitions	Time (min)	$R_{(CoSolvt\%)}$
	X_1	X_2	X_3	X_4	X_5
1	DES1	1:15	1	2	40
2	DES2	1:15	1	2	20
3	DES1	1:25	1	2	20
4	DES2	1:25	1	2	40
5	DES1	1:15	5	2	20
6	DES2	1:15	5	2	40
7	DES1	1:25	5	2	40
8	DES2	1:25	5	2	20
9	DES1	1:15	1	6	20
10	DES2	1:15	1	6	40
11	DES1	1:25	1	6	40
12	DES2	1:25	1	6	20
13	DES1	1:15	5	6	40
14	DES2	1:15	5	6	20
15	DES1	1:25	5	6	20
16	DES2	1:25	5	6	40
17	DES1/DES2	1:20	3	4	30
18	DES1/DES2	1:20	3	4	30
19	DES1/DES2	1:20	3	4	30

Table 3. Independent variable values used in CCRD 2^2 assays with 3 central points, used as independent variables R (CoSolvt%) and extraction repetitions.

Assay	R (CoSolvt%)	Extraction Repetitions
	X_1	X_2
1	16	2
2	44	2
3	16	6
4	44	6
5	10.3	4
6	49.7	4
7	30	1
8	30	7
9	30	4
10	30	4
11	30	4

The main response evaluated for optimization of the process was based on the total yield of cyanidin 3-glucoside (C3G), cyanidin 3-rutinoside (C3R), and total anthocyanin level (TAL), considering only these two major anthocyanins, attempting to obtain models for those responses as well. Statistica 14.0 software was used to analyze the results and plot the response surfaces.

2.4. *Identification and Quantification of Anthocyanins*

For all the samples, anthocyanin separation was carried out on a 250 × 4.6 i.d. mm, 5 μm particle size, C_{18} Shim-pack CLC-ODS column (Shimadzu, Canby, OR, USA), using as the mobile phase a gradient of methanol/5% formic acid (*v*/*v*) from 15:85 to 80:20 in 25 min, the latter proportion being maintained for a further 15 min, at a flow rate of 0.9 mL/min and a column temperature set at 28 °C. The chromatograms were processed at 280 and 520 nm, and the spectra were obtained between 250 and 800 nm [19,30]. The anthocyanins were identified based on the combined information provided by the elution order in the reversed phase column, co-chromatography with standards, UV–VIS, and mass spectra compared to the literature data. Anthocyanins were quantified by HPLC as cyanidin 3-glucoside and cyanidin 3-rutinoside, using an external calibration curve for both with a minimum of six concentration levels. All analyses were performed in triplicate.

2.5. Antioxidant Activity

In order to compare the antioxidant activity of extracts obtained by DES in CCRD trials, ABTS [2,2′-azinobis-(3-ethylbenzothiazoline-6-sulfonate)] assay was performed as described by [31]. Antioxidant activity (AA) of the extracts was expressed in µmol Trolox/g equivalents. Results were then plotted in CCRD again for the same effects, using antioxidant activity as the dependent variable and analyzed by ANOVA with a 90% confidence level.

The oxygen radical absorbance capacity (ORAC) method was also performed as previously described by [32], which consists of measuring the decrease in the fluorescence of a protein as a result of the loss of its conformation when it suffers oxidative damage caused by a source of peroxyl radicals. The measurements were taken in triplicate. The ORAC values, expressed as µM Trolox equivalents (µM TE), were calculated by applying the following Equation (1):

$$\mathrm{ORAC}(\mu\mathrm{M\ TE}) = \frac{\mathrm{C_{trolox}} \cdot \left(\mathrm{AUC_{sample}} - \mathrm{AUC_{blank}}\right) . \mathrm{k}}{(\mathrm{AUC_{trolox}} - \mathrm{AUC_{blank}})} \quad (1)$$

where C_{Trolox} is the concentration (µM) of Trolox (20 µM), k is the sample dilution factor, and AUC is the area below the fluorescence decay curve of the sample, blank, and Trolox, respectively. Results were then plotted in CCRD again for the same effects, using antioxidant activity as the dependent variable and analyzed by ANOVA with a 90% confidence level.

2.6. Anthocyanin Thermal Stability

The determination of the anthocyanin's thermal stability was performed for both best DES and methanolic extracts. The stability at 60 °C and 90 °C was performed as described by De Rosso and Mercadante [33], with modifications. After the extraction using the validated condition from the CCRD, the best DES and methanolic extracts were prepared and pH adjusted to 3.5 for all samples to keep anthocyanins stable (protonated). A food preservative (10 mg potassium sorbate) was added to both extracts to avoid spoilage. Then, each solution absorbance was adjusted to 0.8 through dilution, measured at the maximum absorption wavelength in the visible region for anthocyanins (520 nm) to standardize the assays, ensuring that Lambert–Beer law is considered and the citrate–phosphate buffer was utilized since small variations of pH affects anthocyanin concentration. The solutions were allowed to rest for 3 h to attain equilibrium among the different forms of anthocyanin. The solutions were then distributed in 10 mL centrifuge tubes with a total sample volume of 3 mL.

All tubes were placed in a water bath and heated at 60 °C or 90 °C, removing triplicates of each extract (DES or MeOH) at random times. The samples were immediately cooled in an ice bath for 10 min after removal. After this time, they were placed at room temperature for 50 min before reading the absorbance in a spectrophotometer (520 nm). When the absorbance pointed to half the original value (0.4), the assay was finished and results were plotted, and the emerging curve behavior was analyzed.

2.7. Anthocyanin Photostability

The anthocyanin's photostability was performed for both DES and methanolic extracts in the presence and absence of light. The extracts were prepared as described in Section 2.7.

The solutions were then distributed in 15 mL transparent tubes with a total sample volume of 10 mL.

The tubes were placed both in constant light and constant dark conditions, and samples of each triplicate of each extract (DES or MeOH) were removed at random times over the course of 25 days until absorbance reached half of its initial value (0.4). Samples had their absorbance were read in a spectrophotometer (520 nm) at these intervals. Results were plotted, and the emerging curve behavior was analyzed.

2.8. Disc-Diffusion Antimicrobial Sensitivity Test

The growth method was performed as follows: (1) three colonies, well isolated, of the same morphological type of *Salmonella enterica* subsp. (ATCC 13076) and *Staphylococcus aureus* (ATTCC 19095) were selected from the agar plate with Müeller–Hilton culture medium. Microorganisms were transferred to a tube containing 4–5 mL of culture medium. (2) The culture was incubated in broth at 35 °C until reaching or exceeding the standard McFarland 0.5 solution's turbidity measured in a spectrophotometer at 600 nm. (3) The turbidity of the growing culture was adjusted with a sterile saline solution in order to obtain optical turbidity comparable to that of the 0.5 McFarland standard solution. A sterile cotton swab was then dipped into the adjusted suspension up to 15 min after changing the inoculum suspension's turbidity. The dry surface of the Müeller–Hinton agar plate was inoculated by rubbing the swab across the agar's sterile surface ensuring uniform distribution of the inoculum. A set of antimicrobial discs was placed on the surface of a seeded agar plate, applying samples of the best DES extract, SF (saline), DES (deep eutectic solvent only), and positive control (vancomycin for *S. aureus* and meropenem for *S. enterica*), as well as samples of the optimized extract. The plates were inverted and placed in an oven at 35 °C for up to 15 min after applying the discs. After 16–18 h of incubation, each plate was examined. The diameters of the total inhibition halos were measured by the diameter (mm) of the disc.

3. Results and Discussion

3.1. Screening Process

A chromatographic profile of the methanolic jussara extract from DES screening analyzed the two major anthocyanins: C3G, with a retention time of 13.39 min, and C3R, with a retention time of 13.99 min (Figure S1 of Supporting Information). Quantification of DES and methanolic extract's peak areas to C3G and C3R, respectively, are shown in Figure 1. A confidence level of 95% was used for all analyses.

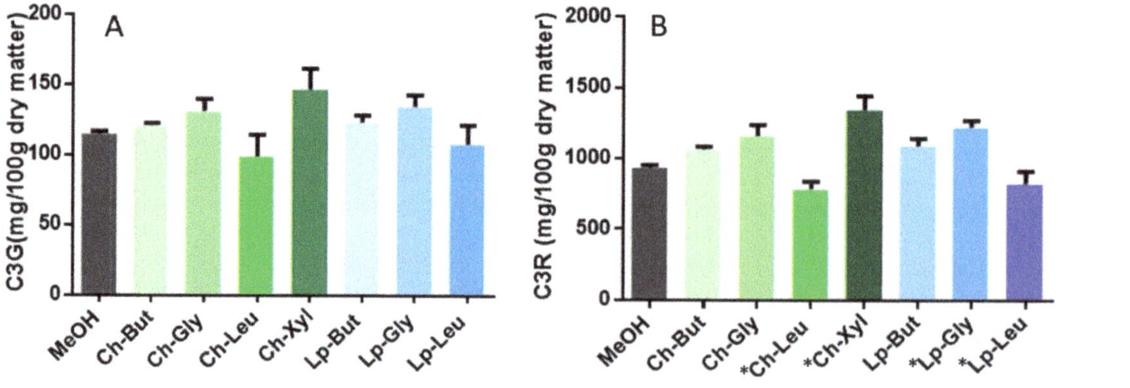

Figure 1. Cyanidin 3-glucoside (**A**) and Cyanidin 3-rutinoside (**B**) concentration (mg 100 g^{-1} dry matter) in different extracts of eutectic solvents tested in lyophilized jussara samples. The * marks DES mixtures with results statistically different from the control (MeOH).

The results of C3G content in DES extracts were not significant (compared to MeOH; Mean: 114.79 ± 3.56 mg/100 g dry matter, ANOVA followed by Dunnett's test) for any of the extracts tested. The results of C3R in the extracts, on the other hand, were significant for Ch-Leu (785.16 ±99.2 mg/100 g dry matter, lower) and Lp-But (1092.41 ± 88.33 mg/100 g dry matter, lower), Ch-Xyl (1343.92 ±172.42 mg/100 g dry matter, higher), and Lp-Gly (1224.55 ± 80.20 mg/100 g dry matter, higher) in comparison to MeOH extract (935.48 ± 22.28 mg/100 g dry matter). C3R is the main anthocyanin of the pulp, and its yield significantly influenced the total anthocyanin level (TAL) of extracts when compared to methanolic extract.

After the sum of both anthocyanins, only the TAL of Ch-Xyl (1490.65 ± 198.27 mg/100 g dry matter) and Lp-Gly (1359.06 ± 92.92 mg/100 g dry matter) showed significant differences compared to the MeOH extract (1050.28 ± 29.32 mg/100 g dry matter). As our samples had 88.7% humidity, the TAL content in the methanolic extract can be expressed as 118.68 ± 3.3 mg/100 g fresh matter, being compatible with the MeOH extracts from the literature. One study [21] obtained 14.84 to 409.85 mg/100 g fresh matter with acidified methanolic extract, using samples gathered in different months. More recently [34], a change was observed in the total monomeric anthocyanins during *E. edulis* ripening. That study found that the anthocyanin content was very positively correlated with the ripening process, with a maximum of 634.26 mg TAL/100 g fresh matter. A progressive increase in the concentration of anthocyanins is believed to occur during ripening, and changes in the color of the fruits are easily observable [35]. Increased anthocyanin levels during ripening have also been previously noted for açai (*Euterpe oleracea*) [35,36].

The medium absolute values of Ch-Xyl anthocyanin extracts were 42% higher than methanolic extract and Lp-Gly was 29% higher. These results clearly lead us to choose Ch-Xyl and Lp-Gly as the best green solvents for further experimental optimization of the extraction process (DES1 and DES2 of FFED, respectively).

3.2. Fractional Factorial Experimental Design (FFED)

After choosing the DES with a higher capacity to extract total anthocyanins from *E. edulis* lyophilized pulp, the optimization of different process conditions was conducted through a fractional factorial experimental design with three central points 2^{5-1}, following the standardized effects described in the methodology section. The extraction process in different experimental conditions of the 2^{5-1} and the yields are shown in Table 4.

Table 4. Experimental values for C3G, C3R, and TAL obtained from fractional experimental design (2^{5-1}) with three central points for Ch-Xyl and Lp-Gly used as independent variables as well as $R_{(S/L)}$, extraction repetitions, time, and $R_{(CoSolvt\%)}$.

Assay	Total C3G Content (mg/100g Dry)	Total C3R (mg/100 g Dry)	TAL (mg/100 g Dry)
	$Y_1 actual$	$Y_2 actual$	$Y_3 actual$
1	110.85	988.37	1099.21
2	81.82	747.47	829.29
3	101.52	933.65	1035.17
4	101.65	913.29	1014.93
5	133.67	1224.93	1358.60
6	116.76	1122.11	1238.87
7	118.17	1092.79	1210.96
8	110.04	997.73	1107.77
9	97.23	891.86	989.09
10	97.67	1059.43	1157.09
11	130.74	1264.24	1394.97
12	111.77	1063.85	1175.61
13	124.15	1174.53	1298.68
14	98.54	921.33	1019.87
15	100.84	961.75	1062.59
16	77.40	769.40	846.80
17	92.46	891.40	983.86

Table 4. *Cont.*

Assay	Total C3G Content (mg/100g Dry)	Total C3R (mg/100 g Dry)	TAL (mg/100 g Dry)
18	94.89	924.33	1019.22
19	95.94	908.96	1004.90

The experimental data shows that total anthocyanins extracted in all factorial design trials ranged from 829.29 to 1394.97 mg/100 g dry matter, indicating that the selected variables and levels could change the process efficiency up to 60%. As mentioned before, the proportion of DES is a qualitative variable and the levels studied were DES1 (−1) and DES2 (+1), and for the central point (0), we prepared a mixture of both DES. The negative effect of the proportion of DES, obtained from the FFED means that when we use DES2 instead of DES1, i.e., when we change the level from −1 to +1, the extraction of anthocyanins is less efficient. The fact that the DES2 variable has a negative effect implies that DES1 (Ch-Xyl) is more efficient in extracting the target bioactive compounds. Co-solvent usage had a positive influence, moving towards level +1 (more co-solvent in the total solvent used) and repetition (more repetitions). The solid/liquid ratio and time had no significant effects in this factional design. Considering these results, only two variables were carried out for the next step: co-solvent and repetitions.

Although the DES choice has had a significant effect, this variable could be fixed with Ch-Xyl for the CCRD (2^2) due to the negative mean effect of the Lp-Gly (DES2) solvent as discussed above. Chlorine chloride and xylitol were present in other tailor-made DES [37,38] for phenolic compounds and anthocyanins [39], where they are described as showing a high affinity for phenolic structure. Analysis obtained with the use of nuclear magnetic resonance spectroscopy by [40] shows that a Ch-Xyl–Water mixture shows great affinity for phenolic compounds compared to many other combinations of DES, including the Ch-Gly tested here. Additionally, Ch-Xyl and water exhibit a more cohesive structure when dissolving quercetin (a phenolic compound) than a citric acid/choline-chloride/water mixture [41]. However, in analyzing extraction of anthocyanins from mulberries, Ch-Gly gave a slightly better anthocyanin extraction than Ch-Xyl and a better extraction than mixtures of chloride–mannitol, chloride–fructose and choline chloride–glucose [39], indicating that food matrix and the type of anthocyanins contained can be important factors when tailoring DES for extraction of these compounds.

3.3. Central Composite Rotatable Design (CCRD)

As the literature indicates that using more than 50% of the water in the mixture can cause problems in DES stability [41,42], the co-solvent effect levels in CCRD were defined as 16% to 49.7%. For repetitions, previous experiments observed that beyond the fourth repetition, the solvent was almost colorless after extraction. A hypothesis was formulated then, assuming that after a hypothetical optimum of repetitions (four), more repetitions could generate TAL loss by excess manipulation. To verify this, a quick assay was done, with fixed variables (R(S/L) 1:15; Time: 2 min, co-solvent: 40%) with different repetitions (three, five, and seven times). Results are shown in Figure S2 of the Supporting Information, confirming a significant loss of efficiency in seven repetitions, compared to five repetitions (ANOVA).

After set repetition effect levels (2, 5 and 7 times), Table 5 shows the total yield obtained for the extraction process in different experimental conditions of the 2^2 with central and axial points, for C3G, C3R, and TAL content from the *E. edulis* used as a response to the construction of the predictive model.

Table 5. Experimental values for C3G, C3R, and TAL extracted by CH-Xyl in the CCRD assays, and relative deviation (%) among experimental and predictive values of TAL.

Assay	C3G (μg/mL)	C3R (μg/mL)	TAL (μg/mL)	Relative Deviation (%)
	Y_1	Y_2	Y_3	Y_3
1	167.4752	1959.889	2127.364	−3.38
2	142.7458	1804.298	1947.044	−12.95
3	175.5200	2050.806	2226.326	20.31
4	182.2589	2016.332	2198.591	19.31
5	160.5043	1860.417	2020.922	−6.15
6	165.3153	1826.223	1991.538	−7.72
7	84.00578	1131.545	1215.551	−27.34
8	156.6489	2012.959	2169.608	29.40
9	179.7469	2078.261	2258.008	−0.83
10	186.2938	2071.040	2257.333	−0.86
11	172.7000	2145.786	2318.486	1.80

The main effects and interactions were estimated for TAL are described in Equation (2).

$$Total\ anthocyanin\ level\ (\mu g/mL) = 2276.84 - 66.2\,(X_1)^2 - 212.6(X_2) - 224.0\,(X_2)^2 \quad (2)$$

where X_1 is co-solvent participation (%) and X_2 is the number of repetitions. ANOVA was used to evaluate the data's adequacy to verify the possibility of obtaining the model considering the different responses; additionally, the R^2 value provided a confidence measure of the model. In this study, an R^2 value of 0.6969 was obtained. The F value 94.9 (F value tabulated: 4.103) for TAL exceeded the 95% confidence level. The TAL response attended the criteria to generate the model and, thus, can be considered predictive; therefore, the response surface was plotted in Figure 2A.

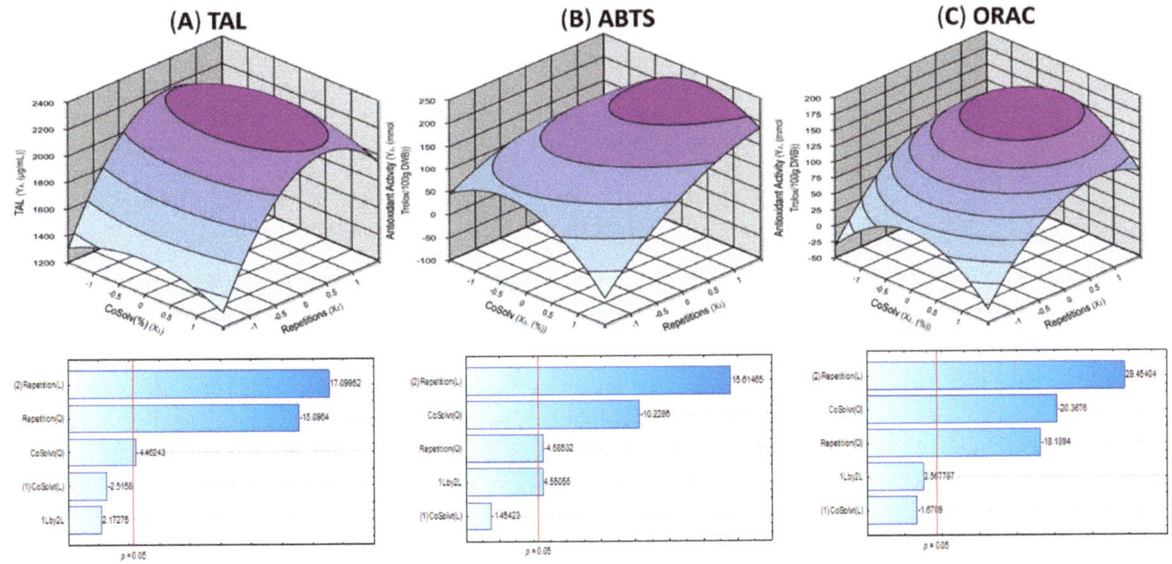

Figure 2. (**A**) Surface model for total anthocyanin level (TAL) (mg/100 g dry matter) from Ch-Xyl extract from *E. edulis* pulp for repetition and co-solvent as main effects and Pareto chart of the results. (**B**) Surface model for ABTS (mmol Trolox 100 g^{-1} dry matter) and Pareto chart of the results. (**C**) Surface model for ORAC assay (mg/100 g dry matter) and Pareto chart of the results.

As observed in the model, repetitions largely influenced the TAL yield of extracts. On the other hand, next to maximum levels, the efficiency is lost; this is attributed to excess manipulation of extracts. The co-solvent had a mild influence on the model, as observed in the Pareto chart of effects. The negative influence when the water participation (%) rises followed other studies, in which it is observed that it can disrupt the DES structure and thus reduce the DES effect in the extraction [25,43]. The best results were obtained around the central point level, with four repetitions and 30% co-solvent participation. The validation assays, performed in triplicate, were reproduced applying the best conditions. The model was effectively validated (mean of TAL was 2350.61 ± 53.05 μg/mL) with low relative deviation (3.10%), evidencing its high confidence in predicting data. These results for co-solvent participation are compatible with others in the literature for DES extraction for phenolic compounds that report best yields at 25–30% water usage [28,44]. Choline cations are highly capable of coordinating the surrounding environment: strong hydrogen bonding-mediated correlations between the hydroxyl group and water or chloride are believed to be formed. In addition, it is possible that the ammonium group drives the formation of a solvating environment, with water, chloride, and hydroxyl moieties approaching it, between the methyl groups [45]. The total anthocyanin content extracted by these assays varied from 1215.51 to 2318.48 μg/mL (assays 7 and 11, respectively), corresponding to 117.301 100 g^{-1} fresh weight and 223.734 100 g^{-1} fresh weight, respectively. Compared to the literature, a careful approach is necessary since the total anthocyanins contained in fruits can vary greatly due to edaphoclimatic variables and maturation stages [46]. The results obtained by [21] also using the experimental design ranged from 25.54 mg 100 g^{-1} fresh weight (methanol) up to 418.52 mg 100 g^{-1} fresh weight (Methanol/HCl 1.5 M). The study used an ultrasonic bath and the best condition with a 1:50 solvent ratio, which is considerably higher than that used in our study (1:15). In a study using the same solvent ratio [47] and 70% ethanolic solvent in ultrasonic bath, with 360 W potency, similar results were found: 190.92 ± 13.48 mg 100 g^{-1} fresh weight to 282.64 ± 11.07 mg 100 g^{-1} fresh weight. When authors tested a higher potency in the bath (900 W), a reduction in the anthocyanin content was observed.

3.4. Antioxidant Activity

After quantifying anthocyanin content in Ch-Xyl extracts for CCRD, the antioxidant activity of the samples was measured by ABTS and ORAC assays. Another CCRD assay was performed, using ABTS and ORAC assays as a response instead of TAL for the very same effects. Results are also shown in Figure 2B,C. Coded values can be found in Table S2A of Supporting Information.

In the ABTS assay, ANOVA resulted in an R^2 value of 0.88 and F value of 20.5, for a 95% confidence level; therefore, it is able to generate a model with a predictive equation (Equation (3)).

$$\textit{Antioxidant activity}\ (\text{mmol Trolox }100/\text{g}) = 189.73 - 36.6\ (X_1)^2 + 46.5(X_2) - 16.4\ (X_2)^2 + 19.3\ (X_1 \times X_2) \quad (3)$$

where X_1 is co-solvent participation (%), and X_2 is the number of repetitions. The graphical model and the Pareto chart of effects are shown in Figure 2B. For the ORAC assay (Table S2B of Supporting Information), ANOVA resulted in an R^2 value of 0.81 for a 95% confidence level and is therefore able to generate a model with a predictive Equation (4):

$$\textit{Antioxidant activity}\ (\text{mmol Trolox }100/\text{g}) = 159.32 - 31.3\ (X_1)^2 + 38.0(X_2) - 27.9\ (X_2)^2 \quad (4)$$

where X_1 is co-solvent participation (%), and X_2 is the number of repetitions.

The precision and exactness of the model were validated by comparing the experimental results with the theoretical data predicted by Equations (2)–(4). As observed, response surfaces of the same extracts show slightly different performances for anthocyanin content and antioxidant activity. It is known that jussara pulp contains multiple hydrophilic compounds such as phenolic compounds [19,21,48,49], and even anthocyanin degradation products, which possess antioxidant properties [50]. Such compounds can explain the dissonant behavior and different maximum values of antioxidant activity when reacting to ABTS and ORAC assays. The results had values between 69.92 and 198.93 for ABTS and between 63.09 and 196.436 mmol Trolox/100 g dwb (dry weight base) for ORAC. Values between 67.7 and 154.4 mmol Trolox/100 g dwb using TEAC (Trolox equivalent antioxidant capacity) and ORAC assays were found for jussara pulp [51]. While analyzing the ripening of the fruit, values between 108.8 (immature) and 207.1 (complete maturation) mmol Trolox/100 g dwb using the ORAC assay were found [52].Considering the evaluated dependent variables in CCRD (TAL and antioxidant activity by the methods ABTS and ORAC), the maximization of anthocyanin content and antioxidant activity are around the central point (applying around four repetitions and 30% co-solvent participation). This way, as presented for the TAL, the validation assays for antioxidant activity were also performed in triplicate for ABTS and ORAC. The result mean of the antioxidant activity was 198.82 ± 10.97 and 155.71 ± 21.77 (mmol Trolox 100 $g^{-1\ dwb)}$, representing a relative deviation from the predicted values of 5.06% and 10.51% for ABTS and ORAC, respectively. This confirms that the models were validated with low relative deviation, evidencing their statistical significance and predictiveness.

3.5. Determination of Anthocyanins Thermal Stability

The changes in anthocyanin concentration from jussara extracts caused by the temperature increase with the time was expected, following a biphasic behavior that was best fitted by a biexponential equation (Equation (5)), considering both evaluated temperatures (60 and 90 °C). Figure 3 shows the degradation kinetics of total anthocyanins extracted with MeOH and with Ch-Xyl, and the calculated kinetic parameters obtained by fitting the data from Equation (5) can be found in Table S3 of Supporting Information.

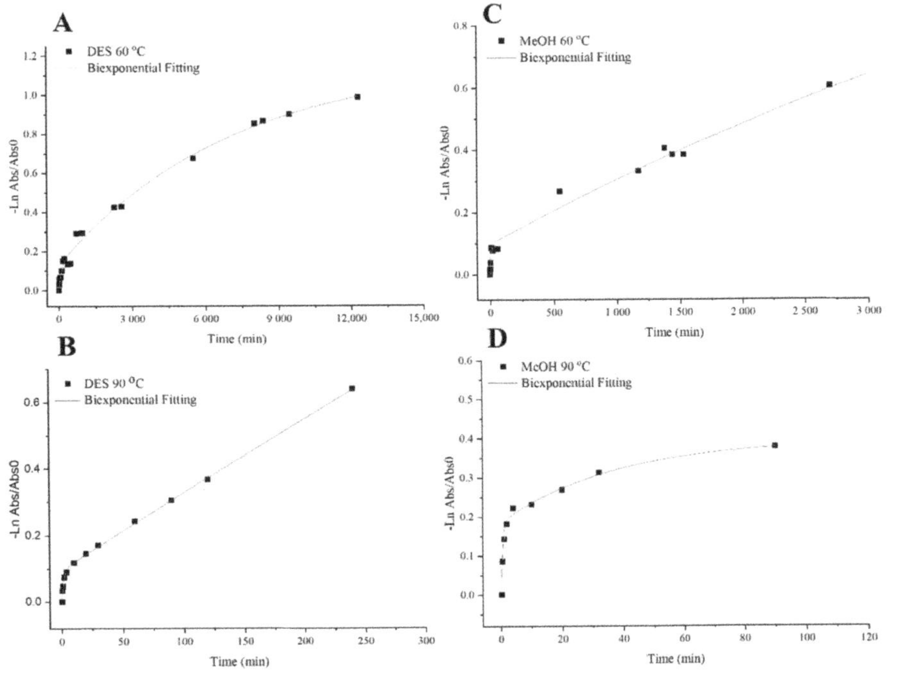

Figure 3. Degradation kinetics of anthocyanins extracted with Ch-Xyl (**A**,**B**) and MeOH (**C**,**D**) at 60 °C and 90 °C. The correlation coefficients of all curves are $R^2 \geq 0.99$.

The biexponential decay behavior, verified for the initial concentrations of anthocyanins, showed the same fast and slow lifetimes at each temperature, thus confirming the occurrence of a shared overall degradation mechanism.

$$y_t = A_1 exp(-\gamma_1 t) + A_2 exp(-\gamma_2 t) + y_\infty \quad (5)$$

y_t and y_∞ are anthocyanin absorbance values at real time and infinite time, respectively. A_1 and A_2 are the pre-exponential factors, whereas γ_1 and γ_2 are the observed rate constants for the fast and slow decays, respectively.

The degradation rate constants (K_d) at 60 °C and 90 °C from anthocyanins present in Ch-Xyl extract are shown in Table 6.

Table 6. Degradation constant (K_d) and half-life ($t_{1/2}$) values for anthocyanins extracted from jussara.

Solvent	T (°C)	T (K)	K_d (min^{-1})	$t_{1/2}$ (min)	E_a (kJ·mol^{-1})
MeOH	60 90	333.15 363.15	0.0002 0.001	3465.7 693.1	55.94
Ch-Xyl	60 90	333.15 363.15	0.00008 0.0025	8664.3 277.3	115.35

The K_d values were estimated from the slope of the semi-natural logarithmic plot of residual activity vs. time, assuming classical Arrhenius linear behavior and using the equation between both temperatures. Similarly, the half-life ($t_{1/2}$) values were calculated using the degradation rate constants, also presented in Table 4. A comparison of the $t_{1/2}$ values of both extracts at different temperatures showed that the results of anthocyanins extracted with Ch-Xyl are more stable than those obtained with MeOH, with half-life 2.5 times greater, in both temperatures.

The activation energy (E_a) for the Ch-Xyl extract (115.35 kJ/mol) was higher compared to the MeOH extracts (53.94 kJ/mol), confirming that the Ch-Xyl utilization provides more thermally stable anthocyanins.

The literature shows that in several studies, the thermal degradation of bioactive compounds, such as enzymes and bioactive peptides, as well as other bioproducts, the kinetics followed a linear adjusted model. Meanwhile, the number of articles evaluating anthocyanins' stability using DES is limited, with different DES combinations and food matrices. Anthocyanins extracted from *Catharanthus roseus* with lactic acid-glucose (5:1) showed better stability in this DES than in acidified ethanol, about three times at 60 °C [9], although the curve showed a first-order model. Previous works from our research group showed that the thermal stability of carotenoids from tomatoes extracted using an ultrasound-assisted approach and 1-Butyl-3-methyl-imidazolium-chloride [BMIM][Cl] followed a first-order model [53]. However, they also showed the biexponential behavior in two studies regarding carotenoid extraction using distinct ionic liquids of orange peel [54] as well as from pupunha (*Bactris gasipaes*) fruit [55]. The higher stabilization ability of Ch-Xyl for anthocyanins may be correlated with the interactions between cyanidin and these molecules. The two components of Ch-Xyl may have intermolecular interactions, mainly hydrogen bonding with the carboxyl and hydroxyl groups of anthocyanins, as was observed for quercetin in Ch-Xyl [40]. This interaction is expected to decrease the movement of solute molecules, reducing its contact time with oxygen at the interface of anthocyanins and air, and consequently reducing oxidative degradation, which is the major degradation mechanism. In a study with anthocyanins extracted from black carrots using choline chloride/citric acid DES [56], the samples also exhibited improved stability.

3.6. Determination of Anthocyanins Photostability

Light is reported to have two different effects on anthocyanins: it accelerates the rate of the thermal reactions and the formation of new molecules, including chalcones [57]. Anthocyanins from jussara pulp changed with the time both in light and dark conditions and followed a first order curve and were best fitted by a linear equation (Equation (6)), considering both situations. Figure 4 and Table 7 present the degradation kinetics of total anthocyanins extracted with MeOH and with Ch-Xyl.

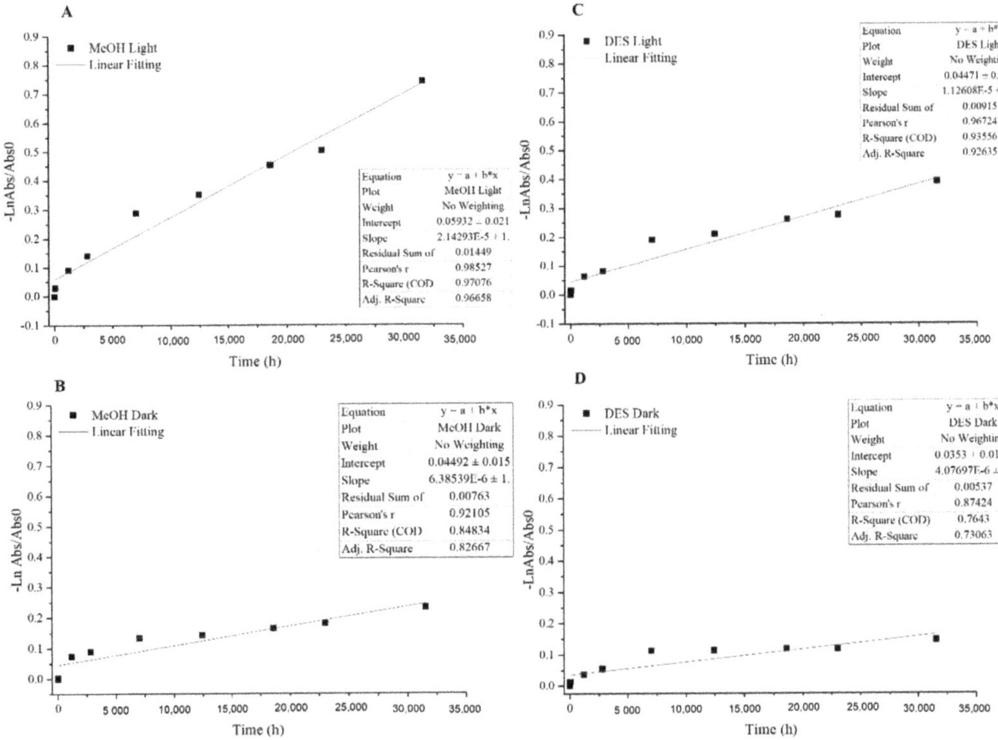

Figure 4. Degradation kinetics of anthocyanins extracted with MeOH (**A**) light and (**B**) Dark and Ch-Xyl (**C**) Light, (**D**) Dark in both conditions. The correlation coefficients of all curves are $R^2 \geq 0.99$.

Table 7. Degradation constant (K_d) and half-life ($t_{1/2}$) values for anthocyanins extracted from jussara.

Solvent	Conditions	K_d (h^{-1})	$t_{1/2}$ (h)
MeOH	Light	0.0015	462.1
	Dark	0.0005	1286.3
Ch-Xyl	Light	0.0008	866.4
	Dark	0.0003	2310.5

From a semi-natural logarithmic plot of the absorbance vs. time, the activation rate constants (K_d) were calculated, and the half-lives were estimated using Equation (6). The half-life ($t_{1/2}$) is defined as the time taken for the absorbance to reach 50% of the initial measurement.

$$t_{1/2} = \frac{-Ln0.5}{K_d} \tag{6}$$

When comparing $t_{1/2}$ values of Ch-Xyl and MeOH, the latter perished significantly ($p < 0.05$, ANOVA) faster—about 1.9 times in dark and 1.5 in light—which indicates that Ch-Xyl extract could have a better storage time than MeOH extract, especially in dark conditions. Anthocyanins from blueberry extract exposed to natural and UV light had the same pattern of degradation after 25 days [58], while they retained about 60% of their initial anthocyanin count in dark conditions and lower than 20% in light conditions. This study obtained 78% (MeOH) and 85% (Ch-Xyl) retention of initial absorbance after 25 days in dark and 43% (MeOH) and 63% (Ch-Xyl) retention of absorbance after 25 days, for jussara extracts exposed to light conditions. Degradation of anthocyanins of purple sweet potato analyzed for 15 days at 25 °C retained more than 80% anthocyanin content both under light and dark conditions [59]. The same was true for anthocyanins of *Hibiscus sabdariffa* for 10 days [60], and similar to results in this study, except for MeOH extract under light exposure, which retained only 63% of its initial absorbance after 15 days.

The strong hydrogen bond formed between Ch-Xyl and anthocyanins can once again explain the slower reaction rate of these extracts. This is pointed to as a possible strategy to deal with anthocyanins' relative instability compared to artificial colorants [39]. Evaluation of the thermal and light stability of anthocyanins from different sources is essential in order to determine their properties and hence apply those pigments, especially the most stable ones, for scaling-up processes and industrial applications. The bioaccessibility of anthocyanins during digestion, usually reported as very low [61], should be further evaluated when used in conjunction with Ch-Xyl solvents, evaluating their stabilization power in this environment to better understand the structure–stability relationships of the anthocyanin/extraction solvent.

3.7. Antimicrobial Activity

In order to evaluate the antimicrobial activity of extracts, a standard disk diffusion test for microbial sensitivity was conducted, and their inhibitory halo was measured (Figure 5 and Table S4 of Supporting Material).

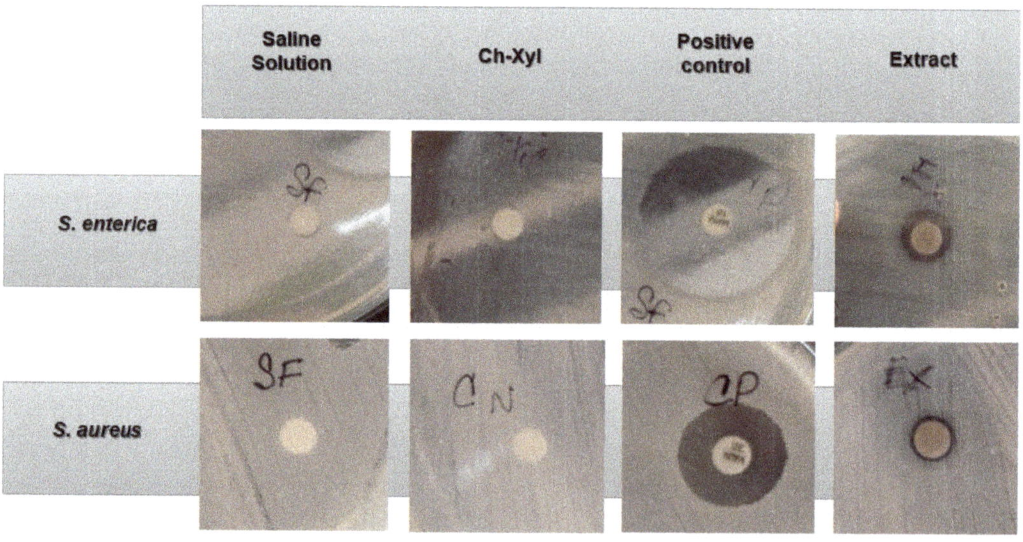

Figure 5. Disc-diffusion antimicrobial sensitivity. Representation of halos of inhibition for *S. enterica* and *S. aureus* in four treatments: saline solution, DES only (Ch-Xyl), positive control (meropen and vancomycin respectively), and optimized extract.

The bacterial pathogens, *Salmonella enterica* and *Staphylococcus aureus*, were tested, and the extract displayed moderate inhibition for both (8.6 mm and 7.6 mm halo respectively). *S. enterica* is one of the most frequent bacterial food-borne pathogens in humans. Salmonella infections range from gastrointestinal infections that are accompanied by inflammation of intestinal epithelia, diarrhea, and vomiting, to typhoid fever, a life-threatening systemic infection [62]. *S. aureus* is indeed found in the skin, hair, and nostrils of warm-blooded animals. This bacterium is a significant cause of nosocomial infections, as well as community-acquired diseases. The extract resistance to this microorganism is a good indicator for stability and avoidance of microbial growth during storage. The results for *S. aureus* are according to [18], where *Euterpe edulis* crude extract displayed slightly positive inhibition against this microorganism. This can be attributed to the antimicrobial activity of anthocyanins and other polyphenols found in the extract, which is usually an antimicrobial agent in plants, and shows that DES extraction did not remove significant antimicrobial components from the crude extract [63]. Coupled with the relevant information of the stability of the extracts, further studies are necessary to determine the minimum inhibitory concentration (MIC) of the extract and to determine the ideal storage concentrations.

Together, these experiments build an additional step in the knowledge of how efficient and stable anthocyanin-rich extracts obtained with DES can be while keeping its antioxidant and antimicrobial properties. Further experiments should be conducted in order to investigate practical applications in a number of cosmetical and food products regarding the stability and interactions in these systems. Reducing the use of VOS is a worldwide effort, and general industry faces additional appeals for "green label" procedures by its consumers. Alternative green solvents fit this necessity well, and natural colorants obtained from jussara are a promising technology that may provide another sustainable economic option for the tropical forest remnants.

Supplementary Materials: The following supporting information can be downloaded at: https://www.mdpi.com/article/10.3390/pr10030615/s1, Table S1. (A) Real and coded values used in the fractional experimental design with three central composites (25-1). (B) Real and coded values used in the Central Composite Rotatable Design (CCRD) 22. Table S2. Real values and experimental values and relative deviation (%) among experimental and predictive values for (A) ABTS and (B) ORAC antioxidant activity for jussara pulp extracted by CH-Xyl in the CCRD assays. Table S3. Kinetic parameters obtained by fitting the experimental data for degradation of anthocyanins during thermal treatment of the extracts obtained with MeOH and DES. Table S4. Determination of halo diameter and susceptibility for extract for strains of *S. enterica* and *S. aureus*. S—sensible. Figure S1. Chromatogram obtained by HPLC-PDA-MS/MS of the jussara methanolic extract from jussara pulp. Cyanidin 3-glucoside, retention time: 13.396 min; cyanidin 3-rutinoside, retention time: 13.991. Figure S2. TAC of Ch-Xyl extracts by repetition. Purple dots represent individual experiments. Extracts with seven repetitions show significantly (ANOVA) less yield of anthocyanins than extracts with five repetitions. They are not statistically different of extracts with three repetitions.

Author Contributions: Conceptualization, V.V.D.R.; methodology, V.V.D.R. and A.R.C.B.; validation, V.V.D.R., A.R.C.B. and N.V.; formal analysis, N.V.; investigation, N.V., V.V.D.R. and A.R.C.B.; resources, V.V.D.R.; writing—original draft preparation, N.V., V.V.D.R. and A.R.C.B.; writing—review and editing, N.V., V.V.D.R. and A.R.C.B.; supervision, V.V.D.R.; project administration, V.V.D.R.; funding acquisition, V.V.D.R. All authors have read and agreed to the published version of the manuscript.

Funding: This work was supported by "Fundação de Amparo a Pesquisa do Estado de São Paulo-FAPESP" for fundings (2016/18910-1) and fellowship (2018/04265-2).

Conflicts of Interest: The authors declare no conflict of interest.

References

1. Anastas, P.; Eghbali, N. Green Chemistry: Principles and Practice. *Chem. Soc. Rev.* **2010**, *39*, 301–312. [CrossRef] [PubMed]
2. Lewis, A.C.; Hopkins, J.R.; Carslaw, D.C.; Hamilton, J.F.; Nelson, B.S.; Stewart, G.; Dernie, J.; Passant, N.; Murrells, T. An increasing role for solvent emissions and implications for future measurements of volatile organic compounds: Solvent emissions of VOCs. *Philos. Trans. R. Soc. A Math. Phys. Eng. Sci.* **2020**, *378*, 20190328. [CrossRef] [PubMed]
3. Beach, E.S.; Cui, Z.; Anastas, P.T. Green Chemistry: A design framework for sustainability. *Energy Environ. Sci.* **2009**, *2*, 1038. [CrossRef]
4. Lenardão, E.J.; Freitag, R.A.; Dabdoub, M.J.; Ferreira Batista, A.C.; Da Cruz Silveira, C. "Green chemistry" Os 12 princípios da química verde e sua inserção nas atividades de ensino e pesquisa. *Quim. Nova* **2003**, *26*, 123–129. [CrossRef]
5. Sheldon, R.A. Green solvents for sustainable organic synthesis: State of the art. *Green Chem.* **2005**, *7*, 267. [CrossRef]
6. Bubalo, M.C.; Vidović, S.; Radojčić Redovniković, I.; Jokić, S. Green solvents for green technologies. *J. Chem. Technol. Biotechnol.* **2015**, *90*, 1631–1639. [CrossRef]
7. Cortez, R.; Luna-Vital, D.A.; Margulis, D.; de Mejia, E.G. Natural Pigments: Stabilization Methods of Anthocyanins for Food Applications. *Compr. Rev. Food Sci. Food Saf.* **2017**, *16*, 180–198. [CrossRef]

8. Radošević, K.; Cvjetko Bubalo, M.; Gaurina Srček, V.; Grgas, D.; Dragičević, T.L.; Redovniković, R.I. Evaluation of toxicity and biodegradability of choline chloride based deep eutectic solvents. *Ecotoxicol. Environ. Saf.* **2015**, *112*, 46–53. [CrossRef]
9. Dai, Y.; Rozema, E.; Verpoorte, R.; Choi, Y.H. Application of natural deep eutectic solvents to the extraction of anthocyanins from Catharanthus roseus with high extractability and stability replacing conventional organic solvents. *J. Chromatogr. A* **2016**, *1434*, 50–56. [CrossRef]
10. Pinheiro, M.C.D.O.; Abrantes, S.D.M.P. Avaliaçao da exposiçao aos corantes artificiais em balas e chicletes por crianças entre 3 e 9 anos estudantes de escolas particulares da Tijuca/Rio de Janeiro. 2012. Available online: http://www.revistaanalytica.com.br/artigos/8.pdf (accessed on 11 July 2014).
11. Ramesh, M.; Muthuraman, A. Flavoring and Coloring Agents: Health Risks and Potential Problems. In *Natural and Artificial Flavoring Agents and Food Dyes*; Elsevier: Amsterdam, The Netherlands, 2018; pp. 1–28. [CrossRef]
12. Blumberg, P.O. Rainbow Risks: 6 Artificial Food Colors You Need to Know about—Reader's Digest. Published 2010. Available online: https://www.rd.com/health/conditions/rainbow-risks-6-artificial-food-colors-you-need-to-know-about (accessed on 17 March 2022).
13. da Silva, J.Z.; dos Reis, M.S. Consumption of *Euterpe edulis* fruit by wildlife: Implications for conservation and management of the southern Brazilian atlantic forest. *An. Acad. Bras. Cienc.* **2019**, *91*, e20180537. [CrossRef]
14. Reis, M.S.; Fantini, A.C.; Nodari, R.O.; Reis, A.; Guerra, M.P.; Mantovani, A. Management and Conservation of Natural Populations in Atlantic Rain Forest: The Case Study of Palm Heart (*Euterpe edulis* Martius). *Biotropica* **2000**, *32*, 894–902. [CrossRef]
15. Silva, P.; Carmo, L.; Silva, G.; Silveira-Diniz, M.; Casemiro, R.; Spoto, M. Physical, chemical, and lipid composition of juçara (*Euterpe edulis* Mart.) pulp. *Alim. Nutr. Braz. J. Food Nutr.* **2013**, *24*, 7–13. Available online: https://www.researchgate.net/publication/284267543_Physical_chemical_and_lipid_composition_of_jucara_Euterpe_edulis_Mart_pulp (accessed on 17 March 2022).
16. Vannuchi, N.; Jamar, G.; Pisani, L.; Braga, A.R.C.; de Rosso, V.V. Chemical composition, bioactive compounds extraction, and observed biological activities from jussara (*Euterpe edulis*): The exotic and endangered Brazilian superfruit. *Compr. Rev. Food Sci. Food Saf.* **2021**, *20*, 3192–3224. [CrossRef]
17. Brito, E.S.; de Araújo, M.C.P.; Alves, E.E.; Carkeet, C.; Clevidence, B.A.; Novotny, J.A. Anthocyanins Present in Selected Tropical Fruits: Acerola, Jambolão, Jussara, and Guajiru. *J. Agric. Food Chem.* **2007**, *55*, 9389–9394. [CrossRef] [PubMed]
18. Favaro, L.; Balcão, V.; Rocha, L.; Silva, E.; Oliveira, J.J.; Vila, M.; Tubino, M.; Favaro, L.I.L.; Balcão, V.M.; Rocha, L.K.H.; et al. Physicochemical characterization of a crude anthocyanin extract from the fruits of jussara (*Euterpe edulis* Martius): Potential for food and pharmaceutical applications. *J. Braz. Chem. Soc.* **2018**, *29*, 2072–2088. [CrossRef]
19. Da Silva, N.A.; Rodrigues, E.; Mercadante, A.Z.; De Rosso, V.V. Phenolic Compounds and Carotenoids from Four Fruits Native from the Brazilian Atlantic Forest. *Agric. Food Chem.* **2014**, *62*, 5072–5084. [CrossRef]
20. Cardoso, L.M.; Dias Novaes, R.; De Castro, C.A. Chemical composition, characterization of anthocyanins and antioxidant potential of *Euterpe edulis* fruits: Applicability on genetic dyslipidemia and hepatic steatosis in mice. *Nutr. Hosp.* **2015**, *32*, 702–709. [CrossRef]
21. Borges, G.D.S.C.; Vieira, F.G.K.; Copetti, C.; Gonzaga, L.V.; Zambiazi, R.C.; Filho, J.M.; Fett, R. Chemical characterization, bioactive compounds, and antioxidant capacity of jussara (*Euterpe edulis*) fruit from the Atlantic Forest in southern Brazil. *Food Res. Int.* **2011**, *44*, 2128–2133. [CrossRef]
22. Vieira, G.S.; Cavalcanti, R.N.; Meireles, M.A.A.; Hubinger, M.D. Chemical and economic evaluation of natural antioxidant extracts obtained by ultrasound-assisted and agitated bed extraction from jussara pulp (*Euterpe edulis*). *J. Food Eng.* **2013**, *119*, 196–204. [CrossRef]
23. Carvalho, A.G.A.; Silva, K.A.; Silva, L.O.; Costa, A.M.M.; Akil, E.; Coelho, M.A.Z.; Torres, A.G. Jussara berry (*Euterpe edulis* M.) oil-in-water emulsions are highly stable: The role of natural antioxidants in the fruit oil. *J. Sci. Food Agric.* **2019**, *99*, 90–99. [CrossRef]
24. AOAC. *Official Methods of Analysis of the Association of Official Analytical Chemists*, 17th ed.; The Association of Official Analytical Chemists International: Gaithersburg, MD, USA, 2000.
25. Gutiérrez, M.C.; Ferrer, M.L.; Mateo, C.R.; del Monte, F. Freeze-drying of aqueous solutions of deep eutectic solvents: A suitable approach to deep eutectic suspensions of self-assembled structures. *Langmuir* **2009**, *25*, 5509–5515. [CrossRef]
26. Bi, W.; Tian, M.; Row, K.H. Evaluation of alcohol-based deep eutectic solvent in extraction and determination of flavonoids with response surface methodology optimization. *J. Chromatogr. A* **2013**, *1285*, 22–30. [CrossRef] [PubMed]
27. Nam, M.W.; Zhao, J.; Lee, M.S.; Jeong, J.H.; Lee, J. Enhanced extraction of bioactive natural products using tailor-made deep eutectic solvents: Application to flavonoid extraction from *Flos sophorae*. *Green Chem.* **2015**, *17*, 1718–1727. [CrossRef]
28. Zhuang, B.; Dou, L.L.; Li, P.; Liu, E.H. Deep eutectic solvents as green media for extraction of flavonoid glycosides and aglycones from Platycladi Cacumen. *J. Pharm. Biomed. Anal.* **2017**, *134*, 214–219. [CrossRef]
29. Torskangerpoll, K.; Andersen, Ø.M. Colour stability of anthocyanins in aqueous solutions at various pH values. *Food Chem.* **2005**, *89*, 427–440. [CrossRef]
30. De Rosso, V.V.; Mercadante, A.Z. Identification and quantification of carotenoids, by HPLC-PDA-MS/MS, from Amazonian fruits. *J. Agric. Food Chem.* **2007**, *55*, 5062–5072. [CrossRef]
31. Do, M.; Rufino, S.M.; Alves, R.E.; de Brito, E.S. *Metodologia Científica: Determinação Da Atividade Antioxidante Total Em Frutas Pela Captura Do Radical Livre ABTS+*; Embrapa Agroindústria Tropical: Fortaleza, Brazil, 2007; Available online: https://www.embrapa.br/busca-de-publicacoes/-/publicacao/426954/metodologia-cientifica-determinacao-da-atividade-antioxidante-total-em-frutas-pela-captura-do-radical-livre-abts (accessed on 17 March 2021).
32. Zulueta, A.; Esteve, M.J.; Frígola, A. ORAC and TEAC assays comparison to measure the antioxidant capacity of food products. *Food Chem.* **2009**, *114*, 310–316. [CrossRef]

33. de Rosso, V.V.; Mercadante, A.Z. Evaluation of colour and stability of anthocyanins from tropical fruits in an isotonic soft drink system. *Innov. Food Sci. Emerg. Technol.* **2007**, *8*, 347–352. [CrossRef]
34. Schulz, M.; Biluca, F.C.; Gonzaga, L.V.; Borges, G.; Vitali, L.; Micke, G.A.; de Gois, J.S.; de Almeida, T.S.; Borges, D.L.G.; Miller, P.R.M.; et al. Bioaccessibility of bioactive compounds and antioxidant potential of juçara fruits (*Euterpe edulis* Martius) subjected to in vitro gastrointestinal digestion. *Food Chem.* **2017**, *228*, 447–454. [CrossRef]
35. Rogez, H.; Pompeu, D.R.; Akwie, S.N.T.; Larondelle, Y. Sigmoidal kinetics of anthocyanin accumulation during fruit ripening: A comparison between açai fruits (*Euterpe oleracea*) and other anthocyanin-rich fruits. *J. Food Compos. Anal.* **2011**, *24*, 796–800. [CrossRef]
36. Gordon, A.; Gil Cruz, A.P.; Cabral, L.M.C.; de Freitas, S.C.; Taxi, C.M.A.D.; Donangelo, C.M.; Mattietto, R.D.A.; Friedrich, M.; da Matta, V.M.; Marx, F. Chemical characterization and evaluation of antioxidant properties of Açaí fruits (*Euterpe oleraceae* Mart.) during ripening. *Food Chem.* **2012**, *133*, 256–263. [CrossRef] [PubMed]
37. Wang, T.; Jiao, J.; Gai, Q.-Y.; Wang, P.; Guo, N.; Niu, L.-L.; Fu, Y.-J. Enhanced and green extraction polyphenols and furanocoumarins from Fig (*Ficus carica* L.) leaves using deep eutectic solvents. *J. Pharm. Biomed. Anal.* **2017**, *145*, 339–345. [CrossRef]
38. Zhao, B.-Y.; Xu, P.; Yang, F.-X.; Wu, H.; Zong, M.-H.; Lou, W.-Y. Biocompatible Deep Eutectic Solvents Based on Choline Chloride: Characterization and Application to the Extraction of Rutin from *Sophora japonica*. *ACS Sustain. Chem. Eng.* **2015**, *3*, 2746–2755. [CrossRef]
39. Guo, N.; Kou, P.; Jiang, Y.-W.; Wang, L.-T.; Niu, L.-J.; Liu, Z.-M.; Fu, Y.-J. Natural deep eutectic solvents couple with integrative extraction technique as an effective approach for mulberry anthocyanin extraction. *Food Chem.* **2019**, *296*, 78–85. [CrossRef] [PubMed]
40. Dai, Y.; van Spronsen, J.; Witkamp, G.J.; Verpoorte, R.; Choi, Y.H. Natural deep eutectic solvents as new potential media for green technology. *Anal. Chim. Acta* **2013**, *766*, 61–68. [CrossRef] [PubMed]
41. López, N.; Delso, I.; Matute, D.; Lafuente, C.; Artal, M. Characterization of xylitol or citric acid:choline chloride:water mixtures: Structure, thermophysical properties, and quercetin solubility. *Food Chem.* **2020**, *306*, 125610. [CrossRef]
42. Dai, Y.; Witkamp, G.-J.; Verpoorte, R.; Choi, Y.H. Tailoring properties of natural deep eutectic solvents with water to facilitate their applications. *Food Chem.* **2015**, *187*, 14–19. [CrossRef]
43. Bubalo, M.C.; Ćurko, N.; Tomašević, M.; Kovačević Ganić, K.; Redovnikovic, I.R. Green extraction of grape skin phenolics by using deep eutectic solvents. *Food Chem.* **2016**, *200*, 159–166. [CrossRef]
44. Triolo, A.; Lo Celso, F.; Brehm, M.; Di Lisio, V.; Russina, O. Liquid structure of a choline chloride-water natural deep eutectic solvent: A molecular dynamics characterization. *J. Mol. Liq.* **2021**, *331*, 115750. [CrossRef]
45. Schulz, M.; Borges, G.; Gonzaga, L.V.; Seraglio, S.K.T.; Olivo, I.S.; Azevedo, M.S.; Nehring, P.; de Gois, J.S.; de Almeida, T.S.; Vitali, L.; et al. Chemical composition, bioactive compounds and antioxidant capacity of juçara fruit (*Euterpe edulis* Martius) during ripening. *Food Res. Int.* **2015**, *77*, 125–131. [CrossRef]
46. Madalão, M.C.M.; Lima, E.M.F.; Benincá, D.B.; Saraiva, S.H.; de Carvalho, R.V.; Silva, P.I. Extraction of bioactive compounds from juçara pulp (*Euterpe edulis* M.) is affected by ultrasonic power and temperature. *Ciência Agrotecnologia* **2021**, *45*, 2021. [CrossRef]
47. Braga, A.R.C.; Murador, D.C.; de Souza Mesquita, L.M.; de Rosso, V.V. Bioavailability of anthocyanins: Gaps in knowledge, challenges and future research. *J. Food Compos. Anal.* **2017**, *68*, 31–40. [CrossRef]
48. de Souza, M.O.; Silva, M.; Silva, M.E.; Oliveira, R.D.P.; Pedrosa, M.L. Diet supplementation with acai (*Euterpe oleracea* Mart.) pulp improves biomarkers of oxidative stress and the serum lipid profile in rats. *Nutrition* **2017**, *26*, 804–810. [CrossRef] [PubMed]
49. Nayak, B.; Berrios, J.D.J.; Powers, J.R.; Tang, J. Thermal Degradation of Anthocyanins from Purple Potato (Cv. Purple Majesty) and Impact on Antioxidant Capacity. *J. Agric. Food Chem.* **2011**, *59*, 11040–11049. [CrossRef] [PubMed]
50. Inada, K.O.P.; Oliveira, A.A.; Revorêdo, T.B.; Martins, A.B.N.; Lacerda, E.C.Q.; Freire, A.S.; Braz, B.F.; Santelli, R.; Torres, A.G.; Perrone, D.; et al. Screening of the chemical composition and occurring antioxidants in jabuticaba (*Myrciaria jaboticaba*) and jussara (*Euterpe edulis*) fruits and their fractions. *J. Funct. Foods* **2015**, *17*, 422–433. [CrossRef]
51. Bicudo, M.O.P.; Ribani, R.H. Anthocyanins, Phenolic Acids and Antioxidant Properties of Juçara Fruits (*Euterpe edulis* M.) Along the On-tree Ripening Process. *Plant Foods Hum. Nutr.* **2014**, *69*, 142–147. [CrossRef]
52. Martins, P.L.G.; de Rosso, V.V.; Larangeira, P.; Martins, G.; De Rosso, V.V. Thermal and light stabilities and antioxidant activity of carotenoids from tomatoes extracted using an ultrasound-assisted completely solvent-free method. *Food Res. Int.* **2016**, *82*, 156–164. [CrossRef]
53. Murador, D.C.; Braga, A.R.C.; Martins, P.L.G.; Mercadante, A.Z.; de Rosso, V.V. Ionic liquid associated with ultrasonic-assisted extraction: A new approach to obtain carotenoids from orange peel. *Food Res. Int.* **2019**, *126*, 108653. [CrossRef]
54. De Souza Mesquita, L.M.; Ventura, S.P.M.; Braga, A.R.C.; Pisani, L.P.; Dias, A.C.R.V.; De Rosso, V.V. Ionic liquid-high performance extractive approach to recover carotenoids from: Bactris gasipaes fruits. *Green Chem.* **2019**, *21*, 2380–2391. [CrossRef]
55. Türker, D.A.; Doğan, M. Ultrasound-assisted natural deep eutectic solvent extraction of anthocyanin from black carrots: Optimization, cytotoxicity, in-vitro bioavailability and stability. *Food Bioprod. Process.* **2022**, *132*, 99–113. [CrossRef]
56. Furtado, P.; Figueiredo, P.; das Neves, H.C.; Pina, F. Photochemical and thermal degradation of anthocyanidins. *J. Photochem. Photobiol. A Chem.* **1993**, *75*, 113–118. [CrossRef]
57. Chen, J.; Ma, X.H.; Yao, G.L.; Zhang, W.T.; Zhao, Y. Microemulsion-based anthocyanin systems: Effect of surfactants, cosurfactants, and its stability. *Int. J. Food Prop.* **2018**, *21*, 1152–1165. [CrossRef]
58. Chen, C.C.; Lin, C.; Chen, M.H.; Chiang, P.Y. Stability and quality of anthocyanin in purple sweet potato extracts. *Foods* **2019**, *8*, 393. [CrossRef] [PubMed]
59. Sipahli, S.; Mohanlall, V.; Mellem, J.J. Stability and degradation kinetics of crude anthocyanin extracts from *H. sabdariffa*. *Food Sci. Technol.* **2017**, *37*, 209–215. [CrossRef]

60. Han, H.; Liu, C.; Gao, W.; Li, Z.; Qin, G.; Qi, S.; Jiang, H.; Li, X.; Liu, M.; Yan, F.; et al. Anthocyanins Are Converted into Anthocyanidins and Phenolic Acids and Effectively Absorbed in the Jejunum and Ileum. *J. Agric. Food Chem.* **2021**, *69*, 992–1002. [CrossRef]
61. Hensel, M. Evolution of pathogenicity islands of *Salmonella enterica*. *Int. J. Med. Microbiol.* **2004**, *294*, 95–102. [CrossRef]
62. Cisowska, A.; Wojnicz, D.; Hendrich, A.B. Anthocyanins as antimicrobial agents of natural plant origin. *Nat. Prod. Commun.* **2011**, *6*, 149–156. [CrossRef]
63. Lima, M.C.; Paiva de Sousa, C.; Fernandez-Prada, C.; Harel, J.; Dubreuil, J.D.; de Souza, E.L. A review of the current evidence of fruit phenolic compounds as potential antimicrobials against pathogenic bacteria. *Microb Pathog.* **2019**, *130*, 259–270. [CrossRef]

processes

Article

Interfacial Microstructure and Mechanical Reliability of Sn-58Bi/ENEPIG Solder Joints

Cheng Chen [1,*], Cheng Wang [1], Huhao Sun [1], Hongbo Yin [1], Xiuli Gao [1], Hengxu Xue [1], Dahai Ni [1], Kan Bian [2] and Qilin Gu [3,*]

[1] China State Shipbuilding Corporation, Limited (CSSC) 723rd Research Institute, Yangzhou 225001, China; nuaawc@163.com (C.W.); sunhuhao2012@163.com (H.S.); yinhongbo2004@126.com (H.Y.); yzgxl@163.com (X.G.); xhx08648@126.com (H.X.); seuseayou@outlook.com (D.N.)
[2] Nanjing Institute of Technology, School of Materials Science and Engineering, Nanjing 211100, China; bk@nuaa.edu.cn
[3] National Engineering Research Center for Special Separation Membrane, Nanjing Tech University, Nanjing 210009, China
* Correspondence: chencheng5670@163.com (C.C.); gu_qilin@nuaa.edu.cn (Q.G.)

Abstract: The 42 wt.% Sn–58 wt.% Bi (Sn-58Bi) Ball Grid Array (BGA) solder balls were mounted to electroless nickel-electroless palladium-immersion gold (ENEPIG) pads by employing the reflow process profile. The effects of reflow cycles and aging time on the interfacial microstructure and growth behavior of intermetallic compounds, as well as the mechanical properties, were investigated. Pd-Au-Sn intermetallic compound (IMC) was formed at the Sn-58Bi/ENEPIG interface. With the increase in reflow cycles and aging time, the IMC grew gradually. After five reflow cycles, the shear strength of the Sn-58Bi/ENEPIG solder joints first decreased and then increased. After 500 h of aging duration under −40 °C, the shear strength of the Sn-58Bi/ENEPIG solder joints decreased by about 12.3%. The fracture mode transferred from ductile fracture to ductile and brittle mixed fracture owing to the fact that the fracture location transferred from the solder matrix to the IMC interface with the increase in reflow cycles and aging time.

Keywords: electroless nickel-electroless palladium-immersion gold; Sn-58Bi; ball grid array; intermetallic compound

Citation: Chen, C.; Wang, C.; Sun, H.; Yin, H.; Gao, X.; Xue, H.; Ni, D.; Bian, K.; Gu, Q. Interfacial Microstructure and Mechanical Reliability of Sn-58Bi/ENEPIG Solder Joints. *Processes* **2022**, *10*, 295. http://doi.org/10.3390/pr10020295

Academic Editors: Prashant K. Sarswat and Zhou Li

Received: 13 October 2021
Accepted: 27 December 2021
Published: 1 February 2022

1. Introduction

The eutectic Sn-37Pb solder is an important material in the field of electronic packing because of its excellent wettability, electronic conductivity and mechanical properties [1,2]. However, due to the threat of lead to the human body and the environment, lead-free solder has become the development direction in the electronic industry with the restriction of hazardous substances (RoHs) [3,4]. Sn–Ag, Sn–Zn, Sn–Ag–Cu and Sn–Bi alloys are developed in order to reach or approach the properties of the eutectic Sn-37Pb solder [5–8]. Eutectic Sn-58Bi solder with a melting point of 138 °C has been considered a promising lead-free solder for low-temperature applications, especially in the fields of MCM (Multichip Module) and 3D heterogeneous integration modules. This solder possesses good creep resistance and high tensile strength [9]. In addition, it can avoid warpage in printed circuit boards (PCBs) and electronic components due to thermal mismatch during the soldering process [10].

On the other hand, during PCB manufacturing, a solderable finish is plated over the Cu substrate. At present, electroless nickel-immersion gold (ENIG) has been widely used as a solderable finish in the electronic industry due to its good wettability [11–13]. During the soldering process, the intermetallic compound (IMC) can be formed at the solder/PCB finish interface, which is essential for mechanical, conductive and thermal properties. As for the ENIG finish, the electroless nickel layer acts as a diffusion barrier against Cu migration, and the immersion gold layer can prevent surface oxidation to improve weldability [14]. However, during the immersion gold plating process, ENIG corrodes the Ni (P) layer and causes cracks, which are called "black pad" phenomena [15,16]. Recently, a new solderable finish called ENEPIG finish has emerged [17,18]. A thin and dense electroless palladium layer is added between the Ni (P) layer and Au layer, which can prevent the oxidation reaction of Ni (P) and the "black pad" phenomenon thoroughly [19]. Additionally, ENEPIG finish shows better weldability, corrosion resistance and reliability [20,21].

Studies on the behavior of IMC chemical reactions with ENIG and ENEPIG have been well documented. Yoon et al. [18] found that needle-shaped (Cu, Ni) $_6Sn_5$ IMC and P-rich Ni layer formation deteriorated the shear strengths of the ENEPIG joints. Tian et al. [22] found that the shear strength of the SAC305/ENEPIG joints was consistently higher than that of the SAC305/ENIG joints due to the slightly slow growth rate of interfacial IMCs. With the increase in thermal shock cycles, the fracture mode of the SAC305/ENIG joints switched from ductile to ductile–brittle mixed fracture mode, while the fracture mode of the SAC305/ENEPIG joints was consistently ductile. Chi et al. [23] found two layers of IMCs with compositions of $(Au_{0.30}Ni_{0.70})(Sn_{0.90}Bi_{0.10})_4$ and Ni_3Sn_4 were formed at the Sn-58Bi solder/ENIG finish interface of the aged joints, and the relationship between ball shear strength(S) and thickness(X) of $(Au_{0.30}Ni_{0.70})(Sn_{0.90}Bi_{0.10})_4$ IMC layer: $S = 7.13-0.33X$ was researched. Chen et al. [24] only detected NiSn IMC at the Sn-58Bi/Ni interface and found that the growth of IMC followed diffusion-controlled kinetics. Kim et al. [25] compared the effects of phosphorous and pure Pd in a thin ENEPIG surface finish on the interfacial reactions and mechanical properties of the Sn-58Bi/thin-ENEPIG solder joints and found that the Sn-58Bi solder with phosphorous in the finish has a higher reliability than that with pure Pd in the finish.

The Sn-58Bi solder joint is applicable for the field of MCM (Multichip Module) and SIP (System in Package) modules. According to the National Military Standard of the People's Republic of China (GJB 8481-2015), General Specification for Microwave Assembly, the working temperature of these modules is in the range of −40 °C to 70 °C. These modules may undergo multiple reflow cycles during rework or multi-gradient assembly [26] and will be in service at temperatures as low as −40 °C for a long time. Therefore, a detailed study on the interfacial reactions and growth behavior of IMC, as well as the shear strength and fracture mode of the Sn-58Bi/ENEPIG solder joints with different reflow cycles and aging at a temperature as low as −40 °C are presented in this work.

2. Materials and Methods

2.1. Experimental Materials

A PCB made of flame-retardant-4 (FR4) with an ENEPIG plating process was used as the substrate. The ENEPIG process included electroless nickel plating, electroless palladium plating, and immersion gold plating, successively. The electroless nickel plating process was a redox reaction under the action of metal catalysis. Hypophosphite was used as a reducing agent in the electroless nickel plating process, and the reaction process includes:

$$[H_2PO_2]^- + H_2O \rightarrow [HPO_3]^{2-} + H^+ + 2H, \quad (1)$$

$$Ni^{2+} + 2H \rightarrow Ni\downarrow + 2H^+, \quad (2)$$

$$2[H_2PO_2]^- + H \rightarrow [HPO_3]^{2-} + H_2O + P + H_2\uparrow. \quad (3)$$

Electroless palladium plating process was also an autocatalytic redox reaction procedure with hypophosphite as a reducing agent. The reaction process includes:

$$[H_2PO_2]^- + H_2O \rightarrow [HPO_3]^{2-} + H^+ + 2H, \quad (4)$$

$$Pd^{2+} + 2H \rightarrow Pd\downarrow + 2H^+, \quad (5)$$

$$2[H_2PO_2]^- + H \rightarrow [HPO_3]^{2-} + H_2O + P + H_2\uparrow. \quad (6)$$

The immersion gold plating process was a displacement process where Au and Pd were replaced to obtain the Au layer. The reaction process was:

$$2Au^{+} + Pd \rightarrow 2Au + Pd^{2+}. \quad (7)$$

The thickness of Ni (P), Pd and Au layers was about 5 μm, 0.1 μm and 0.05 μm, respectively. The PCB pad opening size was 0.4 mm. After being dipped in Sn-58Bi solder paste (HL-6RC-4258, Honglian, Suzhou, China), Sn-58Bi BGA solder balls (QW-SB42-Q500, Qwin, Chongqin, China) of 0.5 mm in diameter were mounted to ENEPIG pads by employing the reflow process profile. After reflowing and cooling to room temperature, the soldering samples were cleaned to remove the flux residuals by manual spraying cleaning agents (VIGON EFM, ZESTRON, Ingolstadt, Germany). Figure 1 shows a schematic diagram of the Sn-58Bi/ENEPIG solder joints. The reflow process was realized by a reflow oven (VXC 421, Rehm, Blauboylen, Germany) with six heating zones and two cooling zones. The temperature of the heating zones was measured by a thermodetector, which is shown in Table 1. The thermodetector displayed the temperature profile for the reflow process of Sn-58Bi/ENEPIG solder joints, and the result is shown in Figure 2. The peak temperature (T_{max}), the melting point (T_e), the preheating time and the cooling rate was about 170 °C, 138 °C, 180–220 s, 60–90 s and 2–5 °C/s, respectively.

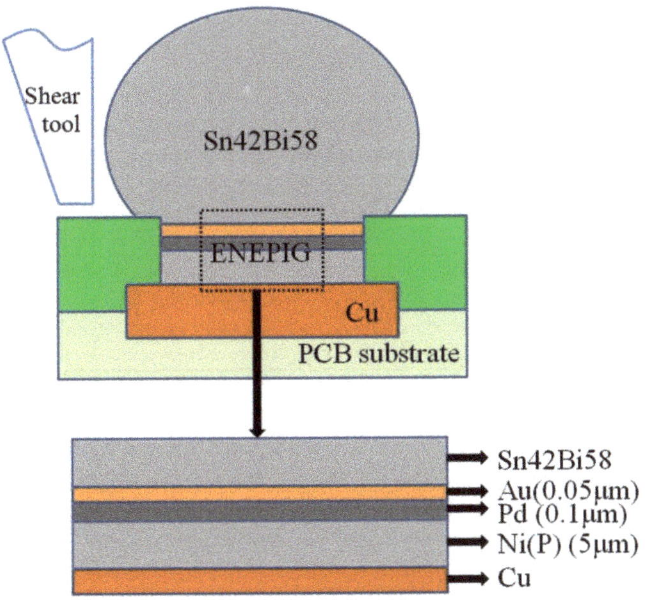

Figure 1. Schematic diagram of the Sn-58Bi/ENEPIG solder joints.

Table 1. Temperature of the heating zones.

Zone / Temperature	1	2	3	4	5	6
Upper furnace/°C	120	130	135	140	160	195
Bottom furnace/°C	120	130	135	140	160	195

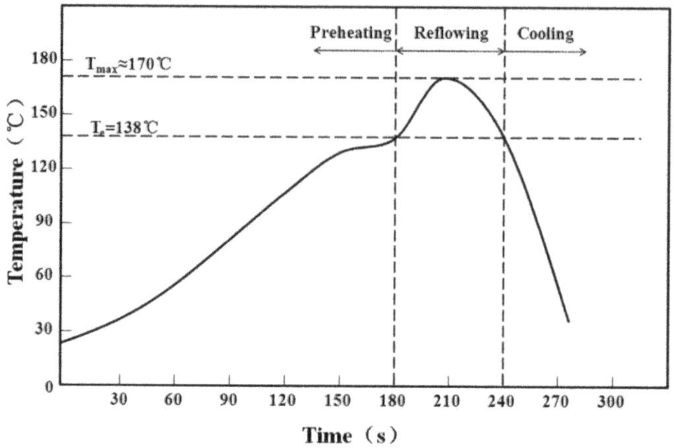

Figure 2. Temperature profile for the reflow process of Sn-58Bi/ENEPIG solder joints.

2.2. *Aging Test*

The aging test at −40 °C ranging from 100 h to 500 h was performed on the reflowed joints by using a rapid temperature change test chamber (SU-662, Espec, Osaka, Japan). In addition, the Sn-58Bi/ENEPIG solder joints were conducted using a reflow oven under different reflow cycles.

2.3. *Microstructure Observation*

After reflowing different cycles, as well as aging under different process parameters, the samples were mounted, inserted, ground and polished successively to observe the cross-sectional microstructure. The interfacial microstructure of the aged solder joints was observed by a scanning electron microscope (SEM, S-4800, Hitachi, Tokyo, Japan) and the composition of IMC, solder and finish was analyzed by energy-dispersive spectroscopy (EDS). The interfacial IMC thickness was calculated using the following equation:

$$L_{IMC} = (N_{IMC}/N_{SEM}) \times L_{SEM} \quad (8)$$

where L_{IMC} was the interfacial IMC thickness, L_{SEM} was the height of the SEM image, and N_{IMC} and N_{SEM} were the number of pixels for the interfacial IMC and the entire SEM image, respectively. The values of N_{IMC} and N_{SEM} were obtained using Image J software. For each test condition, at least five SEM images taken at different locations were utilized to calculate the average value of the interfacial IMC thickness.

2.4. *Ball Shear Test and Fracture Morphology Analysis*

The ball shear strength test was performed using a global bond tester (DAGE4000, Nordson, Aylesbury, UK). The shear speed was 500 μm/s, and the shear height was set at about 50 μm. After the ball shear test, fracture morphology and composition were analyzed by SEM and EDS.

3. Results and Discussion

3.1. *Interfacial Reactions*

Figure 3 shows the cross-sectional image of Sn-58Bi/ENEPIG solder joints after reflowing and the enlarged view of the white square area is presented in Figure 4. In combination with the Sn-58Bi phase diagram, the Sn-58Bi solder ball contained an eutectic lamellar microstructure of β-Sn solid solution (white color) and Bi-rich phase (gray color) [22]. A continuous flake-type IMC layer with a thickness of 1–3 μm was formed at the Sn-58Bi/ENEPIG interface. In addition, it can be seen that a P-rich Ni layer was beneath the interfacial IMC layer. EDS was further carried out to analyze the interfacial reactions between the Sn-58Bi solder ball and ENEPIG finish. According to Figures 5 and 6 results, Pd and Au elements in the ENEPIG finish with the Sn element in Sn-58Bi solder were formed at the Sn-58Bi/ENEPIG interface. Additionally, it can be seen that during the reflow process, most of the Pd element was dissolved into the molten solder to react at the interface. The remaining thin Pd layer prevented the diffusion of the Ni layer, which could be evidenced by the results in Figure 5. The chemical composition of the interfacial IMC was 1.82 at. % Ni, 14.78 at. % Pd, 74.36 at. % Sn, and 9.04 at. % Au. As can be seen, the atomic proportion of Pd and Au is not 1:1 in the IMC, which means Au has replaced some of the Pd atoms in $PdSn_4$. At the same time, the atomic proportion of (Pd, Au) and Sn is 1:3.12. Considering the acceptable measurement error and the possible compound that three elements can form, we infer that it is most likely $(Pd, Au)Sn_4$. The results were coincident with those of a previous study [25].

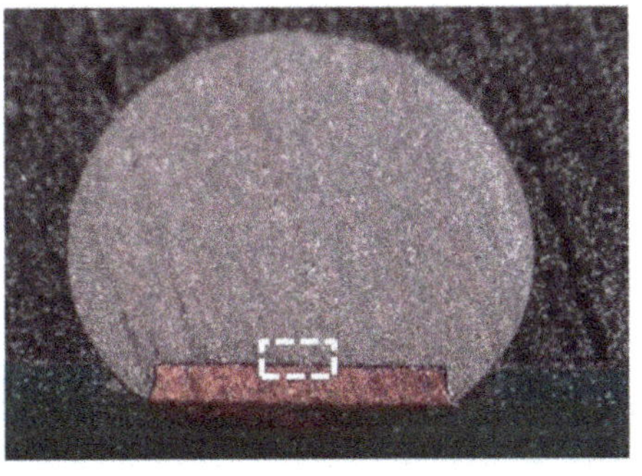

Figure 3. Cross-sectional images of Sn-58Bi/ENEPIG solder joints after reflowing.

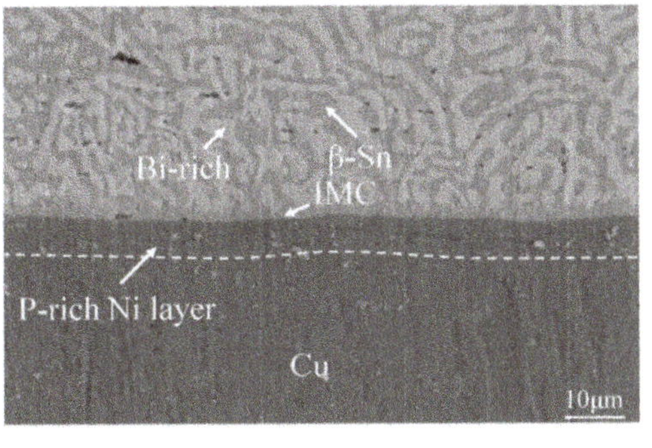

Figure 4. Cross-sectional SEM images of the white square in Figure 3.

3.2. Growth Behavior of Interfacial IMC and Mechanical Properties for Different Reflow Cycles

The reflow process of Sn-58Bi/ENEPIG solder joints was carried out for one to five cycles, and the surface morphology of solder joints is shown in Figure 7. It can be seen that after one reflow cycle, the solder joints were bright and had good wettability. After two reflow cycles, the surface of the solder joints was wrinkled. With the increase in reflow cycles, the wrinkle became obvious, the solder joint surface became black and the oxidation phenomenon was more serious. SEM images of Sn-58Bi/ENEPIG joints with reflow cycles ranging from one to five are shown in Figure 8. It was found that the thickness of the IMC layer gradually increased with the increase in reflow cycles. The thickness of the IMC layer formed at the interface for reflow cycles ranging from one to five was 1.7, 2.3, 3.2, 3.8 and 4.3 μm, respectively, as shown in Figure 9. By fitting the results, the relationship between reflow cycles (X) and thickness (Y, μm) of the IMC layer can be described as:

Y = 0.67X + 1.05. In addition, with the increase in reflow cycles, the Bi rich phase and β-Sn phase showed a coarse growth trend.

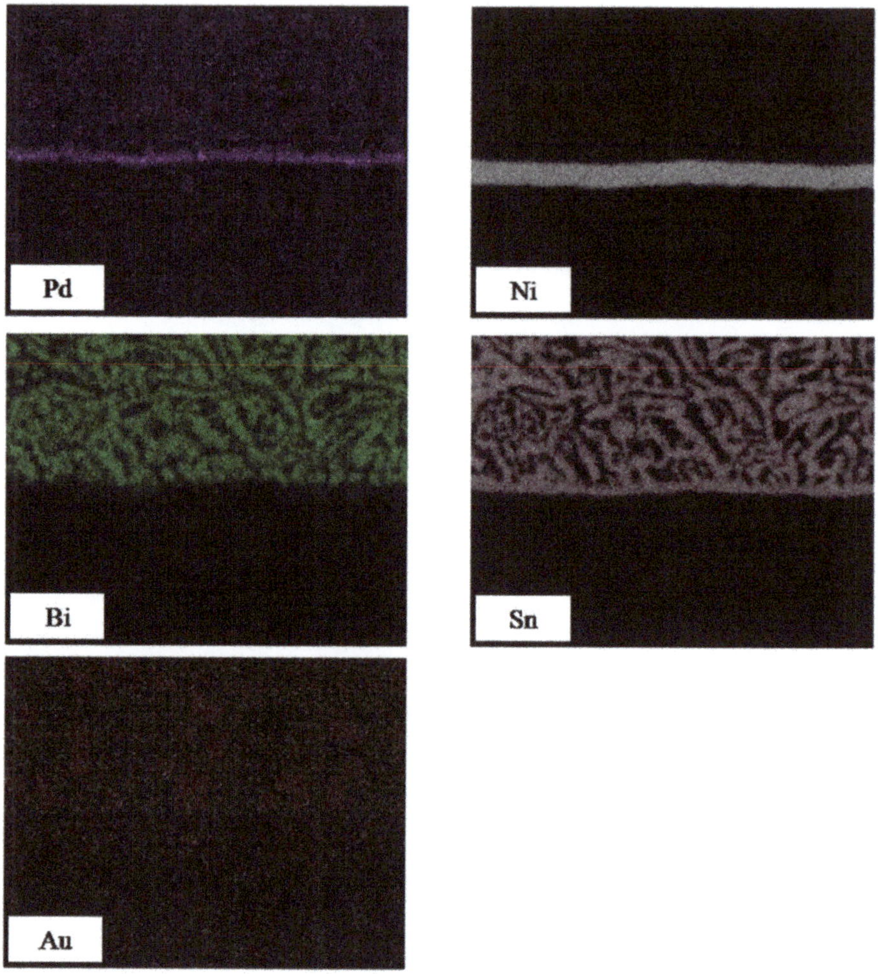

Figure 5. EDS results of Sn-58Bi/ENEPIG solder joints after reflowing.

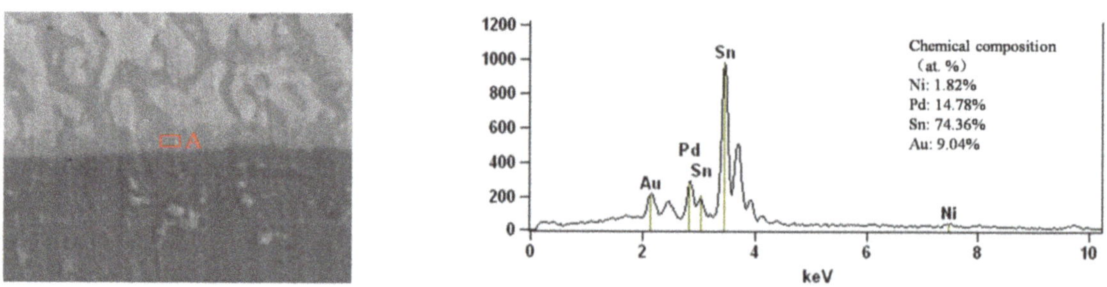

Figure 6. EDS result of the interfacial IMC in area A of enlarged Figure 4.

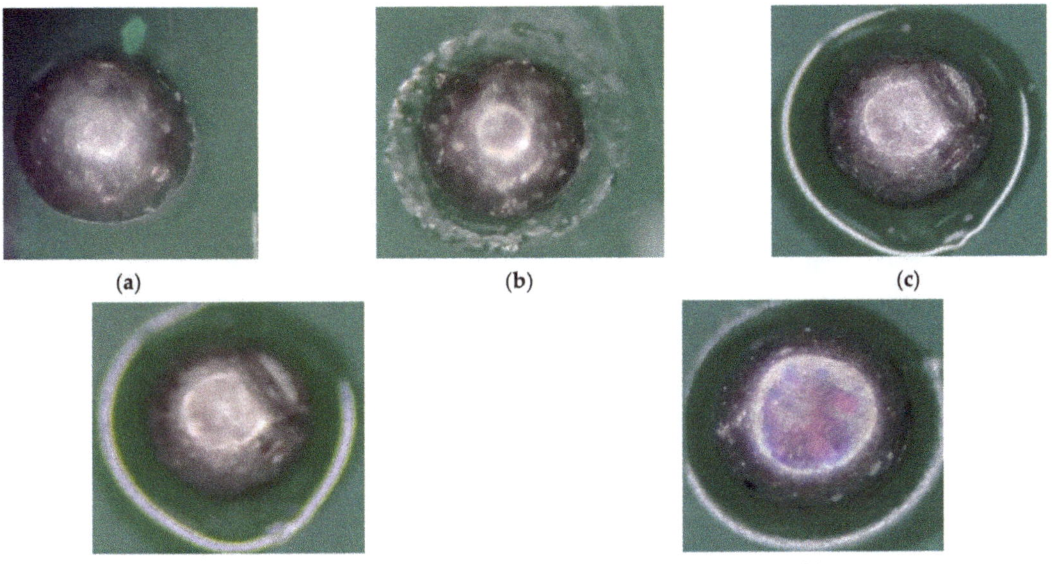

(a) (b) (c) (d) (e)

Figure 7. Surface morphology of solder joints with different reflow cycles: (**a**) one reflow cycle, (**b**) two reflow cycles, (**c**) three reflow cycles, (**d**) four reflow cycles and (**e**) five reflow cycles.

The ball shear test on solder joints with different reflow cycles was carried out. The ball shear strength of the solder joints for reflow cycles ranging from one to five was 86.15, 81.85, 66.00, 76.20 and 84.79 MPa, respectively. With the increase in reflow cycles, the ball shear strength first decreased and then increased. The shear strength of solder joints reached the lowest value of 66.00 MPa after three reflow cycles. However, the increase in reflow cycles had little effect on the shear strength of the solder joints. In addition, the shear strength of the solder joints was slightly higher than that reported in reference [25]. In the literature [25], the shear strength of the solder joints was about 80 MPa after one reflow cycle, while the shear strength of the solder joints in this work was 86.15 MPa. This was due to the different shear speeds of the solder joints in the measurement. Specifically, in the literature [25], the shear speed of the solder joints was 200 μm/s, while in this work, the shear speed of the solder joints was 500 μm/s. According to the National Military Standard of the People's Republic of China (GJB 7677-2012), Test Methods for Ball Grid Array (BGA), the shear strength of the solder joints increases with the increase of shear test speed. This can well explain the slightly higher shear strength of the solder joints in this work. The ball shear tested samples were analyzed by SEM and the result is shown in Figure 10. It was found that after two reflow cycles, the fracture surface consisted of a solder matrix. After three reflow cycles, the fracture surface consisted completely of an IMC layer, which also explained its lowest shear strength. After five reflow cycles, the fracture surface consisted of solder matrix and IMC layer, and the shear strength was improved compared with that of three reflow cycles. According to the analysis of SEM images, the fracture mode of one and two reflow cycles was ductile fracture, while the fracture of three reflow cycles was brittle fracture, and the fracture of four and five reflow cycles was mixed with ductile and brittle fracture. Therefore, in practical application, the number of reflow cycles shall not exceed three.

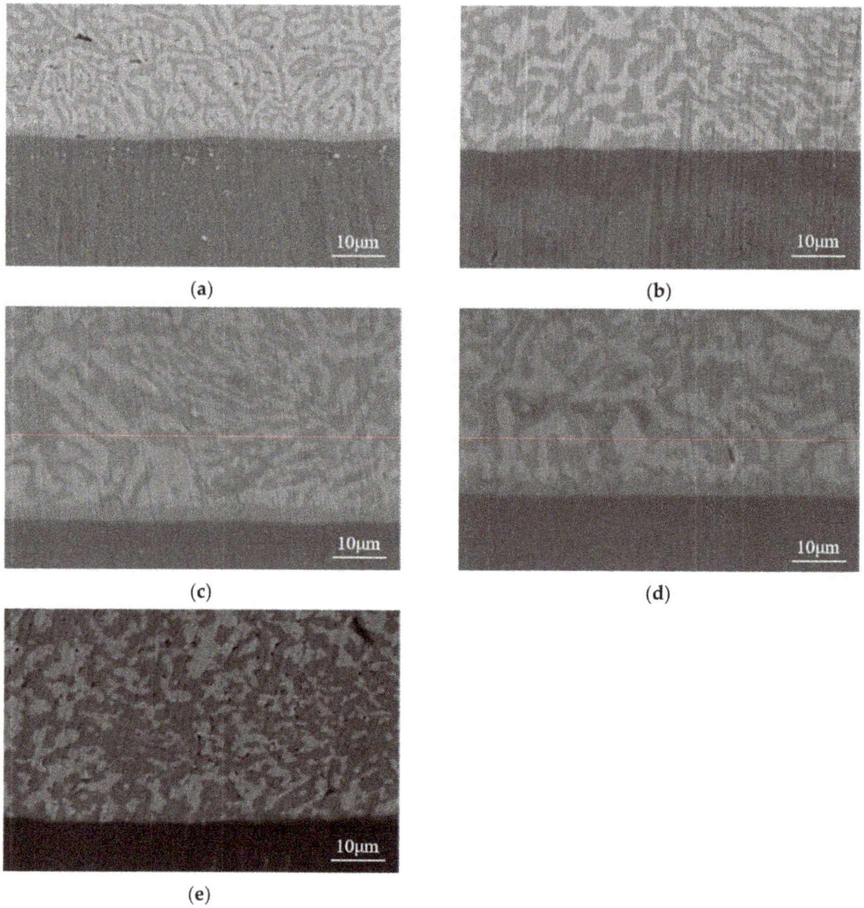

Figure 8. Cross-sectional SEM images of Sn-58Bi/ENEPIG joints with different reflow cycles: (**a**) one reflow cycle (L_{IMC} 1.7 μm), (**b**) two reflow cycles (L_{IMC} 2.3 μm), (**c**) three reflow cycles (L_{IMC} 3.2 μm), (**d**) four reflow cycles (L_{IMC} 3.8 μm) and (**e**) five reflow cycles (L_{IMC} 4.3 μm).

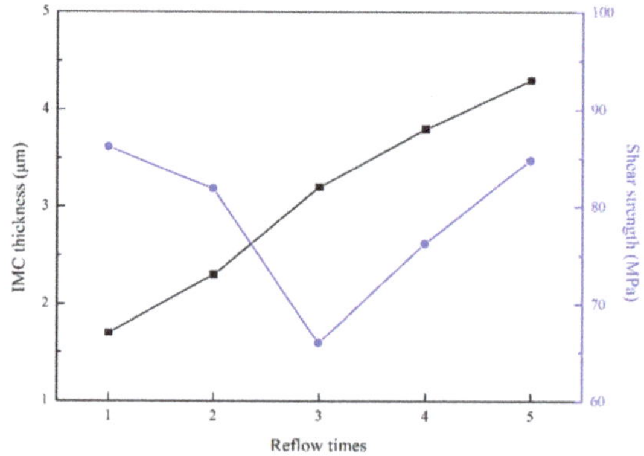

Figure 9. IMC thickness and shear strength of Sn-58Bi/ENEPIG joints as a function of reflow cycles.

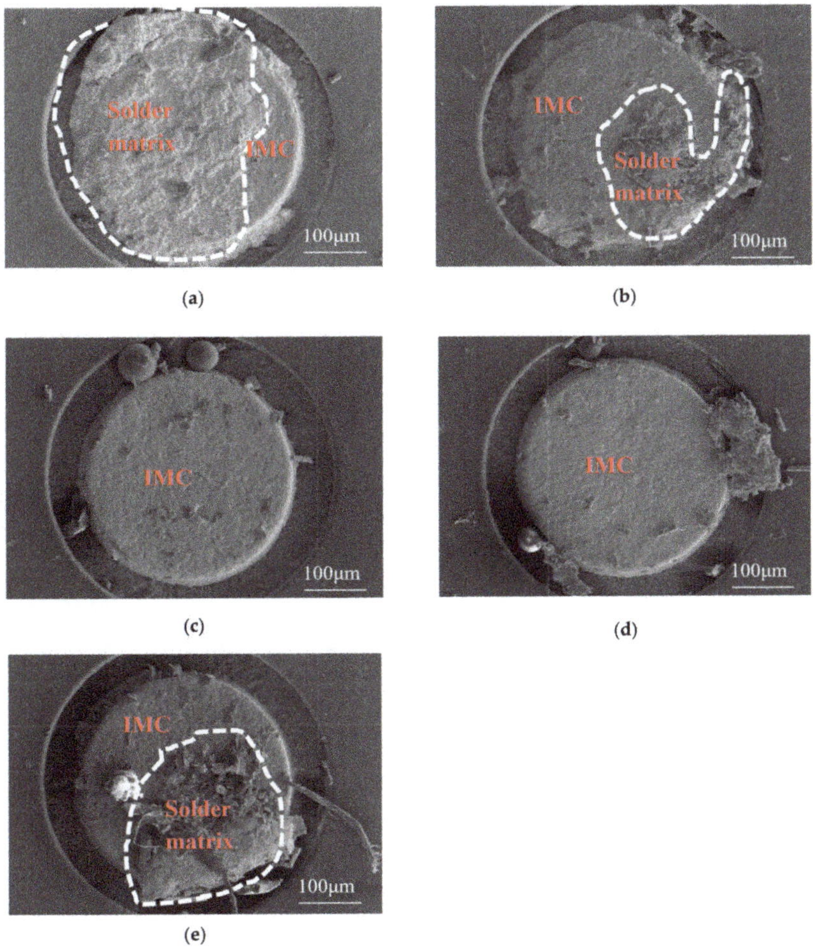

Figure 10. Fracture surfaces of Sn-58Bi/ENEPIG joints with different reflow cycles: (**a**) one reflow cycle, (**b**) two reflow cycles, (**c**) three reflow cycles, (**d**) four reflow cycles and (**e**) five reflow cycles.

3.3. Interfacial Microstructure and Mechanical Properties for Different Aging Times

Figure 11 show the surface morphology and cross-sectional SEM micrographs of the Sn-58Bi/ENEPIG joints aged at −40 °C for 100 h, 200 h, 300 h and 500 h, respectively. Figure 12 shows the relationship between IMC thickness and shear strength of Sn-58Bi/ENEPIG joints with aging time ranging from 100 h to 500 h. The thickness of the IMC layer formed at the interface aged at −40 °C, ranging from 100 to 500 h, was 1.8, 2.0, 2.2 and 2.3 μm, respectively. When the aging time was less than 300 h, the high driving force of IMC growth was provided by high temperature and a large amount of Pd, Au and Sn atoms. The thickness increased from 1.8 μm to 2.2 μm with a higher growth rate of 2×10^{-3} μm/h. When the aging time was more than 300 h, the atom diffusion became slower because of the reduction of Pd, Au and Sn atoms. Under the same aging time, the driving force of IMC growth was lower because of the change in atomic diffusion rate. The thickness increased from 2.2 μm to 2.3 μm, with a growth rate of 0.5×10^{-3} μm/h. In addition, there was no obvious increase in the Bi rich phase or β-Sn phase. Figure 13 shows the fracture surfaces of Sn-58Bi/ENEPIG joints. The ball shear strength of solder joints aged at −40 °C, ranging from 100 h to 500 h, was 83.76, 81.85, 75.72 and 73.49 MPa, respectively. It was found from

Figure 12 that the ball shear strength decreased by about 12.3% with the increase of aging time, which means the extension of aging time has little effect on the strength change of Sn-58Bi/ENEPIG joints and the joints show good mechanical properties at −40 °C. It can be seen from the fracture morphology that the fracture surface consisted of the solder matrix aged for 100 h, and when the aging time increased to 200 h, the fracture surface consisted of the combination of the solder matrix and the IMC layer. Therefore, when the aging time was 100 h, the fracture mode was ductile fracture. When the aging time increased to more than 200 h, the fracture mode was ductile and brittle mixed fracture. As observed from the SEM image of fracture with magnification of 5000 times, the fracture surface of Sn-58Bi/ENEPIG joints was blocked by the Bi-rich phase and β-Sn phase. When the joints were aged for 200 h, the structure of some Bi-rich phases and β-Sn phases was triangular. The fracture surface of Sn-58Bi/ENEPIG joints aged at 300 h and 500 h contained slender dendritic, large flake structures, as well as IMC particles, and the large flake structures coarsen when the aging time increased to 500 h. According to the fracture morphology, it can be concluded that the decrease in ball shear strength was mainly caused by the transition of the fracture location from the solder matrix to the IMC layer.

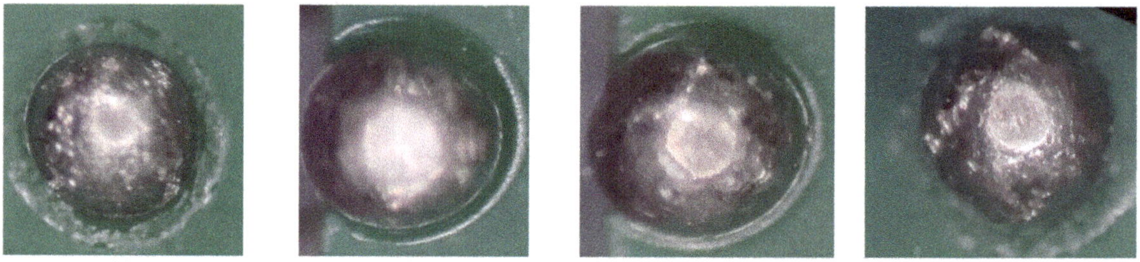

(a) (b) (c) (d)

Figure 11. Surface morphology of solder joints with different aging times: (**a**) 100 h, (**b**) 200 h, (**c**) 300 h and (**d**) 500 h.

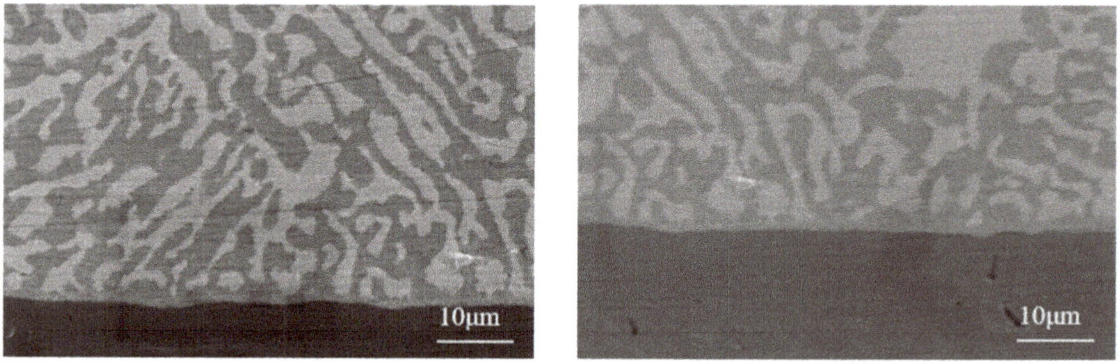

(a) (b)

Figure 11. *Cont.*

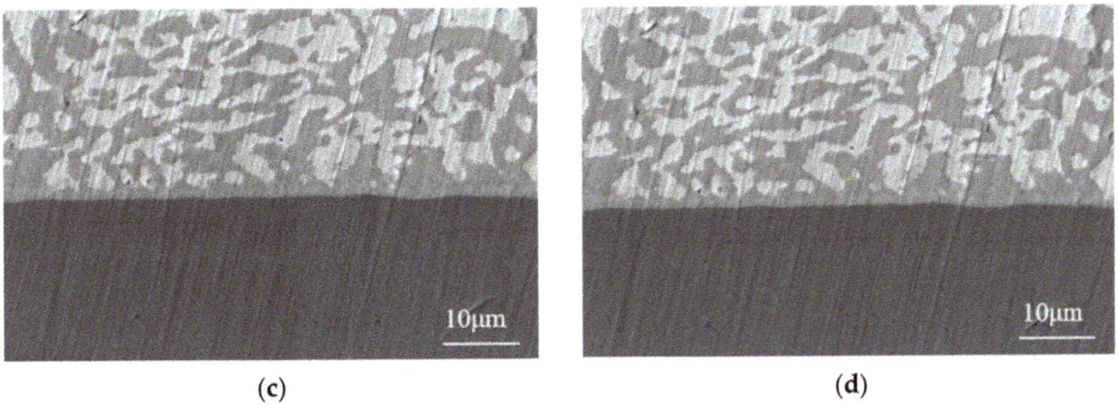

Figure 11. Cross-sectional SEM micrographs of solder joints after aging at −40 °C: (**a**) 100 h (L_{IMC} 1.8 μm), (**b**) 200 h (L_{IMC} 2.0 μm), (**c**) 300 h (L_{IMC} 2.2 μm) and (**d**) 500 h (L_{IMC} 2.3 μm).

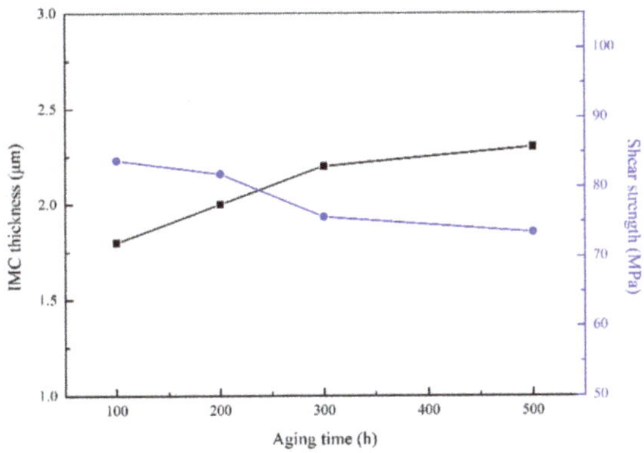

Figure 12. IMC thickness and shear strength of Sn-58Bi/ENEPIG joints with aging time ranging from 100 h to 500 h.

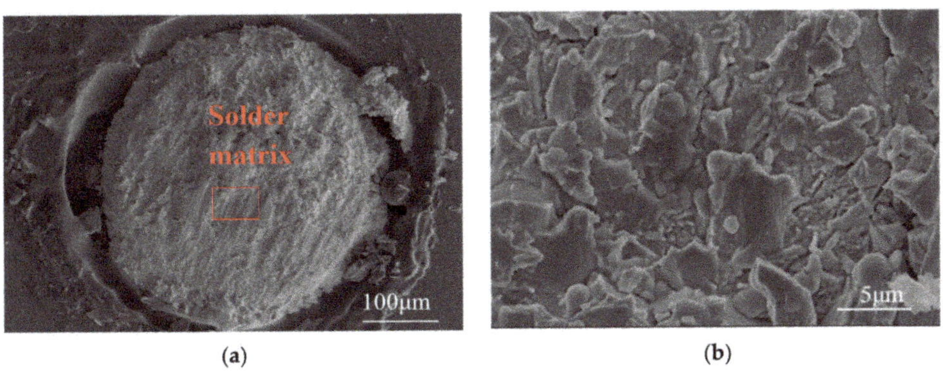

Figure 13. *Cont.*

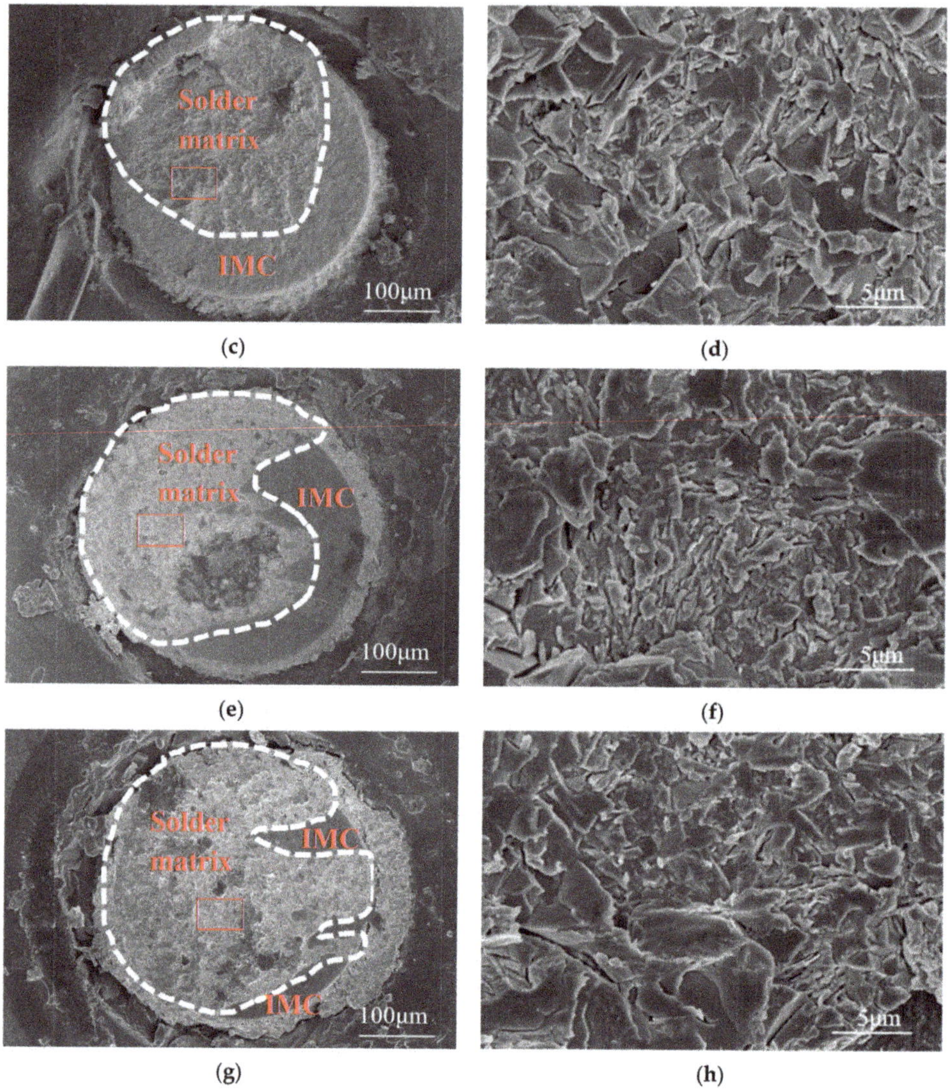

Figure 13. Fracture surfaces of Sn-58Bi/ENEPIG joints with various aging times: (**a**,**b**) 100 h, (**c**,**d**) 200 h, (**e**,**f**) 300 h and (**g**,**h**) 500 h. (**b**) Enlarged view in the frame of (**a**); (**d**) Enlarged view in the frame of (**c**); (**f**) Enlarged view in the frame of (**e**); (**h**) Enlarged view in the frame of (**g**).

4. Conclusions

In this work, the interfacial microstructure and mechanical reliability of the Sn-58Bi/ENEPIG solder joints were systematically studied. The Sn-58Bi solder ball contained an eutectic lamellar microstructure of a β-Sn solid solution and Bi-rich phase. Pd–Au–Sn IMC was formed at the Sn-58Bi/ENEPIG interface. Pd and Au elements in ENEPIG finish with Sn element in Sn-58Bi solder were formed at the Sn-58Bi/ENEPIG interface, and most of the Pd element was dissolved into the molten solder to react at the interface. The remaining thin Pd layer prevented the diffusion of the Ni layer. With the increase in reflow cycles, the surface of the solder joints was wrinkled and blackened, and the oxidation phenomenon was more serious. In addition, the IMC layer grew gradually, and by fitting the data, the relationship between reflow cycles (X) and thickness (Y, μm) of the IMC layer was established as: Y = 0.67X + 1.05 IMC. In the ball shear test, the shear strength first decreased and then increased. After three reflow cycles, the shear strength of solder joints was the lowest. The IMC layer increased to a certain extent, and the ball shear strength slightly decreased after aging at −40 °C for up to 500 h because the low temperature limited the growth of IMC. According to the fracture morphology, the fracture mode transferred from the ductile fracture to a mixture of ductile and brittle fracture when the reflow exceeded three cycles and the aging time exceeded 200 h, owing to the fact that the fracture location transferred from the solder matrix to the IMC interface. This work provides a vital understanding of the interfacial microstructure and mechanical reliability of the Sn-58Bi/ENEPIG solder joints at low temperatures.

Author Contributions: Conceptualization, C.C. and C.W.; methodology, C.C., H.S. and H.X.; software, H.X. and D.N.; validation, C.C. and Q.G.; formal analysis, C.C., H.Y. and X.G.; investigation, C.C., H.S. and Q.G.; resources, C.C., H.Y. and D.N.; data curation, C.C. and C.W.; writing—original draft preparation, C.C., C.W., H.S. and H.X.; writing—review and editing, C.C., C.W., H.S., H.X. and K.B.; visualization, C.C., D.N. and Q.G.; supervision, C.C., H.Y. and Q.G.; project administration, C.C. and Q.G. All authors have read and agreed to the published version of the manuscript.

Funding: This research received no external funding.

Institutional Review Board Statement: Not applicable.

Informed Consent Statement: Not applicable.

Data Availability Statement: Not applicable.

Conflicts of Interest: The authors declare no conflict of interest.

References

1. Adetunji, O.R.; Ashimolowo, R.A.; Aiyedun, P.O.; Adesusi, O.M.; Adeyemi, H.O.; Oloyede, O.R. Tensile, hardness and microstructural properties of Sn-Pb solder alloys. *Mater. Today Proc.* **2021**, *41*, 321–325. [CrossRef]
2. Lee, J.H.; Kim, Y.S. Kinetics of intermetallic formation at Sn-37Pb/Cu interface during reflow soldering. *J. Electron. Mater.* **2002**, *35*, 576–583. [CrossRef]
3. Chen, W.X.; Xue, S.B.; Wang, H. Wetting properties and interfacial microstructures of Sn–Zn– xGa solders on Cu substrate. *Mater. Des.* **2010**, *31*, 2196–2200. [CrossRef]
4. Silva, B.L.; da Silva, V.C.E.; Garcia, A.; Spinelli, J.E. Effects of Solidification Thermal Parameters on Microstructure and Mechanical Properties of Sn-Bi Solder Alloys. *J. Electron. Mater.* **2017**, *46*, 1754–1769. [CrossRef]
5. Zhang, L.; Xue, S.B.; Gao, L.L.; Zeng, G.; Yu, S.L.; Sheng, Z. Thermal Fatigue Behavior of SnAgCu Soldered Joints in Fine Pitch Devices. *J. Rare Met. Mater. Eng.* **2010**, *39*, 382–387.
6. Ye, H.; Xue, S.B.; Luo, J.D.; Li, Y. Properties and interfacial microstructure of Sn–Zn–Ga solder joint with rare earth Pr addition. *Mater. Des.* **2013**, *46*, 816–823. [CrossRef]
7. Gain, A.K.; Zhang, L.C. Thermal aging effects on microstructure, elastic property and damping characteristic of a eutectic Sn–3.5Ag solder. *J. Mater. Sci. Mater. Electron.* **2018**, *29*, 14519–14527. [CrossRef]
8. Zhou, S.; Mokhtari, O.; Rafique, M.G.; Shunmugasamy, V.C.; Mansoor, B.; Nishikawa, H. Improvement in the mechanical properties of eutectic Sn58Bi alloy by 0.5 and 1 wt% Zn addition before and after thermal aging. *J. Alloys Compd.* **2018**, *765*, 1243–1252. [CrossRef]
9. Shalaby, R.M. Effect of silver and indium addition on mechanical properties and indentation creep behavior of rapidly solidified Bi–Sn based lead-free solder alloys. *Mater. Sci. Eng. A* **2013**, *560*, 86–95. [CrossRef]
10. Kang, H.; Rajendran, S.H.; Jung, J.P. Low Melting Temperature Sn-Bi Solder: Effect of Alloying and Nanoparticle Addition on the Microstructural, Thermal, Interfacial Bonding, and Mechanical Characteristics. *Metals* **2021**, *11*, 364. [CrossRef]
11. Shen, C.; Hai, Z.; Zhao, C.; Zhang, J.; Evans, J.L.; Bozack, M.J.; Suhling, J.C. Packaging Reliability Effect of ENIG and ENEPIG Surface Finishes in Board Level Thermal Test under Long-Term Aging and Cycling. *Metals* **2017**, *10*, 451. [CrossRef]
12. Liu, B.; Tian, Y.; Liu, W.; Wu, W.; Wang, C. TEM observation of interfacial compounds of SnAgCu/ENIG solder bump after laser soldering and subsequent hot air reflows. *Mater. Lett.* **2016**, *163*, 254–257. [CrossRef]
13. Sauli, Z.; Retnasamy, V.; Fuad, F.A.A.; Ehkan, P.; Ab Aziz, M.H. Relationship between Controllable Process Parameters on Bump Height in ENIG. *Appl. Mech. Mater.* **2013**, *404*, 62–66. [CrossRef]
14. Ho, C.E.; Fan, C.W.; Hsieh, W.Z. Pronounced effects of Ni(P) thickness on the interfacial reaction and highimpact resistance of the solder/Au/Pd(P)/Ni(P)/Cu reactive system. *Surf. Coat. Technol.* **2014**, *259*, 244–251. [CrossRef]
15. Wu, Y.S.; Lee, P.T.; Huang, Y.H.; Kuo, T.T.; Ho, C.E. Interfacial microstructure and mechanical reliability of the Sn-Ag-Cu/Au/Pd(xP)/Ni(P) reactive system: P content effects. *Surf. Coat. Technol.* **2018**, *350*, 874–879. [CrossRef]

16. Braga, M.H.; Vizdal, J.; Kroupa, A.; Ferreira, J.; Soares, D.; Malheiros, L.F. The experimental study of the Bi–Sn, Bi–Zn and Bi–Sn–Zn systems. *Calphad* **2007**, *31*, 468–478. [CrossRef]
17. Kim, J.; Jung, S.B.; Yoon, J.W. Effects of crystalline and amorphous Pd layers on initial interfacial reactions at Sn-3.0Ag-0.5Cu/thin-Au/Pd/Ni(P) solder joints. *Appl. Surf. Sci.* **2020**, *503*, 144339. [CrossRef]
18. Yoon, J.W.; Back, J.H.; Jung, S.B. Comparative study of ENEPIG and thin ENEPIG as surface finishes for SAC305 solder joints. *J. Mater. Sci. Mater. Electron.* **2018**, *29*, 4724–4731. [CrossRef]
19. Lee, C.Y.; Yang, S.P.; Yang, C.H.; Lu, M.K.; Kuo, T.T.; Ho, C.E. Influence of Pd(P) thickness on the Pd-free solder reaction between eutectic Sn-Ag alloy and Au/Pd(P)/Ni(P)/Cu multilayer. *Surf. Coat. Technol.* **2020**, *395*, 125879. [CrossRef]
20. Yoon, J.W.; Lee, C.B.; Jung, S.B. Interfacial Reactions Between Sn-58 mass% Bi Eutectic Solder and (Cu, Electroless Ni-P/Cu) Substrate. *Mater. Trans.* **2002**, *43*, 1821–1826. [CrossRef]
21. Pun, K.P.; Dhaka, N.S.; Cheung, C.W.; Chan, A.H. Effect of ENEPIG metallization for solid-state gold-gold diffusion bonds. *Microelectron. Reliab.* **2017**, *78*, 339–348. [CrossRef]
22. Chi, C.C.; Tsao, L.C.; Tsao, C.W.; Chuang, T.H. Intermetallic Reactions in Reflowed and Aged Sn-58Bi BGA Packages with Au/Ni/Cu Pads. *J. Mater. Eng. Perform.* **2008**, *17*, 134–140. [CrossRef]
23. Tian, R.; Tian, Y.; Huang, Y.; Yang, D.; Chen, C.; Sun, H. Comparative study between the Sn–Ag–Cu/ENIG and Sn–Ag–Cu/ENEPIG solder joints under extreme temperature thermal shock. *J. Mater. Sci. Mater. Electron.* **2021**, *32*, 6890–6899. [CrossRef]
24. Chen, C.; Ho, C.E.; Lin, A.H.; Luo, G.L.; Kao, C.R. Long-term aging study on the solid-state reaction between 58Bi42Sn solder and Ni substrate. *J. Electron. Mater.* **2000**, *29*, 1200–1206. [CrossRef]
25. Kim, J.; Jung, S.B.; Yoon, J.W. Effects of a phosphorous-containing Pd layer in a thin-ENEPIG surface finish on the interfacial reactions and mechanical strength of a Sn–58Bi solder joint. *J. Alloys Compd.* **2020**, *820*, 153396. [CrossRef]
26. Wu, W.; Lin, W.H. Study on Bonding Reliability of Mixed Multichip Microwave Modules. *Electron. Process Technol.* **2017**, *38*, 141–143.

processes MDPI

Article

Evaluation of Landsat 8-like Land Surface Temperature by Fusing Landsat 8 and MODIS Land Surface Temperature Product

Shenglin Li [1], Jinglei Wang [1,*], Dacheng Li [2], Zhongxin Ran [3] and Bo Yang [1]

1 Farmland Irrigation Research Institute, Chinese Academy of Agricultural Sciences, Xinxiang 453002, China; shenglinlirs@163.com (S.L.); yangbo01@caas.cn (B.Y.)
2 Department of Surveying and Mapping, Taiyuan University of Technology, Taiyuan 030024, China; lidacheng@tyut.edu.cn
3 Hanjiang River Bureau of Hydrology and Water Resources Survey, Xiangyang 441000, China; ranzhongxin123@126.com
* Correspondence: firiwjl@126.com

Abstract: High-spatiotemporal-resolution land surface temperature (LST) is a crucial parameter in various environmental monitoring. However, due to the limitation of sensor trade-off between the spatial and temporal resolutions, such data are still unavailable. Therefore, the generation and verification of such data are of great value. The spatiotemporal fusion algorithm, which can be used to improve the spatiotemporal resolution, is widely used in Landsat and MODIS data to generate Landsat-like images, but there is less exploration of combining long-time series MODIS LST and Landsat 8 LST product to generate Landsat 8-like LST. The purpose of this study is to evaluate the accuracy of the long-time series Landsat 8 LST product and the Landsat 8-like LST generated by spatiotemporal fusion. In this study, based on the Landsat 8 LST product and MODIS LST product, Landsat 8-like LST is generated using Spatial and Temporal Adaptive Reflectance Fusion Model (STARFM), Enhanced STARFM (ESTARFM), and the Flexible Spatiotemporal DAta Fusion (FSDAF) algorithm, and tested and verified in the research area located in Gansu Province, China. In this process, Landsat 8 LST product was verified based on ground measurements, and the fusion results were comprehensively evaluated based on ground measurements and actual Landsat 8 LST images. Ground measurements verification indicated that Landsat 8 LST product was highly consistent with ground measurements. The Root Mean Square Error (RMSE) was 2.862 K, and the coefficient of determination R^2 was 0.952 at All stations. Good fusion results can be obtained for the three spatiotemporal algorithms, and the ground measurements verified at All stations show that R^2 was more significant than 0.911. ESTARFM had the best fusion result (R^2 = 0.915, RMSE = 3.661 K), which was better than STARFM (R^2 = 0.911, RMSE = 3.746 K) and FSDAF (R^2 = 0.912, RMSE = 3.786 K). Based on the actual Landsat 8 LST images verification, the fusion images were highly consistent with actual Landsat 8 LST images. The average RMSE of fusion images about STARFM, ESTARFM, and FSDAF were 2.608 K, 2.245 K, and 2.565 K, respectively, and ESTARFM is better than STARFM and FSDAF in most cases. Combining the above verification, the fusion results of the three algorithms were reliable and ESTARFM had the highest fusion accuracy.

Keywords: Landsat 8 LST product; MODIS; spatiotemporal fusion; validation

Citation: Li, S.; Wang, J.; Li, D.; Ran, Z.; Yang, B. Evaluation of Landsat 8-like Land Surface Temperature by Fusing Landsat 8 and MODIS Land Surface Temperature Product. *Processes* **2021**, *9*, 2262. https://doi.org/10.3390/pr9122262

Academic Editors: Roberta Campardelli and Paolo Trucillo

Received: 9 November 2021
Accepted: 14 December 2021
Published: 15 December 2021

1. Introduction

Land surface temperature (LST) is a key variable of the surface energy process, hydrological balance, and climate change [1–5]. Accurate measurement and the estimation of LST are of great significance to many research fields, such as weather forecasting, ocean circulation, drought monitoring, and energy balance [6]. At present, thermal infrared remote sensing is almost the only method for regional monitoring of LST.

However, due to the mutual restriction between spatial and temporal resolution, no satellite system can provide a thermal band or LST data with high spatial and temporal resolution at the same time [7]. Sensors with good spatial resolution show poor temporal resolution and vice versa [8]. For example, MODIS provides LST at a resolution of 1 km per day, while polar-orbiting Landsat 8 100 m every 16 days. In order to meet the fine-scale monitoring, it is necessary to combine these two types of data to obtain high spatiotemporal resolution LST.

To solve this problem, methods such as downscaling [9,10], sharpening [11,12], image fusion [13–15], and disaggregation [7,8] were proposed to improve the LST resolution. These methods can be divided into two categories: kernel-driven method and fusion method. The former downscales the LST through the auxiliary data of the multispectral sensor, while the latter can predict the LST of fine resolution by integrating different sensors' time change information and neighborhood information. Among them, the kernel-driven method only improves the spatial resolution without improving the temporal resolution of the sensor at the same time [15], while the fusion method improves both the temporal and spatial resolution and has been developed rapidly in the past decade [16].

The most widely used spatiotemporal fusion method is perhaps the Spatial and Temporal Adaptive Reflectance Fusion Model (STARFM) [17], which has been widely used [18–20]. This method assumes that land cover types and systematic errors do not change between the start time and target time, and a coarse pixel contains only one land cover type, Landsat reflectance images of the prediction period is generated using MODIS and Landsat reflectance images of the same period combined with MODIS images of the prediction period. Due to the above assumptions, STARFM is not feasible in heterogeneous regions. Therefore, STARFM has been improved for more complex surface conditions, resulting in Spatial Temporal Adaptive Algorithm for mapping Reflectance Change (STAARCH) [21] and Enhanced STARFM (ESTARFM) [22], which respectively improves the performance of STARFM in the land cover type change and disturbance existence, and STARFM's accuracy in heterogeneous areas. Huang et al. [23,24] proposed dictionary-pair learning methods. Although the pixels of land change can be better predicted, the shape of the ground objects cannot be maintained. Aiming at the limitations of existing spatiotemporal fusion algorithms, Zhu et al. proposed the Flexible Spatiotemporal DAta Fusion (FSDAF) [25] algorithm, which combines temporal and spatial predictions to capture gradually and suddenly land cover type changes with the need for only one high spatial resolution image. The current research and application of the FSDAF algorithm is mainly based on surface reflectance data [26–29].

Although these spatiotemporal fusion methods were originally designed for reflectance, they have also been widely used in temperature fusion. Li et al. [30] and Ma et al. [31] used the STARFM and ESTARFM algorithm to generate thermal band data for the study of evapotranspiration. Yang et al. [32] generated ASTER surface temperature based on the ESTARFM algorithm and verified it based on the ground measurements. According to the STARFM framework, Huang [33] considered the thermal effects of adjacent pixels and used a bilateral filtering algorithm to construct a new weighting function for the fusion model to monitor the urban thermal environment. Weng et al. [15] proposed the Spatial-temporal Adaptive Data Fusion Algorithm for Temperature mapping (SADFAT) algorithm for LST prediction based on the STARFM algorithm framework, considering the annual temperature cycle (ATC) and thermal landscape heterogeneity. These STARFM-like algorithms are mainly applied to two sensors. For temperature data fusion of three or more sensors, Wu et al. [13] proposed the Spatiotemporal Integrated Temperature Fusion Model (STITFM) algorithm based on the STARFM algorithm, analyzed the fusion of geostationary satellite, Landsat, and MODIS LST. However, similar to STARFM, STITFM is poorly adaptable to areas with complex surface heterogeneity and land cover type changes. Quan et al. [14] proposed a temperature fusion algorithm—Blend Spatiotemporal Temperatures (BLEST), which can be used for heterogeneous regions and land cover type changes by integrating the characteristics of existing spatiotemporal fusion algorithms [15,17,21,25], and combined Landsat, MODIS, and geostationary satellite to generate Landsat-like LST at 1 h interval.

Although the spatiotemporal fusion algorithm is widely used in the generation of high spatiotemporal resolution temperature data, however, there are still some issues that need to be explored. On one hand, previous studies mainly focused on the improvement and application of algorithms and lacked ground verification of long-time sequence fusion results. The reliability of the fusion results requires long-term sequence verification. On the other hand, in previous studies, researchers have inverted Landsat 8 LST based on thermal band data, and the steps are cumbersome. Using the standard Landsat 8 LST product can solve this problem, but there is no relevant report so far, so the reliability of using it in fusion needs to be studied. In response to the above problems, this study used MODIS LST product and Landsat 8 LST product to participate in the fusion, and used three common spatiotemporal fusion algorithms, i.e., STARFM, ESTARFM, and FSDAF, to generate long-time sequence data of 28 scenes in the study area from 2013 to 2016, and verified and analyzed the fusion results based on ground measurements LST. Since Landsat 8 LST product is rarely used and verified at present, the accuracy of the Landsat 8 LST product is also evaluated in order to have a more objective analysis of fusion results. Satellite data and ground measurement data are introduced in Section 2. In Section 3, the temperature inversion of the Landsat 8 LST product, the theoretical basis of the spatiotemporal fusion method, the fusion scheme, and the evaluation strategy are described. The evaluation results are detailed in Section 4. Finally, the discussion and conclusions of this study are summarized in Sections 5 and 6.

2. Study Area and Data

2.1. Study Area

The study area is located in Zhangye City, Gansu Province, China, which belongs to the midstream of the Heihe River Basin, with a geographic range of 38°43′15.60″–39°02′16.8″ N, 100°17′45.59″–100°39′21.59″ E. The climate in this area is arid, with an average annual temperature of 7 °C, and the area has obvious characteristics of the four seasons. The average temperature in summer (June to August) is 20 °C; the winter is cold (December to February) with an average temperature of −8 °C; the temperature rises faster in spring (March to May) with an average temperature of 9 °C; and in autumn (September to November), the temperature drops quickly with an average temperature of 7 °C. Here July is the hottest and January is the coldest. The area is flat and the primary land cover is farmland, Gobi, sandy land, forest, water area, and buildings, most of which are farmland and Gobi. The main crops grown in farmland are spring wheat, corn, barley, flax, and so on. The location of the study area and stations is shown in Figure 1.

2.2. Data

2.2.1. MODIS and Landsat 8 Data

The Landsat 8 LST product from Landsat 8 collection 2 level-2 data (downloaded from https://earthexplorer.usgs.gov/ accessed on 9 November 2021). in this study has an overpass time, which is approximately 10:55 a.m. at local solar time, and a revisiting period of 16 days. The spatial resolution of the TIR channel is 100 m (note: downloaded images are provided with 30 m), with 30 pieces of Landsat 8 scene data of Path 133, Row 33 without cloud collected in the study area from 2013 to 2016. The Landsat 8 LST product provides a basis for the wide application of the Landsat8 thermal band. The corresponding daily MODIS datasets (downloaded from https://ladsweb.modaps.eosdis.nasa.gov/search/ accessed on 9 November 2021) include MOD11A1 (1 km MODIS LST product) and MOD05_L2. MOD11A1 was retrieved by a generalized split-window algorithm [34], and in most cases, the accuracy was reported within 2.0 K [35]. The MOD05_L2 product contains 1 and 5 km atmospheric column water vapor (CWV) [36]; in this study, 1 km CWV data generated by the near-infrared algorithm are used. The overpass time is approximately 11:40 a.m. at the local solar time for MOD11A1 and MOD05_L2 in this study area.

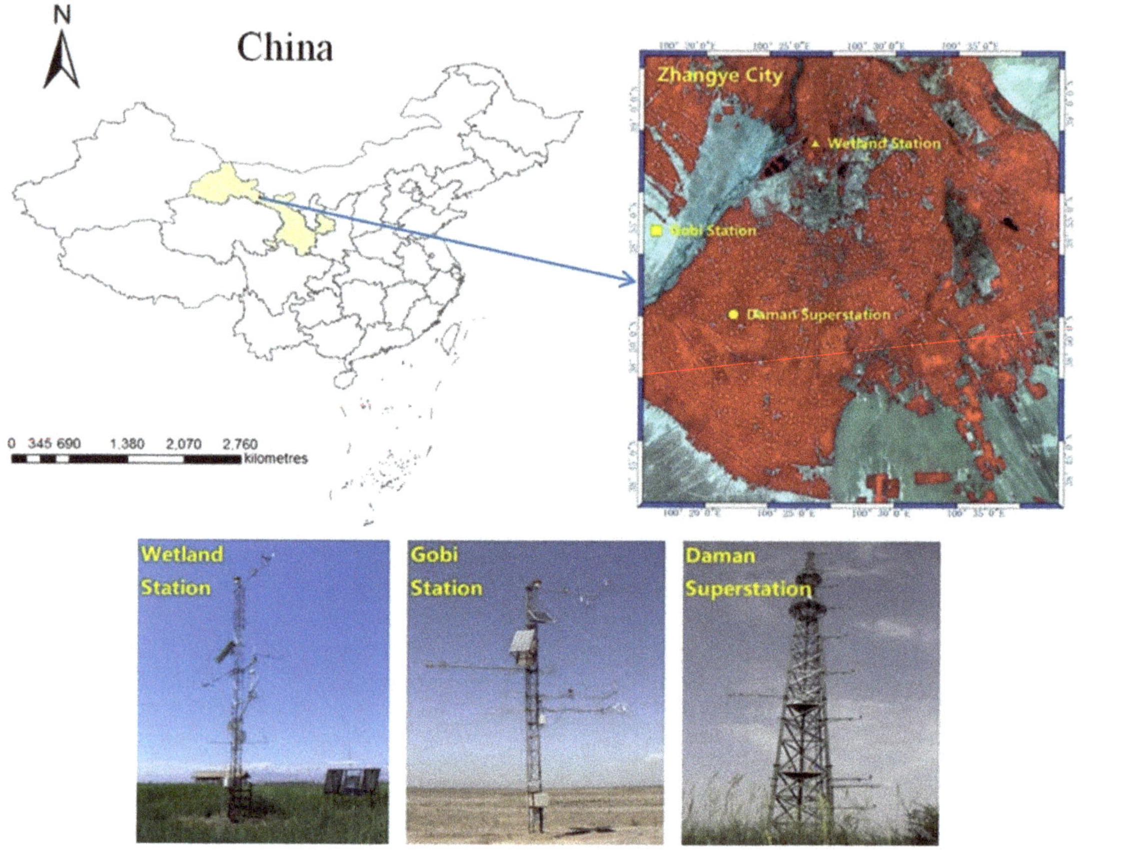

Figure 1. Location of the study area and stations' spatial distribution.

Before implementing the fusion method, both the Landsat 8 and MODIS data are registered to the same coordinate system and resampled to the exact spatial resolution (using a nearest-neighbor method). Since Landsat 8 LST product is provided at 30 m, the MODIS is resampled to 30 m resolution (note: the actual resolution of Landsat 8 LST is 100 m).

2.2.2. Ground Measurements LST

The ground measurements are from HiWATER: Dataset of hydrometeorological observation network [37,38]. The observation data of three automatic weather stations-Zhangye wetland station, Bajitan Gobi station, and Daman superstation-are used in this study. The details of the three verification stations are shown in Table 1. The observation frequency of the automatic weather station is 30 s, and the output is average data of 10 min. According to the Landsat 8 satellite transit time, data within 10 min are linearly interpolated to estimate the measured value of the satellite transit time. Surface radiometric temperature T_r is observed by SI-111 radiometers (8–14 um) installed in the automatic weather station. The true land surface temperature T_S can be obtained by Equation (1):

$$B(T_S) = \left(B(T_r) - (1-\varepsilon)L_{sky}\right)/\varepsilon \quad (1)$$

where B is the Planck function, ε the surface emissivity of the SI-111 channel calculated by ASTER GEDv3, and L_{sky} the atmospheric downward radiation measured by the SI-111.

Table 1. Verifying station information.

Station Name	Geolocation	Altitude/m	Land Cover	Duration
Wetland	100.4464° E 35.9751° N	1460	Reed	June 2012–Present
Gobi	100.3042° E 38.9150° N	1562	Gobi	May 2012–April 2015
Daman	100.3722° E 38.8555° N	1556	Maize	May 2012–Present

3. Methodology

3.1. Landsat 8 LST Retrieval

The global standard Landsat 8 LST product is used in this study, which is based on the inversion of TIR10 and has been used since mid-2020. Firstly, the temperature retrieval process involves calculating ground radiance L_s using emissivity, atmospheric parameters, and the observed Landsat 8 radiance for band 10, and then inverting the Planck function.

$$L_s = B(T_s) = \frac{\frac{L-L^{\uparrow}}{\tau} - (1-\varepsilon)L^{\downarrow}}{\varepsilon} \tag{2}$$

$$B(T_s) = \frac{C_1\lambda^{-5}}{\left(e^{\frac{C_2}{\lambda T_s}} - 1\right)} \tag{3}$$

where T_s is the LST, $B(T_s)$ the Planck function at temperature T_s; L represents the Top of the Atmosphere (TOA) radiance received by the TIR sensor, ε the surface emissivity, $L^{\uparrow}$, $L^{\downarrow}$ and τ upwelling atmospheric radiance, downwelling atmospheric radiance, and atmospheric transmittance, respectively, and λ the effective band wavelength; C_1 and C_2 are Planck's radiation constants with value 1.19104×10^8 W·μm^4·m^{-2}·sr^{-1} and 14,387.7 μm·K, respectively.

Theoretically, the surface temperature T_s can be obtained by inverting the Planck function in Formula (3) and combining L_s in Formula (2) as follows:

$$T_s = \frac{C_2}{\lambda ln\left(1 + \frac{C_1\lambda^{-5}}{L_s}\right)} \tag{4}$$

However, this formulation will increasingly become inaccurate for a sensor's spectral response that deviates from delta function behavior. Instead, use a lookup table (LUT) to calculate Landsat 8 LST. Calculate the expected radiances of the spectral response of Landsat 8 sensor in the temperature range of 0.01 K, which encompasses all expected Earthlike temperatures (typically 150–380 K). This results in a table of values of radiances versus temperatures for Landsat 8 sensor. Landsat 8 LST can be retrieved by using the ground radiance and the table. For a more detailed description of the Landsat 8 LST product, please refer to Malakar et al. (2018) [39].

3.2. Introduction of Spatial Temporal Fusion Models

The STARFM [17] algorithm is the most widely used spatiotemporal fusion algorithm and uses a pair of high/low-resolution images at the reference time and a low-resolution image at the target time to fuse. The moving window and similar pixels are introduced during fusion to reduce the influence of the pixel boundary of low spatial resolution images, using similar pixels in the window and comprehensively considering the spatial weight, spectral weight and temporal weight to calculate the value of the center pixel in the window, and finally predicting the high-resolution image of the target time through the movement of the moving window.

The ESTARFM [22] algorithm, which is improved on the basis of STARFM, introduces a conversion coefficient to improve the accuracy of STARFM in heterogeneous regions. When the ESTARFM model is fused, it is necessary to input two pairs of high/low-resolution images at the reference time and a low-resolution image at the target time. Similar to STARFM, ESTARFM still uses the moving window technology, which uses similar pixels in the window and comprehensively considers the weight function to solve the value of the center pixel to realize the prediction of the high-resolution image at the target time.

The FSDAF [25] algorithm is an unmixing-based fusion model, which can predict the gradual changes and mutations of land cover types. The input data of the algorithm are the same as those of STARFM. First, the time change information is obtained through the pure pixels of the low-resolution images at time t_1 and t_2 that have not changed the ground cover, and then the time-predicted high-resolution image is obtained. The spatially-predicted high-resolution image is obtained by increasing the image resolution through the thin-plate spline interpolation function so as to obtain the residual R of each lower resolution pixel on the prediction date. Finally, the distribution weight function is used to assign the residuals, and this function generates the final target image by using similar pixels similar to STARFM and ESTARFM to reduce the blocking effect.

3.3. Fusion Scheme

The following two kinds of data need to be input for temperature fusion: reference images, namely Landsat 8 LST and MODIS LST images of the same date, and MODIS LST image of the predicted date. When the predicted date is determined, the selection of the reference images has a great impact on the prediction results. Generally, the date adjacent to the predicted date is used as the reference image date under which circumstance the temperature change, surface coverage type change, and sensor difference change between reference and predicted images are small, which is helpful to improve the prediction accuracy. A pair of reference images are required for STARFM and FSDAF, however, for the predicted date, there are two adjacent dates, involving two pairs of reference images. The dates of the reference images are selected by comparing the coefficient of determination (R^2) between the MODIS LST images of the two dates and the MODIS LST images at the predicted date, and the date with higher correlation is used as the reference image date. Two pairs of reference images are required for ESTARFM, and the adjacent dates of the predicted date are directly selected. The fusion scheme is shown in Figure 2. T1 represents the reference image date of STARFM and FSDAF, T1 and T3 ESTARFM, and T2 represents the fusion image date.

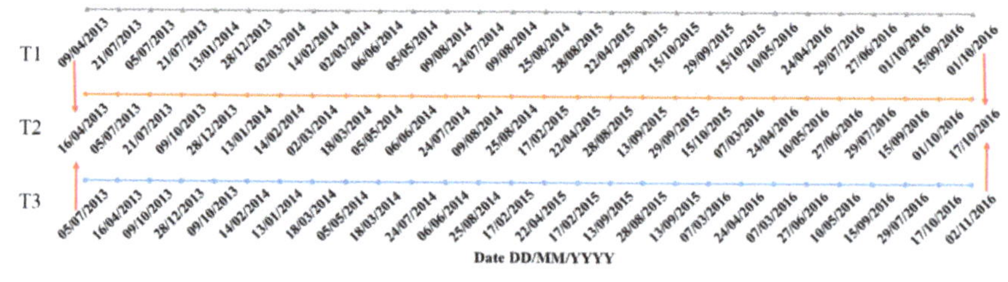

Figure 2. The scheme for estimation of Landsat 8-like LST by STARFM, ESTARFM, and FSDAF.

3.4. Verification and Evaluation

The Landsat 8 LST product is verified based on ground measurements, and the fusion results are comprehensively evaluated based on the ground measurements and the actual Landsat 8 LST images. The prediction accuracy is evaluated with the coefficient of determination (R^2), mean difference (bias), mean absolute error (MAE), and root mean square error (RMSE). The statistical metrics are given as follows:

$$bias = \frac{1}{m}\sum_{i=1}^{m}(P_i - O_i) \tag{5}$$

$$MAE = \frac{1}{m}\sum_{i=1}^{m}|P_i - O_i| \tag{6}$$

$$RMSE = \sqrt{\frac{\sum_{i=1}^{m}(P_i - O_i)^2}{m}} \tag{7}$$

$$R^2 = \left(\frac{\sum(P_i - \overline{P})(O_i - \overline{O})}{\sqrt{\sum(P_i - \overline{P})^2(O_i - \overline{O})^2}}\right)^2 \tag{8}$$

where m is the number of observations, O_i and P_i are the ith verified and estimated LST, respectively.

4. Results

4.1. Test Landsat 8 LST with Ground Measurements

The Landsat 8 LST product became available in mid-2020. At present, few applications and verifications are available for this product, and when generating Landsat-like LST based on the spatiotemporal fusion algorithm, many scholars [13–15,40] used single-channel (SC) algorithm proposed by Jiménez-Muñoz et al. [41–43] to estimates the Landsat LST. Therefore, it is necessary to verify the Landsat 8 LST product before fusion. In this study, Landsat 8 LST product was verified based on ground measurements and compared with the SC algorithm. In the SC algorithm for Landsat 8, which is not described here, the emissivity comes from the emissivity band in Landsat 8 Collection 2 Level-2 data, and the water vapor comes from the MOD05_L2 product. For the detailed algorithm flow, please refer to Jiménez-Muñoz et al. (2014) [43].

It is necessary to analyze the heterogeneity of the ground station before using ground measurements LST. Generally, 3 × 3 pixel LST around the station is counted due to the Landsat 8 LST product is provided with the resolution resampled to 30 m, but the actual resolution is 100 m, so 9 × 9 pixel (270 m) around the station was counted in this study, and the average standard deviation of all 30 Landsat 8 LST scene images was calculated as the basis for site heterogeneity analysis. The statistical results show that the average standard deviations of the Wetland station, Gobi station, and Daman superstation are 0.864 K, 0.411 K, and 0.485 K, respectively, which are all less than 1 K. It meets the requirement that the uncertainty of the ideal surface temperature verification site should be within 1 K, as pointed out by Wan et al. [44].

According to the ground measurements at different sites, the bias, MAE, and RMSE of the SC LST and LST product are listed in Table 2. At the Wetland and Gobi station, LST product RMSE are 2.916 K and 3.776 K, respectively, which are smaller than that of SC LST (3.219 K and 3.851 K, respectively). At Daman superstation, the two types of LST RMSE are close to 2 K with 2.157 K of LST product and 2.007 K of SC LST. All stations represent the comprehensive verification results of the three stations; the LST accuracy of these two algorithms in All stations is within 3 K, and the accuracy of the LST product is slightly better than that of the SC LST. The comparison between SC LST and LST product is shown in Figure 3, the two types of LST are highly linearly related to ground-measured LST with R^2 greater than 0.953, and the R^2 of LST product and SC LST are very close at each site. The histogram distribution of the LST differences between retrieved and measured LST shows that for the two types of LST, most biases are between −3 K and 3 K, and the overall bias of LST product and SC LST at All stations are −0.760 K and −1.284 K, respectively. LST product has reliable accuracy and is suitable for input data in spatiotemporal fusion based on the above verification.

Table 2. The bias (K), MAE (K), and RMSE (K) of retrieved LST over stations.

Station Name	LST Product			SC		
	Bias	MAE	RMSE	Bias	MAE	RMSE
Wetland	−1.602	2.195	2.916	−2.100	2.477	3.219
Gobi	−1.254	2.748	3.776	−1.682	2.805	3.851
Daman	0.470	1.688	2.157	−0.139	1.499	2.007
All	−0.760	2.160	2.862	−1.284	2.207	2.967

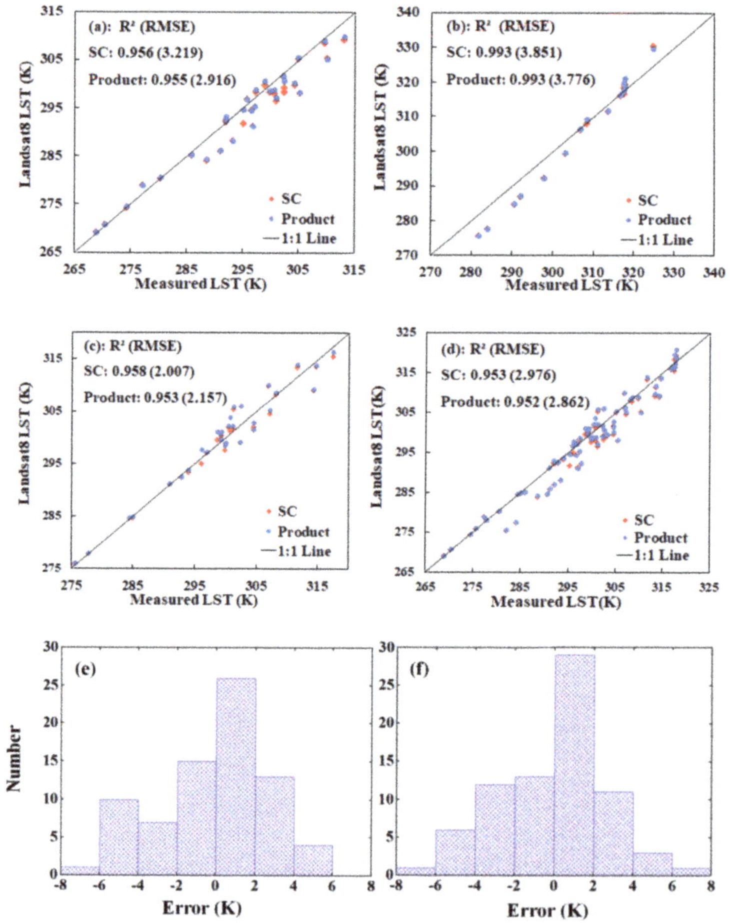

Figure 3. Scatter plot and error histogram of retrieved LST and measured LST. (**a**): Wetland station; (**b**): Gobi station; (**c**): Daman superstation; (**d**): All stations; (**e**): LST product error histogram at All stations; (**f**): SC error histogram at All stations.

4.2. Test Fusion Data with Ground Measurements and Actual Landsat 8 LST Product

Twenty-eight Landsat 8-like LST scene images are generated according to the fusion scheme of Figure 2. The comparison of the ground measurements and fusion results at the site is shown in Figures 4 and 5. The fusion results of the three algorithms are highly consistent with the ground measurements, and R^2 is greater than 0.853 (Figure 4). In addition to the Wetland station, ESTARFM has a higher correlation with measured LST at the remaining stations compared with STARFM and FSDAF. From the perspective of All stations, the R^2 of STARFM, ESTARFM, and FSDAF is 0.911, 0.915, and 0.912, respectively. The R^2 of Gobi station is larger than that of Wetland station and Daman superstation, for the surface coverage of Gobi Station is almost unchanged over time.

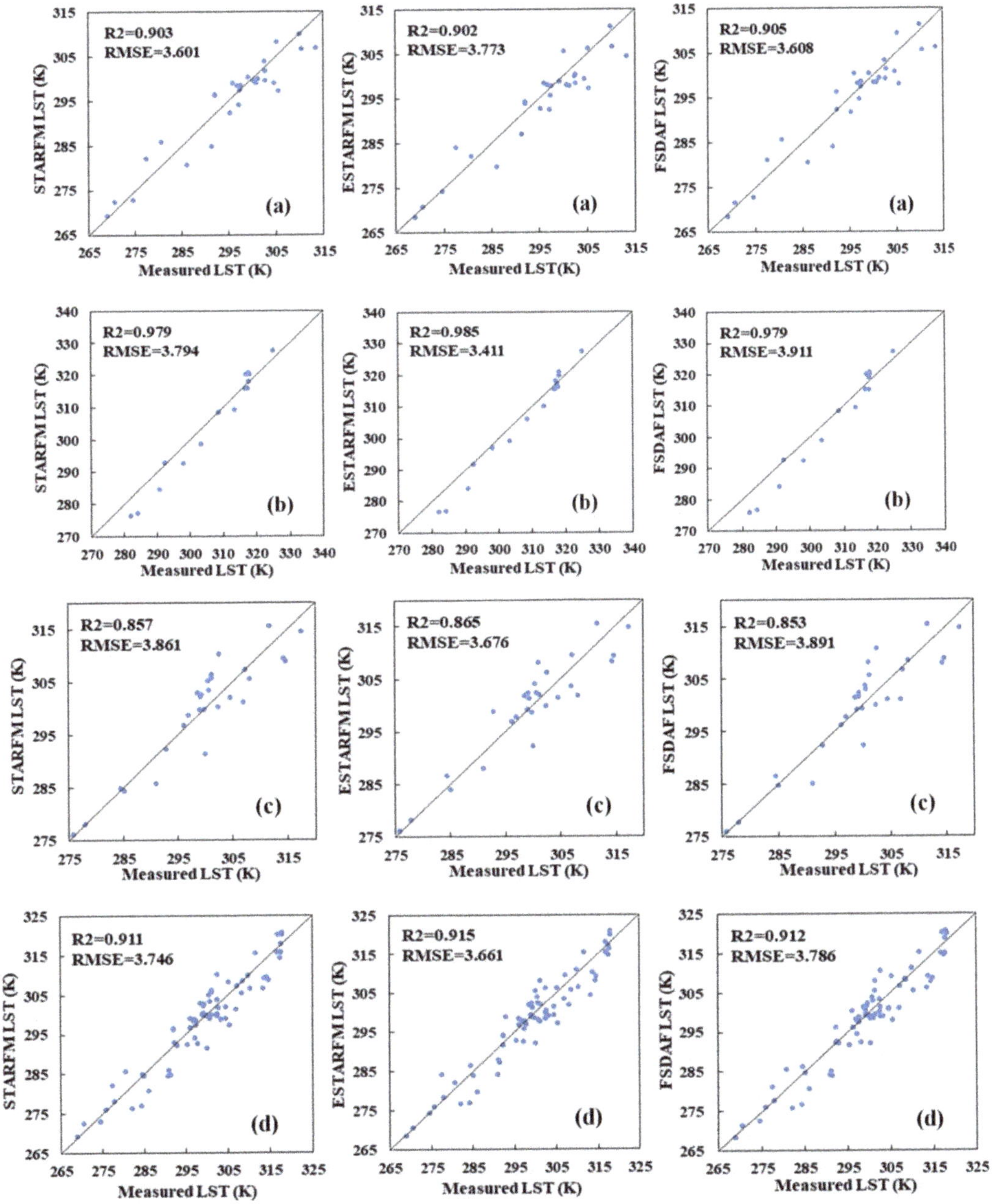

Figure 4. *Cont.*

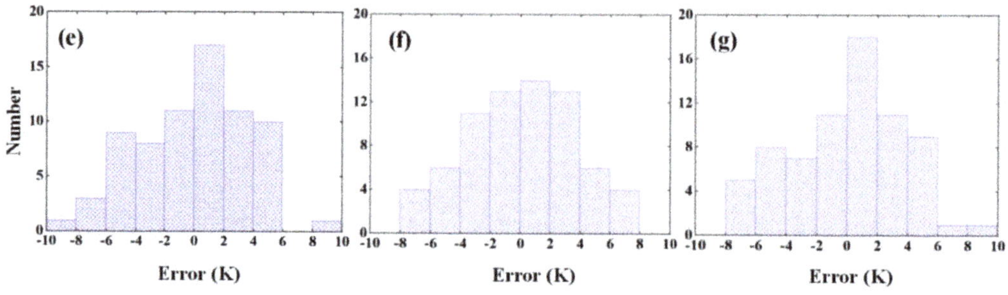

Figure 4. Scatter plot and error histogram of fused LST and measured LST. (**a**): Wetland station; (**b**): Gobi station; (**c**): Daman superstation; (**d**): All stations; (**e**): STARFM error histogram at All stations; (**f**): ESTARFM error histogram at All stations; (**g**): FSDAF error histogram at All stations.

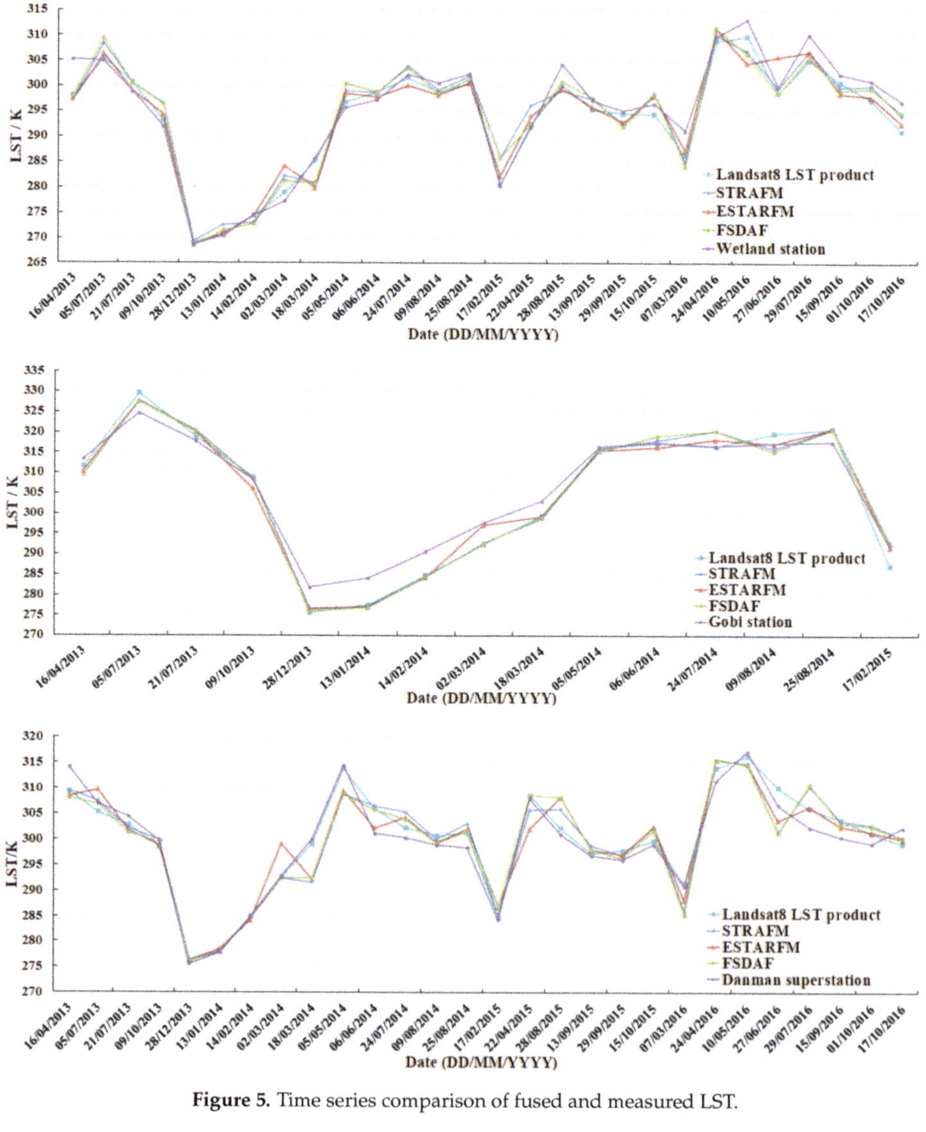

Figure 5. Time series comparison of fused and measured LST.

The results of the correlation analysis between the fusion LST and the ground measurements are shown in Table 3. By analyzing evaluation indicators from All stations, it is found that ESTARFM (MAE = 2.965 K, RMSE = 3.661 K) is less than STARFM (MAE = 3.064 K, RMSE = 3.746 K) and FSDAF (MAE = 3.065 K, RMSE = 3.786 K). Based on the analysis of a single site, except for the Wetland station, ESTARFM performed better than STARFM and FSDAF on the other two sites. For bias, although ESTARFM (bias = −0.607 K) is greater than STARFM (bias = −0.274 K) and FSDAF (bias = −0.470 K), it is closer to Landsat 8 LST product (bias = −0.760 K), STARFM and FSDAF show too small deviations, which may be caused by error propagation in the fusion. The histogram distribution of the LST differences between the fused and measured LST shows most biases are between −4 K and 4 K (Figure 4e–g). Comprehensive analysis shows that ESTARFM fusion results are slightly better than STARFM and FSDAF at the site.

Table 3. The bias (K), MAE (K), and RMSE (K) of fused LST over stations.

Station Name	STARFM			ESTARFM			FSDAF		
	Bias	MAE	RMSE	Bias	MAE	RMSE	Bias	MAE	RMSE
Wetland	−0.356	2.943	3.601	−1.041	2.995	3.773	−0.635	2.988	3.608
Gobi	−1.310	3.197	3.794	−1.407	2.724	3.411	−1.429	3.336	3.911
Daman	0.362	3.114	3.861	0.256	3.064	3.676	0.209	2.997	3.891
All	−0.274	3.064	3.746	−0.607	2.965	3.661	−0.470	3.065	3.786

In order to analyze the fusion results of the three algorithms in more detail, the predicted Landsat 8-like LST images are evaluated from the correlation analysis and visual interpretation based on actual Landsat 8 LST images as the verification data. In the study area, some MODIS LST images have default data (the value is 0), and errors will be brought into the fusion results. Therefore, when evaluating the images of the study area, these fused images with errors will be eliminated.

The correlation analysis results between fused images and the actual Landsat 8 LST images are shown in Figure 6. It can be seen that the Landsat 8 LST images generated by all three fusion methods are very similar to the actual images, and R^2 is higher than 0.769. In most cases, the R^2 of ESTARFM is larger and RMSE is smaller compared with STARFM and FSDAF. After calculating, the average R^2 of the fusion data of STARFM, ESTARFM, and FSDAF were 0.883, 0.898, 0.888, and the RMSE were 2.608 K, 2.245 K, and 2.565 K, respectively. ESTARFM evaluation index is better than STARFM and FSDAF. However, the RMSE of ESTARFM on 5 July 2013 was 3.846 K, which was significantly greater than that of STARFM and FSDAF. The main reason is that compared with STARFM and FSDAF, the image of 16 April 2013 was added for ESTARFM in the fusion process. The reason for the errors were introduced is that large land cover difference between 16 April 2013 (Figure 7a) and 5 June 2013 (Figure 7b), which should be paid attention to when using ESTARFM algorithm.

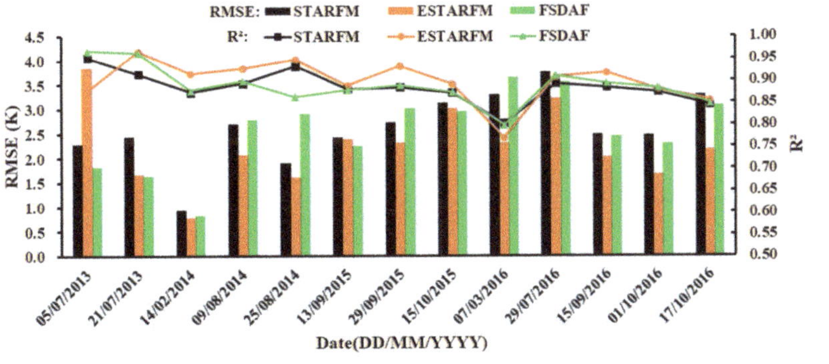

Figure 6. The R^2 and RMSE (K) of fused LST images and actual LST images.

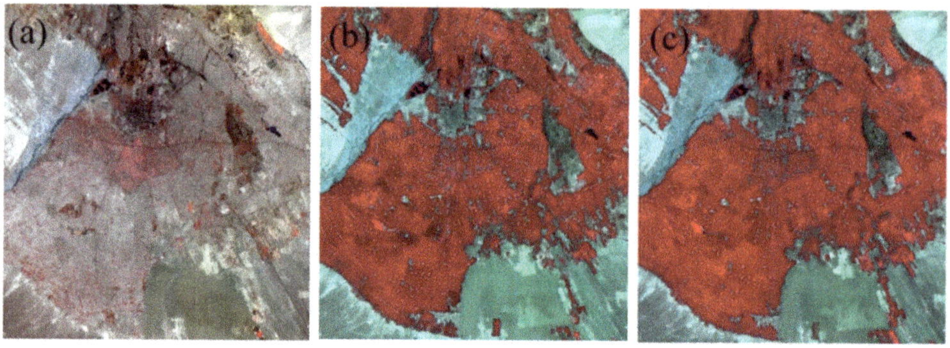

Figure 7. Landsat 8 standard false color image (**a**): 16 April 2013; (**b**): 5 June 2013; (**c**): 21 June 2013.

The visual interpretation results between the fused images and the actual Landsat 8 LST images on 9 August 2014 are obtained from Figures 8 and 9. From the overall image (Figure 8b–d), the fusion result of ESTARFM and FSDAF is clearer than that of STARFM. In the subset (Figure 8e–h), the spatial detail of ESTARFM is the closest to the actual Landsat 8 LST, followed by FSDAF, and the STARFM result is the worst. However, for the fusion results of all three algorithms, the temperature was underestimated compared with the actual Landsat 8 LST. After analysis, it can be inferred that the cause is radiometric and geometric inconsistencies between the two sensors. The low-resolution data use simulated MODIS LST (resampled to 1000m based on Landsat 8 LST) in Figure 9, there is no underestimation in the subset, and the overall fusion accuracy has been improved. It can be seen that ESTRAFM still has the best fusion effect, which is the closest to the actual Landsat 8 LST, followed by FSDAF, and then STARFM.

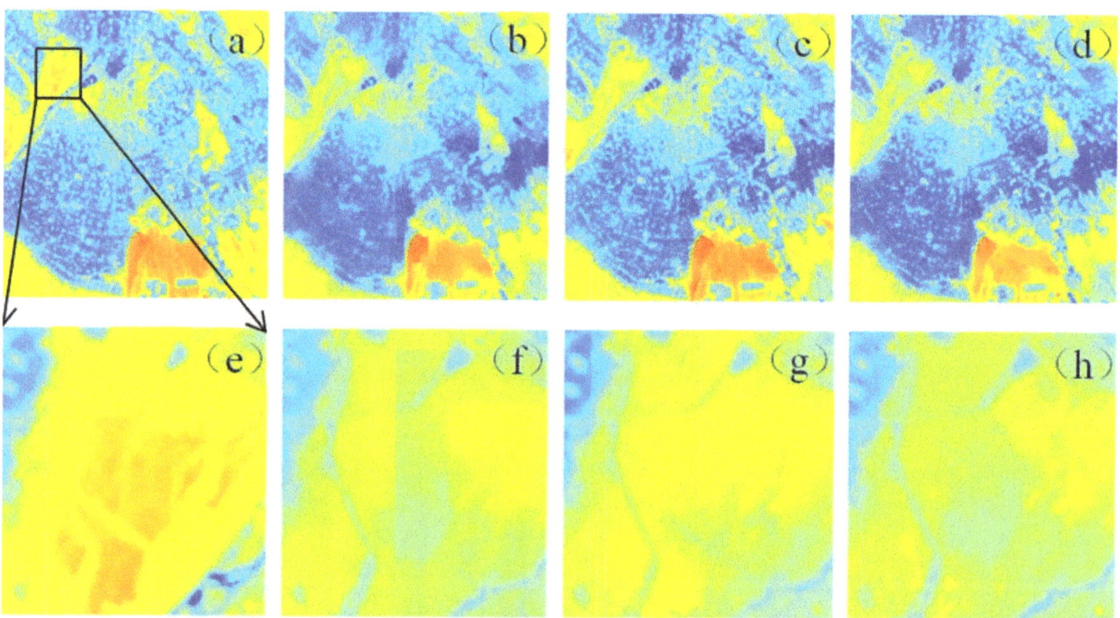

Figure 8. *Cont.*

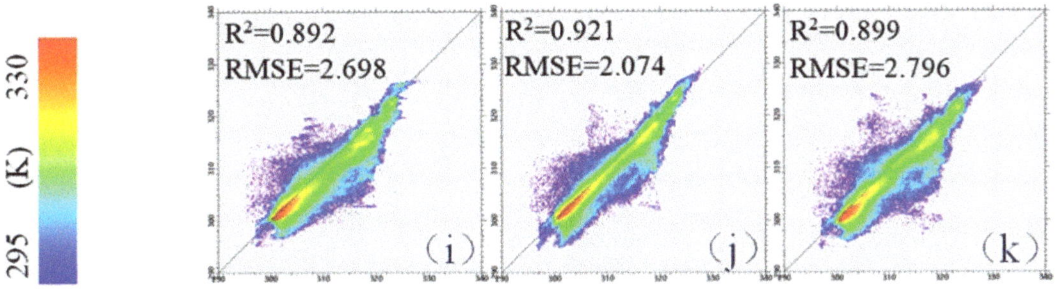

Figure 8. Results of fusing MODIS LST and Landsat 8 LST on 9 August 2014. (**a**): Actual Landsat 8 LST; (**b**): STARFM LST; (**c**): ESTARFM LST; (**d**): FSDAF LST; (**e–h**): the subsets of (**a–d**) corresponding to the blacks quare area in (**a**). (**i–k**): scatter plot between (**a–d**).

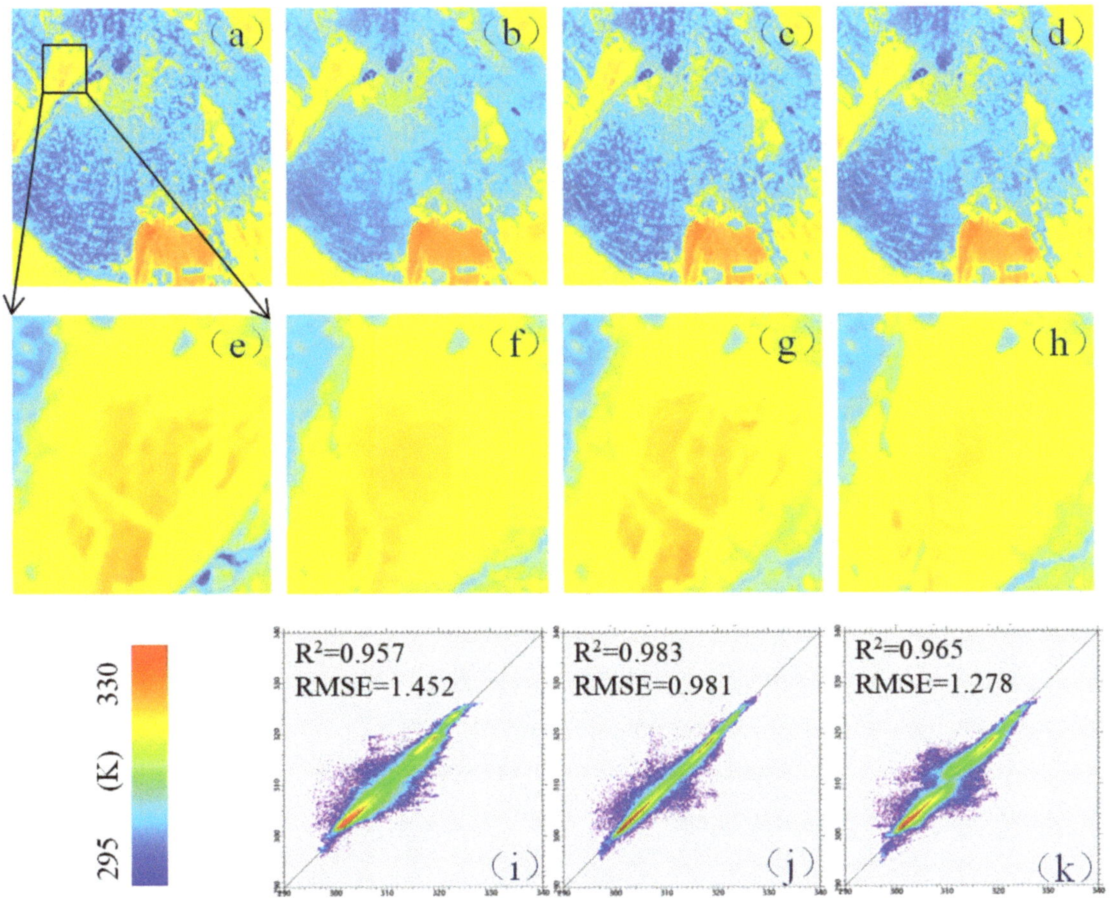

Figure 9. Results of fusing simulated MODIS LST and Landsat 8 LST on 9 August 2014. (**a**): Actual Landsat 8 LST; (**b**): STARFM LST; (**c**): ESTARFM LST; (**d**): FSDAF LST; (**e–h**): the subsets of (**a–d**) corresponding to the black square area in (**a**). (**i–k**): scatter plot between (**a**) and (**b–d**).

From the analysis of the above results, ESTARFM shows the best performance in most cases, whether it is verified from the ground site or based on the actual Landsat 8 LST image. However, when there is a large land cover difference between the reference and predicted images, errors will be introduced instead, and attention needs to be paid when performing ESTARFM fusion. The STARFM fusion images are blurry, and the spatial details are not as good as FSDAF. It is worth noting that the three algorithms will be affected by the difference between high- and low-resolution data, which can be seen by comparing Figures 8 and 9.

5. Discussion

In this study, the Landsat 8 LST product and the results of spatiotemporal fusion are analyzed and verified for a long time series. The accuracy of Landsat 8 LST product is better than that of SC LST, and it has a strong correlation with ground measurements. STARFM, ESTARFM, and FSDAF algorithms can all be used to generate Landsat 8-like LST that is highly correlated with the actual Landsat 8 LST. However, in the application of these methods, there are still problems that need to be solved.

5.1. Lack of Landsat 8 LST Product Verification in High Water Vapor Content

In this article, the Landsat 8 LST product in the study area was verified. From Figure 3, it can be seen that R^2 at All stations is 0.952, and most of the deviations are between −3 K and 3 K, which have high accuracy. However, the atmospheric water vapor content in this study area is relatively low, less than 2 g/cm^2 on most days. Figure 10 shows the average value of atmospheric water vapor content in the study area (from MOD05_L2). The case of high water vapor content has not been verified, and it is necessary to consider conducting comparative verification in areas with high water vapor content in future studies.

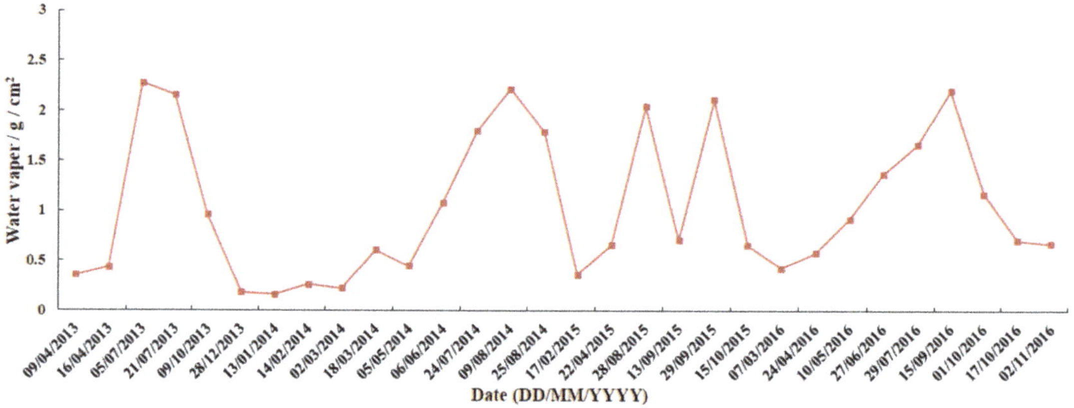

Figure 10. Atmospheric water vapor content in the study area.

5.2. Differences in High- and Low-Resolution Data

From the comparison of Figures 8 and 9, it can be seen that replacing the actual MODIS LST with the simulated MODIS LST will improve the accuracy of the fusion results and have better spatial details. This is mainly because the simulated MODIS LST is based on Landsat 8 LST resampling, which is more consistent with the actual Landsat 8 LST and avoids errors caused by different differences between sensors, satellite angles, and temperature retrieval algorithms. The fusion results of simulated and actual MODIS LST are verified based on ground measurements (Figure 11). The fusion accuracy of the three algorithms has been improved when using simulated MODIS LST, and the RMSE is reduced by about 0.5 K, so improving the consistency of the Landsat 8 LST and MODIS LST images is a problem that needs to be explored.

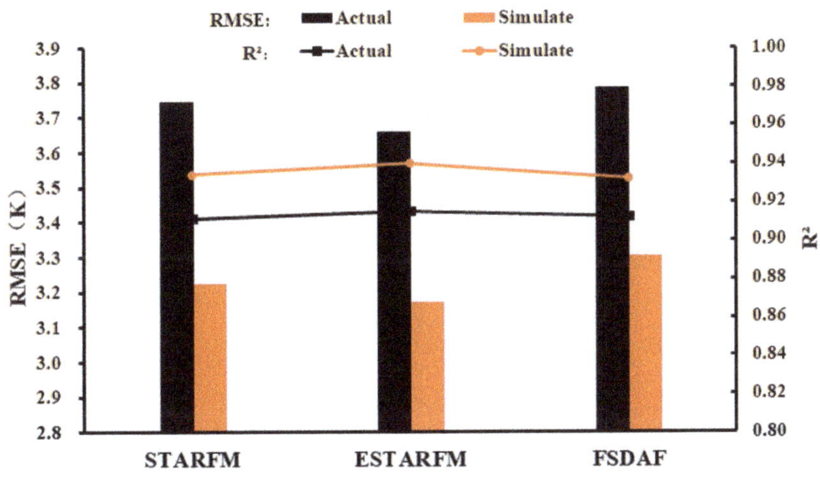

Figure 11. Comparison of simulation and actual fusion results over stations.

5.3. Uncertainty of STARFM, ESTARFM, and FSDAF Input Parameters

The three spatiotemporal fusion algorithms contain the following two important parameters: the number of land cover types and the size of the search window. The former is set to 4, while the latter 2 × 2 MODIS pixel in this study. However, the number of land cover types and spatial heterogeneity in the study area will change over time, especially in the study of long time series. It is necessary to adjust the amount of land cover and the size of the search window according to the actual situation of the study area. How to automatically optimize these two parameters during fusion is a question that needs to be studied in the next step.

5.4. Selection of Reference Images

Only one high-resolution reference image is required for STARFM and FSDAF, while two are required for ESTARFM. Errors may be introduced by the additional reference images. For example, when the situation shown in Figure 7 occurs, the difference in surface coverage between the two high-resolution reference images is large, which seriously affects the ESTARFM fusion result. Therefore, it is necessary to fully consider the conditions of the reference image when using the ESTARFM fusion algorithm.

5.5. MODIS LST Noise

There is noise in MODIS LST data, and the noise error will be introduced into the fusion result. Due to the influence of noise, only 13 scene fusion images are used in the accuracy evaluation based on the actual images (Figure 6). Therefore, noise removal should be performed on MODS LST before fusion to ensure the reliability of the fusion image.

In this study, the three algorithms of STARFM, ESTARFM, and FSDSAF are compared and verified. The original application scenarios of these algorithms are the fusion of surface reflectance data, and the temperature change characteristics are not considered. At the same time, some researchers combined the annual temperature cycle (ATC) [14,15] model with the spatiotemporal fusion algorithm. However, considering the ATC model cannot describe specific daily surface temperature characteristics, which leads to the uncertainty of the model, this kind of fusion algorithm is not validated in this article.

6. Conclusions

In this study, we applied the Landsat 8 LST product to the fusion, and evaluated the accuracy of the Landsat 8 LST product and the fusion results of STARFM, ESTARFM, and FSDAF.

We used ground measurements LST to verify Landsat 8 LST product. The results indicated that Landsat 8 LST product is highly correlated with ground measurements, and accuracy over All stations of LST product (bias = −0.760 K, RMSE = 2.862 K) is better than that of SC algorithm (bias = −1.284 K, RMSE = 2.967 K). The use of this standard product avoids the LST inversion before the fusion and makes the Landsat 8 thermal band more widely used.

The three algorithms of STARFM, ESTARFM, and FSDAF were comprehensively evaluated based on the ground measurements and actual Landsat 8 LST images. The verification results based on the ground measurements show that in All stations, the R^2 of the three algorithms is greater than 0.911 and the MAE is less than 3.065 K, the smallest with MAE of ESTARFM, which is 2.965 K, followed by STARFM 3.064 K and then FSDAF 3.065 K. Based on the actual Landsat 8 LST images as the verification data, the R^2 of the three algorithms for the 13 scene fusion images is all greater than 0.769, which has a high correlation with the actual Landsat 8 LST images. The evaluation index of MAE and RMSE (the average of the 13 scene fusion images) shows ESTARFM (MAE = 1.771 K, RMSE = 2.245 K) is smaller than STARFM (MAE = 2.129 K, RMSE = 2.608 K) and FSDAF (MAE = 2.077 K, RMSE = 2.565 K). In the 13 scene fusion images, ESTARFM is better than STARFM and FSDAF in most cases (Figure 6). From the perspective of the spatial details of the fusion results (Figures 8 and 9), ESTARFM has the best fusion effect, followed by FSDAF, and the STARFM fusion results are relatively fuzzy. The results of integrating ground measurements verification and actual Landsat 8 images verification indicated the reliability of the three algorithms and proved that ESTARFM performs best in this study.

Author Contributions: S.L. and J.W. conceived and designed the experiments; S.L. performed the experiments; D.L. and Z.R. contributed with data processing and analysis; B.Y. checked the experimental results; S.L. wrote the paper; and J.W. revised the manuscript. All authors have read and agreed to the published version of the manuscript.

Funding: This research received no external funding.

Institutional Review Board Statement: Not applicable.

Informed Consent Statement: Not applicable.

Data Availability Statement: Not applicable.

Conflicts of Interest: The authors declare no conflict of interest.

References

1. Li, Z.L.; Tang, B.H.; Wu, H.; Ren, H.; Yan, G.; Wan, Z.; Trigo, I.F.; Sobrino, J.A. Satellite-derived land surface temperature: Current status and perspectives. *Remote Sens. Environ.* **2013**, *131*, 14–37. [CrossRef]
2. Eleftheriou, D.; Kiachidis, K.; Kalmintzis, G.; Kalea, A.; Bantasis, C.; Koumadoraki, P.; Spathara, M.E.; Tsolaki, A.; Tzampazidou, M.I.; Gemitzi, A. Determination of annual and seasonal daytime and nighttime trends of MODIS LST over Greece-climate change implications. *Sci. Total Environ.* **2018**, *616*, 937–947. [CrossRef]
3. Shamir, E.; Georgakakos, K.P. MODIS Land Surface Temperature as an index of surface air temperature for operational snowpack estimation. *Remote Sens. Environ.* **2014**, *152*, 83–98. [CrossRef]
4. Anderson, M.C.; Kustas, W.P.; Norman, J.M.; Hain, C.R.; Mecikalski, J.R.; Schultz, L. Mapping daily evapotranspiration at field to global scales using geostationary and polar orbiting satellite imagery. *Hydrol. Earth Syst. Sci. Discuss* **2010**, *7*, 5957–5990.
5. Coll, C.; Valor, E.; Galve, J.M.; Mira, M.; Bisquert, M.; García-Santos, V.; Caselles, E.; Caselles, V. Long-term accuracy assessment of land surface temperatures derived from the Advanced Along-Track Scanning Radiometer. *Remote Sens. Environ.* **2012**, *116*, 211–225. [CrossRef]
6. Li, Z.L.; Duan, S.B.; Tang, B.H.; Wu, H.; Ren, H.Z.; Yan, G.J.; Tang, R.L.; Leng, P. Review of methods for land surface temperature derived from thermal infrared remotely sensed data. *J. Remote Sens.* **2016**, *20*, 899–920.
7. Zhan, W.; Chen, Y.; Zhou, J.; Wang, J.; Liu, W.; Voogt, J.; Zhu, X.; Quan, J.; Li, J. Disaggregation of remotely sensed land surface temperature: Literature survey, taxonomy, issues, and caveats. *Remote Sens. Environ.* **2013**, *131*, 119–139. [CrossRef]
8. Zhan, W.; Huang, F.; Quan, J.; Zhu, X.; Gao, L.; Zhou, J.; Ju, W. Disaggregation of remotely sensed land surface temperature: A new dynamic methodology. *J. Geophys. Res.-Atmos.* **2016**, *121*, 538–554. [CrossRef]
9. Liu, D.; Pu, R. Downscaling Thermal Infrared Radiance for Subpixel Land Surface Temperature Retrieval. *Sensors* **2008**, *8*, 2695–2706. [CrossRef]

10. Stathopoulou, M.; Cartalis, C. Downscaling AVHRR land surface temperatures for improved surface urban heat island intensity estimation. *Remote Sens. Environ.* **2009**, *113*, 2592–2605. [CrossRef]
11. Agam, N.; Kustas, W.P.; Anderson, M.C.; Li, F.; Neale, C.M. A vegetation index based technique for spatial sharpening of thermal imagery. *Remote Sens. Environ.* **2007**, *107*, 545–558. [CrossRef]
12. Dominguez, A.; Kleissl, J.; Luvall, J.; Rickman, D. High-resolution urban thermal sharpener (HUTS). *Remote Sens. Environ.* **2011**, *115*, 1772–1780. [CrossRef]
13. Wu, P.; Shen, H.; Zhang, L.; Göttsche, F.-M. Integrated fusion of multi-scale polar-orbiting and geostationary satellite observations for the mapping of high spatial and temporal resolution land surface temperature. *Remote Sens. Environ.* **2015**, *156*, 169–181. [CrossRef]
14. Quan, J.L.; Zhan, W.F.; Ma, T.D.; Du, Y.Y.; Guo, Z.; Qin, B.Y. An integrated model for generating hourly Landsat-like land surface temperatures over heterogeneous landscapes. *Remote Sens. Environ.* **2018**, *206*, 403–423. [CrossRef]
15. Weng, Q.; Fu, P.; Gao, F. Generating daily land surface temperature at Landsat resolution by fusing Landsat and MODIS data. *Remote Sens. Environ.* **2014**, *145*, 55–67. [CrossRef]
16. Zhu, X.; Cai, F.; Tian, J.; Williams, T.K.-A. Spatiotemporal Fusion of Multisource Remote Sensing Data: Literature Survey, Taxonomy, Principles, Applications, and Future Directions. *Remote Sens.* **2018**, *10*, 527. [CrossRef]
17. Gao, F.; Masek, J.; Schwaller, M.; Hall, F. On the Blending of the Landsat and MODIS Surface Reflectance: Predicting Daily Landsat Surface Reflectance. *IEEE Trans. Geosci. Remote Sens.* **2006**, *44*, 2207–2218.
18. Gao, F.; Kustas, W.; Anderson, M. A Data Mining Approach for Sharpening Thermal Satellite Imagery over Land. *Remote Sens.* **2012**, *4*, 3287–3319. [CrossRef]
19. Emelyanova, I.V.; McVicar, T.R.; van Niel, T.G.; Li, L.T.; van Dijk, A.I.J.M. Assessing the accuracy of blending Landsat-MODIS surface reflectances in two landscapes with contrasting spatial and temporal dynamics: A framework for algorithm selection. *Remote Sens. Environ.* **2013**, *133*, 193–209. [CrossRef]
20. Gevaert, C.M.; García-Haroa, F.J. A comparison of STARFM and an unmixing-based algorithm for Landsat and MODIS data fusion. *Remote Sens. Environ.* **2015**, *156*, 34–44. [CrossRef]
21. Hilker, T.; Wulder, M.; Coops, N.; Linke, J.; McDermid, G.; Masek, J.G.; Gao, F.; White, J. A new data fusion model for high spatial- and temporal-resolution mapping of forest disturbance based on Landsat and MODIS. *Remote Sens. Environ.* **2009**, *113*, 1613–1627. [CrossRef]
22. Zhu, X.; Chen, J.; Gao, F.; Chen, X.; Masek, J.G. An enhanced spatial and temporal adaptive reflectance fusion model for complex heterogeneous regions. *Remote Sens. Environ.* **2010**, *114*, 2610–2623. [CrossRef]
23. Huang, B.; Song, H. Spatiotemporal Reflectance Fusion via Sparse Representation. *IEEE Trans. Geosci. Remote Sens.* **2012**, *50*, 3707–3716. [CrossRef]
24. Song, H.; Huang, B. Spatiotemporal Satellite Image Fusion through One-Pair Image Learning. *IEEE Trans. Geosci. Remote Sens.* **2013**, *51*, 1883–1896. [CrossRef]
25. Zhu, X.; Helmer, E.H.; Gao, F.; Liu, D.; Chen, J.; Lefsky, M.A. A flexible spatiotemporal method for fusing satellite images with different resolutions. *Remote Sens. Environ.* **2016**, *172*, 165–177. [CrossRef]
26. Li, X.; Foody, G.M.; Boyd, D.S.; Ge, Y.; Zhang, Y.; Du, Y.; Ling, F. SFSDAF: An enhanced FSDAF that incorporates sub-pixel class fraction change information for spatio-temporal image fusion. *Remote Sens. Environ.* **2019**, *237*, 111537. [CrossRef]
27. Guo, D.; Shi, W.; Hao, M.; Zhu, X. FSDAF 2.0: Improving the performance of retrieving land cover changes and preserving spatial details. *Remote Sens. Environ.* **2020**, *248*, 111973. [CrossRef]
28. Tang, J.; Zeng, J.; Zhang, L.; Zhang, R.; Li, J.; Li, X.; Zou, J.; Zeng, Y.; Xu, Z.; Wang, Q.; et al. A modified flexible spatiotemporal data fusion model. *Front. Earth Sci.* **2020**, *14*, 601–614. [CrossRef]
29. Shi, C.; Wang, X.; Zhang, M.; Liang, X.; Niu, L.; Han, H.; Zhu, X. A comprehensive and automated fusion method: The enhanced flexible spatiotemporal data fusion model for monitoring dynamic changes of land surface. *Appl. Sci.* **2019**, *9*, 3693. [CrossRef]
30. Li, Y.; Huang, C.; Hou, J.; Gu, J.; Zhu, G.; Li, X. Mapping daily evapotranspiration based on spatiotemporal fusion of ASTER and MODIS images over irrigated agricultural areas in the Heihe River Basin, Northwest China. *Agric. Forest Meteorol.* **2017**, *244*, 82–97. [CrossRef]
31. Ma, Y.; Liu, S.M.; Song, L.; Xu, Z.; Liu, Y.; Xu, T.; Zhu, Z. Estimation of daily evapotranspiration and irrigation water efficiency at a Landsat-like scale for an arid irrigation area using multi-source remote sensing data. *Remote Sens. Environ.* **2018**, *216*, 715–734. [CrossRef]
32. Yang, G.; Weng, Q.; Pu, R.; Gao, F.; Sun, C.; Li, H.; Zhao, C. Evaluation of ASTER-Like Daily Land Surface Temperature by Fusing ASTER and MODIS Data during the HiWATER-MUSOEXE. *Remote Sens.* **2016**, *8*, 75. [CrossRef]
33. Huang, B.; Wang, J.; Song, H.; Fu, D.; Wong, K. Generating High Spatiotemporal Resolution Land Surface Temperature for Urban Heat Island Monitoring. *IEEE Geosci. Remote Sens. Lett.* **2013**, *10*, 1011–1015. [CrossRef]
34. Wan, Z.M.; Dozier, J. A generalized split-window algorithm for retrieving landsurface temperature from space. *IEEE Trans. Geosci. Remote Sens.* **1996**, *34*, 892–905.
35. Wan, Z. New refinements and validation of the collection-6 MODIS land-surface temperature/emissivity product. *Remote Sens. Environ.* **2014**, *140*, 36–45. [CrossRef]
36. Gao, B.; Kaufman, Y.J. *MODIS Atmosphere L2 Water Vapor Product; NASA MODIS Adaptive Processing System*; Goddard Space Flight Center: Greenbelt, MD, USA, 2017.

37. Liu, S.; Li, X.; Xu, Z.; Che, T.; Xiao, Q.; Ma, M.; Liu, Q.; Jin, R.; Guo, J.; Wang, L.; et al. The Heihe Integrated Observatory Network: A basin-scale land surface processes observatory in China. *Vadose Zone J.* **2018**, *17*, 1–21. [CrossRef]
38. Liu, S.M.; Xu, Z.W.; Wang, W.Z.; Bai, J.; Jia, Z.; Zhu, M.; Wang, J.M. A comparison of eddy-covariance and large aperture scintillometer measurements with respect to the energy balance closure problem. *Hydrol. Earth Syst. Sci. Discuss.* **2011**, *15*, 1291–1306. [CrossRef]
39. Malakar, N.K.; Hulley, G.C.; Hook, S.J.; Laraby, K.; Cook, M.; Schott, J.R. An Operational Land Surface Temperature Product for Landsat Thermal Data: Methodology and Validation. *IEEE Trans. Geosci. Remote Sens.* **2018**, *56*, 5717–5735. [CrossRef]
40. Xia, H.; Chen, Y.; Li, Y.; Quan, J. Combining kernel-driven and fusion-based methods to generate daily high-spatial-resolution land surface temperatures. *Remote Sens. Environ.* **2019**, *224*, 259–274. [CrossRef]
41. Jiménez-Muñoz, J.C.; Sobrino, J.A. A generalized single-channel method for retrieving land surface temperature from remote sensing data. *J. Geophys. Res.-Atmos.* **2003**, *108*, 4688. [CrossRef]
42. Jiménez-Muñoz, J.C.; Cristóbal, J.; Sobrino, J.A.; Soria, G.; Ninyerola, M.; Pons, X. Revision of the single-channel algorithm for land surface temperature retrieval from Landsat thermal-infrared data. *IEEE Trans. Geosci. Remote Sens.* **2008**, *47*, 339–349. [CrossRef]
43. Jiménez-Muñoz, J.C.; Sobrino, J.A.; Skokovic, D.; Mattar, C.; Cristóbal, J. Land Surface Temperature Retrieval Methods from Landsat-8 Thermal Infrared Sensor Data. *IEEE Geosci. Remote Sens. Lett.* **2014**, *11*, 1840–1843. [CrossRef]
44. Wan, Z. New refinements and validation of the MODIS land-surface temperature/emissivity products. *Remote Sens. Environ.* **2008**, *112*, 59–74. [CrossRef]

processes

MDPI

Article

Optimization of PCL Polymeric Films as Potential Matrices for the Loading of Alpha-Tocopherol by a Combination of Innovative Green Processes

Emanuela Drago [1], Roberta Campardelli [1,*], Iolanda De Marco [2,*] and Patrizia Perego [1]

1 Department of Civil, Chemical and Environmental Engineering (DICCA), University of Genoa, Via Opera Pia 15, 16145 Genova, Italy; emanuela.drago@edu.unige.it (E.D.); p.perego@unige.it (P.P.)
2 Department of Industrial Engineering, University of Salerno, Via Giovanni Paolo II, 132, 84084 Fisciano, Italy
* Correspondence: roberta.campardelli@unige.it (R.C.); idemarco@unisa.it (I.D.M.)

Abstract: Active food packaging represents an innovative way to conceive food packages. The innovation lies in using natural-based and biodegradable materials to produce a system intended to interact with the food product to preserve its quality and shelf-life. Compared to traditional plastics, active packaging is designed and regulated to release substances in a controlled manner, mainly antimicrobial and antioxidant compounds. Conventional technologies are not suitable for treating these natural substances; therefore, the research for innovative and green techniques represents a challenge in this field. The aim of this work is to compare two different polymeric structures: nanofibrous films obtained by electrospinning and continuous films obtained by solvent casting, to identify the best solution and process conditions for subjecting the samples to the supercritical fluids impregnation process (SFI). The supports optimized were functionalized by impregnating alpha-tocopherol using the SFI process. In particular, the different morphologies of the samples both before and after the supercritical impregnation process were initially studied, identifying the limits and possible solutions to obtain an optimization of the constructs to be impregnated with this innovative green technology in the packaging field.

Keywords: supercritical fluid impregnation; electrospinning; solvent casting; antioxidant packaging; materials optimization

Citation: Drago, E.; Campardelli, R.; Marco, I.D.; Perego, P. Optimization of PCL Polymeric Films as Potential Matrices for the Loading of Alpha-Tocopherol by a Combination of Innovative Green Processes. *Processes* **2021**, *9*, 2244. https://doi.org/10.3390/pr9122244

Academic Editor: Reza Tahergorabi

Received: 23 November 2021
Accepted: 10 December 2021
Published: 13 December 2021

1. Introduction

In recent years, the food packaging sector has been subject to an important re-evaluation derived by the current environmental needs to find alternatives to traditional packaging. Materials of petrochemical derivation have dominated this field for decades due to their numerous advantages in terms of low cost and ease of production. It facilitates the marketing of a wide range of food products through lightweight, highly flexible, elastic, and performing packaging systems [1]. All these advantages have led to a constant increase, over the years, in the production of mainly disposable packaging, which, given their nonbiodegradability and disposal difficulties, have inexorably accumulated not only in landfills but also in the environment, bringing to a global environmental pollution problem [2,3]. For this reason, the scientific community is increasingly aiming at finding new biodegradable materials that can gradually replace plastics. This research seems to have found a solution in using biopolymers synthesized from natural sources, such as biomass and microorganisms [4–7], or even extracted from agri-food wastes, such as polysaccharides and proteins [8–10].

Polycaprolactone (PCL) is a biopolymer of fossil origin, but it is biodegradable. This polyester is attractive because of its excellent properties of biocompatibility, nontoxicity, and thermal stability. For these reasons, it has already found wide use in the biomedical field, also as a drug delivery system approved by the Food and Drug Administration

(FDA) [11–14]. This polymer has been used in the field of food packaging through the production of active packaging [15,16], which are systems capable of interacting with packaged food products by releasing or absorbing controlled and regulated substances [17], in order to preserve the quality and shelf-life of the products [18]. The use of natural substances, such as essential oils, natural extracts, vitamins, etc., for the fabrication of active packaging with mainly antimicrobial and antioxidant action has been reported [19–22]. However, traditional packaging production techniques are no longer an obvious choice when working with biopolymers and natural substances, as they have several drawbacks. For example, in extrusion [23], high process temperatures are required for the melting of polymers, large quantities of often toxic and polluting organic solvents are necessary, and the incorporation of bioactive substances into the polymers themselves is difficult, as they can be degraded by the high process temperatures. In some cases, the natural ingredients are deposited as external coatings of the packaging film; in this case, a low penetration into the polymer structure and significant degradation of the active compounds by external agents have been observed [24].

In order to try to overcome these drawbacks, innovative green techniques that are more suitable for the processing of biopolymers and bioactive molecules have been recently proposed in the literature. In particular, electrospinning is a technique capturing more and more attention, as it allows to work at low temperatures, with the use of green solvents, and it can provide micro and nanostructured polymer matrix [25,26]. It is a technique that uses only the electric field as a driving force to produce fibers arranged in a random way that gives rise to a porous structure with a large surface-to-volume ratio. This characteristic makes electrospun mats suitable for drug delivery systems or active packaging films [27]. Another method, already used on an industrial scale, is solvent casting, a low-cost, versatile, and green technique, which consists of the solubilization of the polymer of interest in an appropriate solvent. Specifically, the solution is poured into a mold, and the evaporation of the solvent leads to the production of thin, isotropic, and transparent films [28]. These techniques have been extensively investigated in the literature for biopolymers processing [29,30]. The impregnation process assisted by supercritical fluids (SFI) is an interesting method for loading active ingredients inside polymeric preformed structures [31]. SFI works by exploiting the high solvent power of CO_2 in supercritical conditions ($scCO_2$) and its high diffusivity to solubilize the active ingredient of interest and diffuse it within the polymer matrix, without the need to use other solvents or high temperatures. Indeed, carbon dioxide reaches the supercritical state in mild conditions ($Tc = 31.1$ °C, $pc = 7.38$ MPa) [32,33]. It is a process that allows high penetration of the compound within the polymeric structure [34].

SFI has been applied in the literature for active compound impregnation into PCL [35]. Most of the time, these impregnation studies have been conducted using PCL granules [36,37]. PCL films have been rarely processed with SFI due to the foaming activity of CO_2 against this polymer; indeed, the most used polymers matrix for the SFI process are poly(lactic acid) (PLA) [38], acrylic-based hydrogels [39], cellulose acetate [40], and poly(methyl methacrylate) (PMMA) [41]. In some cases, the use of polymeric blend with PCL was demonstrated to improve the structure resistance to the CO_2 foaming effect [42]. Process conditions are particularly important to allow good impregnation without significantly modifying the polymeric structures. For these reasons, PCL was generally processed at low temperatures from 35 to 40 °C and moderate pressure from 110 to 200 bar [35,43].

Therefore, the first part of this paper concerns the optimization of SFI of different polymeric PCL-based structures produced by both electrospinning and solvent casting methods. PCL blended with polyethylene glycol (PEG) has been also studied. PEG was selected to improve the handling of the films produced, especially by solvent casting [28], and for its antifogging properties reported in the literature due to its hydrophilic nature [44]. Then, in the second part of this work, SFI of alpha-tocopherol, selected as a model natural active compound, has been performed [45,46]. The best combination of process conditions has been researched to provide a possible process layout for PCL-loaded alpha-tocopherol films production that allowed good loading of the active compounds, avoiding their degradation, high penetration inside the polymeric film, and preservation of the polymeric structure and functionality.

2. Materials, Equipment, and Methods

2.1. Materials

Polycaprolactone (PCL, Mn 80,000), alpha-tocopherol (purity > 97%), and polyethylene glycol (PEG 400) were purchased by Sigma-Aldrich, Milan, Italy; acetone and acetic acid glacial (extra pure) were provided by Carlo Erba Reagents S.A.S, Val-de-Reuil, France; carbon dioxide (CO_2, purity 99%) was purchased by Morlando Group S.R.L., Sant'Antimo, Naples, Italy.

2.2. Production of PCL-Based Matrices

The polymeric matrices were produced both by electrospinning and solvent casting following the scheme reported in Figure 1. The electrospinning used (Spinbow, Bologna, Italy) is composed of a high voltage generator (PCM, Spellman, New York, NY, USA), a syringe pump (KDS-100, KD Scientific, Holliston, MA, USA) which feeds an 18 Gauge injector needle with a blunt tip to which the anode is connected, while the cathode consists of a flat plate collector discharged to the ground where the fibers are deposited. The solutions to be processed were prepared by dissolving the PCL with a fixed concentration of 20% w/w with respect to the solvent consisting of a mixture of acetone and acetic acid in a ratio of 3:7 v/v. PEG was added to the same solution as a plasticizer at different concentrations (from 0% to 10% w/w with respect to the polymer content). The solutions were stirred at 60 °C for about 1 h, until the complete solubilization of the polymer, after which, once cooled to room temperature, different quantities of solution were processed to obtain films of different thicknesses, with a voltage of 19 kV, a flow rate of 1.7 mL/h, and a needle-collector distance of 16.5 cm.

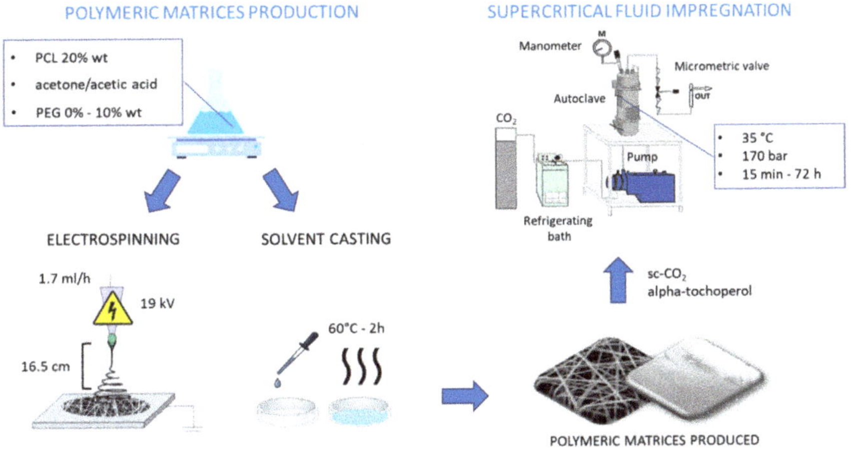

Figure 1. Scheme of the experimental steps based on different combinations of green processes: production of the polymeric matrices by electrospinning and solvent casting (left); supercritical impregnation of the supports (right).

Similarly, for the solvent casting process, the PCL concentration was set at 20% w/w while the amount of PEG was always varied from 0 to 10% in weight with respect to the PCL content. A variable amount of the solutions was poured into Petri dishes with a diameter of 9 cm, placed in an oven at 60 °C for two days, and then left in the desiccator for at least 30 min before being removed from the plates.

2.3. Supercritical Fluid Impregnation

The feasibility tests of supercritical impregnation and the effective loading of alpha-tocopherol into the polymeric supports were performed in a homemade lab-scale plant. It consists of a 100 mL stainless steel autoclave (NWA GmbH, Ahlen, Germany) into which the CO_2, first cooled and then compressed, is fed by a piston pump (Milton Roy, model Milroyal B, Pont-Saint-Pierre, France). An internal impeller, axially mounted, ensures homogeneous mixing. The operating pressure inside the autoclave is monitored by a digital manometer (Parker, Minneapolis, MN, USA), while the thermal control is performed by a PID controller (Watlow, mod. 93, St. Louis, MO, USA). Finally, the CO_2 is returned in gaseous form in atmospheric conditions by depressurization through a micrometric valve (Hoke, mod. 1315G4Y, Spartanburg, SC, USA).

Before proceeding with the supercritical impregnation, a feasibility study was conducted on films of PCL and PCL + PEG at different concentrations, obtained by electrospinning and solvent casting, placed in contact with $scCO_2$. In particular, a fixed surface (about 2 cm^2) of each sample at different thicknesses was weighed and placed inside the autoclave. The tests were carried out at a temperature of 35 °C and a pressure of 170 bar, testing different contact times up to 72 h. At the end of each test, the CO_2 was depressurized to environmental conditions, and the surface and weight of each sample were re-measured. Once the best combinations of the PEG percentage, thickness, and contact times were identified, the supercritical fluid impregnation experiments were performed; the process was carried out by placing a quantity of alpha-tocopherol inside a container mounted axially on the impeller to let it enter in contact with CO_2 once the process has started, keeping the pressure and temperature conditions, reported above, fixed, and varying the impregnation times. Figure 1 reassumed each step of the work, starting from the production of polymeric matrices through electrospinning and solvent casting, followed by the supercritical impregnation of alpha-tocopherol on the supports fabricated.

2.4. Analytical Methods

The morphology of the samples was observed both before and after the supercritical process through Field Emission Scanning Electron Microscopy (FESEM, LEO 1525, Carl Zeiss SMT AG, Jena, Germany) coating the samples with a gold–palladium layer. The mean diameter of the fibers was evaluated using the image processing software ImageJ, analyzing no less than 300 fibers per sample.

The effective loading of alpha-tocopherol on the supports was evaluated by gravimetric method carried out on all samples for each impregnation time. In detail, the initial weight of the samples before the supercritical process and the final weight of the impregnated samples were evaluated. The effective loading was then calculated as the difference between the final and the initial weight.

The migration tests of alpha-tocopherol were carried out on the samples produced both by electrospinning and by solvent casting after the supercritical impregnation process. The samples with known weight were immersed, in static, in a defined volume of EtOH 50% to simulate the behavior of food products with lipidic content. The samples were placed under accelerated mass transfer conditions at 40 °C, analyzing the release fluid by UV-vis at 295 nm at fixed time intervals until a constant value was reached. Migration kinetic was expressed as the mg of alpha-tocopherol released with respect to the mg of alpha-tocopherol loaded by impregnation.

3. Results

3.1. Feasibility Study of the SFI Process and PCL Supports Optimization

Electrospinning and solvent casting techniques were used to produce different polymeric films, as shown in Figure 2, to support the supercritical impregnation of alpha-tocopherol. To prepare the films, different amounts of polymeric solution were processed by both techniques to obtain supports of different thicknesses, according to the method described in Section 2.2. In particular, nanofibrous samples (fiber mean diameter of 850 nm) with a thickness of 30 to 300 μm were obtained by electrospinning, processing from 5 to 10 mL of a 20% *w*/*w* solution of PCL. Samples with a continuous structure, with a thickness of 50 to 200 μm, were produced by solvent casting, pouring on the casting mold from 5 to 10 mL of the 20% PCL solution. The percentage of PEG, in the case of both techniques, was varied. The presence of PEG was instrumental in the case of solvent casting, as it facilitates the removal of the samples from the mold.

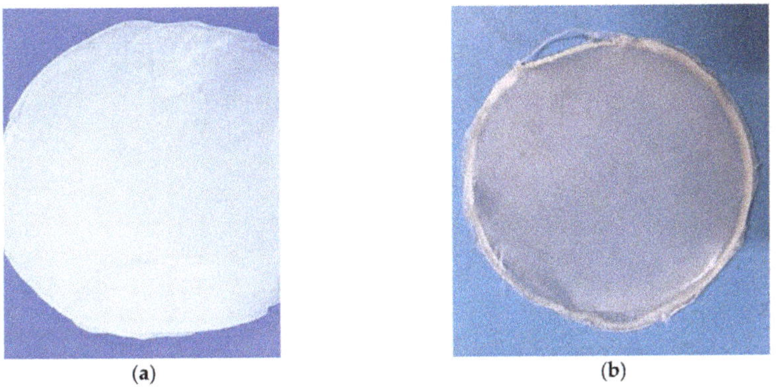

(**a**) (**b**)

Figure 2. PCL 20% wt + PEG 10% wt films produced by (**a**) electrospinning and (**b**) solvent casting.

The feasibility study of supercritical impregnation was conducted by first investigating the behavior of the two different materials placed in contact with supercritical CO_2 (in the absence of alpha-tocopherol). When in contact with $scCO_2$, PCL is subjected to the foaming phenomenon; this causes the formation of pores inside the structure in which the active ingredient can then be impregnated. Unfortunately, however, $scCO_2$ generates a depression of the melting temperature (T_m) in semicrystalline polymers. Campardelli et al. [37] observed that PCL is processable with $scCO_2$ at 40 °C/150 bar or 35 °C/200 bar. For this reason, the first part of the work concerned optimizing the solution composition and operating parameters of film production to obtain a structure that can be maintained after contact with $scCO_2$. In particular, the $scCO_2$ contact tests were conducted on samples with different thicknesses and different PEG content (from 0 to 10% *w*/*w*), at 35 °C, 170 bar, varying the contact times with the $scCO_2$ between 15 min and 72 h.

Table 1 summarizes and schematizes the experimental tests of contact with $scCO_2$, which indicate, for both techniques, the effect obtained and the parameters that most influenced the process.

Table 1. Summarized results of tests conducted on all supports in contact with $scCO_2$ at 35 °C and 170 bar.

Thickness (μm)	% PEG	Contact Time (h)	Macroscopic Aspect	Formation of Pores
		ELECTROSPUN SUPPORTS		
<100	0–2.5–5–10	15 min–72 h	Nonintact structure	yes
>100	0–2.5–5	<2 h	Nonintact structure	yes
>100	5–10	<2 h	Intact structure	yes
		SOLVENT CASTING SUPPORTS		
<100	0–2.5–5–10	15 min–72 h	Nonintact structure	yes
>100	0	<24 h	Nonintact structure	yes
>100	2.5–5	<24 h	Semi-intact structure	yes
>100	10	<24 h	Intact structure	no

In particular, for very thin samples, with a thickness lower than 100 µm, placed in contact with $scCO_2$, it was not possible to maintain the structure of the films, neither for those produced by electrospinning nor for those made by solvent casting; indeed, the effect of T_m depression was predominant for each contact time. Furthermore, the presence of PEG has also been found to have an important role in improving the structure of the samples, since, on the one hand, it acts to improve the handling and, on the other hand, it has been shown to act as a protective barrier of the polymer from the action of $scCO_2$. This is especially the case for supports produced by solvent casting, where its effect is particularly evident and allows to work for longer contact times than electrospun supports, for which, instead, the limiting factor is mainly the contact time. In fact, in the case of casted films, for all the PEG contents, the porous structure was maintained, as shown in Figure 3, in which the FESEM images of the samples obtained by casting before and after the contact with $scCO_2$ are reported. Indeed, after the supercritical process, the structure of the supports obtained by casting has holes, which decrease in size as the PEG content increases. The diameters of the holes, measured by image analysis software, are reported in Table 2.

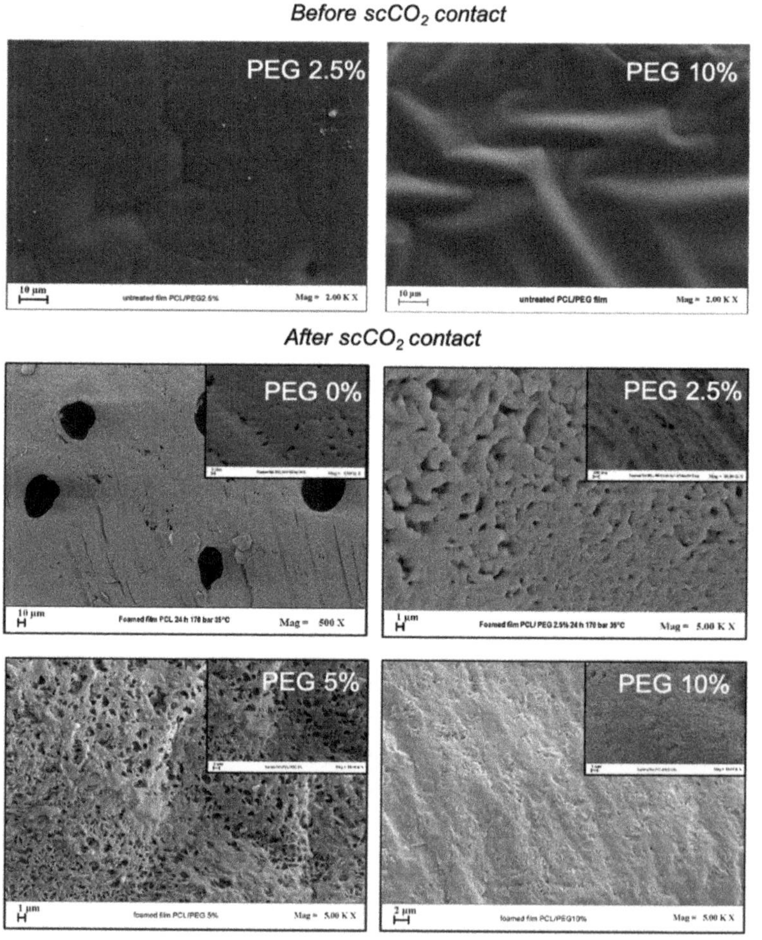

Figure 3. FESEM images of casted film of PCL + PEG 2.5% and PCL + PEG 10% before the $scCO_2$ process; and FESEM images after the contact with $scCO_2$ for 24 h at 170 bar and 35 °C as the PEG percentage varies.

Table 2. The pore size of the casted films after contact with $scCO_2$ for 24 h at 170 bar and 35 °C.

PEG Content (%)	Pore Size (µm)	Standard Deviation (µm)
0	1.22	0.28
2.5	0.85	0.22
5	0.46	0.09
10	0.15	0.04

In Figure 4, the FESEM images of the samples obtained by electrospinning before the contact with $scCO_2$ and after the contact referring to the best conditions investigated are reported.

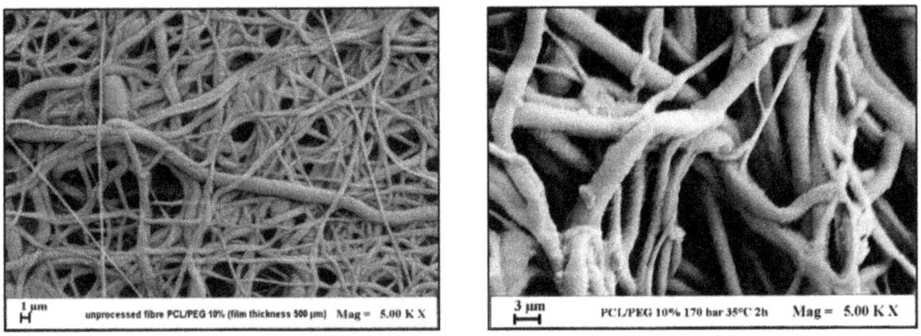

(a) (b)

Figure 4. FESEM images of the electrospun films of PCL + PEG 10% with a thickness of 500 µm: (**a**) before the contact with $scCO_2$ and (**b**) after the 2 h of contact.

Considering that the FESEM images reported in Figure 4 are taken at the same magnification, it is evident that the contact of the fibers with $scCO_2$ causes the swelling of the fibers without compromising the structure of the sample. The optimization of the supports, intended as maintaining the macrostructure of the samples, was therefore obtained by working with samples with a thickness greater than 100 µm, with a PEG content equal to 10%, in particular for casting supports, and for contact times with the supercritical fluid not longer than 2 h for the electrospun supports. For lower thicknesses, lower quantities of PEG, and higher contact times, the structure is damaged, as seen in Figure 5, which shows the results before (a) and after (b) optimizing the materials' production.

(a) (b)

Figure 5. Macroscopic structure of electrospun PCL films after contact with $scCO_2$: (**a**) not optimized and (**b**) optimized.

3.2. Loading of Alpha-Tocopherol and Study of the Supercritical Impregnation Kinetics Performed on the Optimized Supports

Having identified the limits and the best conditions for subjecting both types of polymeric supports to the supercritical process, the systematic study of alpha-tocopherol loading was started. The impregnation process was carried out on the samples (support size of 2 cm^2) at 170 bar and 35 °C at different impregnation times, depending on the support production technique, considering the process optimization performed in Section 3.1.

In particular, concerning the samples produced by casting, it was seen that the concentration of PEG is a determining parameter for the preservation of the macrostructure. It was, therefore, decided to investigate the kinetics of impregnation as the concentration of PEG varies, from 0% to 10%, initially setting an impregnation time equal to 24 h, with time identified as the maximum above which even the samples produced with the maximum concentration of PEG and high thicknesses are damaged. The main results obtained are shown in Table 3.

Table 3. Kinetics of impregnating alpha-tocopherol in PCL films by casting varying % of PEG [set at impregnation time 24 h, at 250 °C and 170 bar].

% PEG	Alpha-Tocopherol Impregnated (mg_{TOC}/mg_{FILM})	Effect
0	1.234	Higher impregnation but a significant modification of the structure
2.5	0.998	Impregnation and foaming
5	0.959	Impregnation and foaming
10	0.828	Lower impregnation but preservation of the structure

Specifically, loading the antioxidant into the supports without PEG, a higher effect of CO_2 was observed: larger pores were formed, a higher amount of alpha-tocopherol was charged into the sample, but the macroscopic structure was damaged. In the samples with PEG values equal to 2.5% and 5%, both the impregnation of the antioxidant compound and the formation of pores were observed, while for PEG values equal to 10%, the impregnation was prevalent with respect to the modification of the structure.

On the other hand, alpha-tocopherol impregnation results showed lower loadings as the percentage of PEG increases, going from 1.234 mg_{TOC}/mg_{FILM} loaded on the sample of only PCL to 0.828 mg_{TOC}/mg_{FILM} loaded on samples of PCL + PEG 10% with the same contact time of 24 h. This trend can be explained by considering the morphology study carried out through FESEM analysis, reported in Figure 6. Indeed, as the concentration of PEG increases, the size of the pores decreases. Considering that the structures formed are closed-cell structures, the larger pores offer a greater exposed surface in which the alpha-tocopherol can be impregnated; therefore, the larger the pores, the higher the amount of alpha-tocopherol charged.

In the case of the samples obtained by electrospinning, the effect of PEG% on alpha-tocopherol impregnation was not evaluated since PCL + PEG 10% was the only condition that allows preserving the structure of the polymeric matrix for a contact time lower than 2 h. The alpha-tocopherol impregnation kinetics was then studied, varying the impregnation time, using PCL + PEG 10% films for both electrospun and casting films. The results are reported in Table 3. For the films obtained by casting, the impregnation kinetics was studied from 2 to 24 h. In the case of electrospun mats, the impregnation time varied in the range from 15 min to 2 h, considering the previous feasibility study reported in Section 3.1. From Table 4, it can be seen that the fibrous supports allow a more effective impregnation of the active compound than those obtained by casting with the same contact time; the alpha-tocopherol amount, which is loaded on the fibers in 2 h, is loaded on the film obtained by casting in 6 h. Therefore, the porous structure of the electrospun sample, once optimized, the composition of the fibers with PEG at 10% *w/w* and the impregnation time can produce the same active compound loading in a lower time.

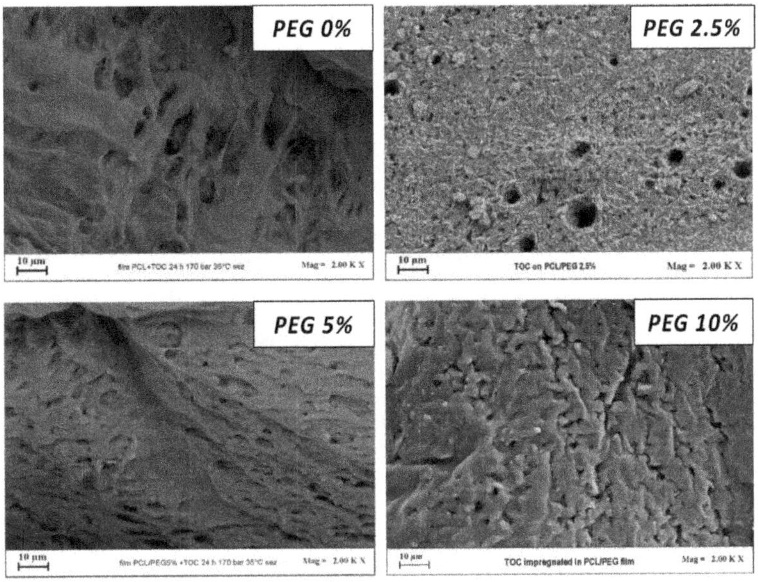

Figure 6. FESEM images of the casted samples at different PEG percentages after an impregnation time of 24 h, at 35 °C and 170 bar.

Table 4. Alpha-tocopherol loading on PCL + 10% PEG films produced by casting and electrospinning technique at different impregnation times, at 35 °C and 170 bar.

Impregnation Time	Alpha-Tocopherol Impregnated (mg_{TOC}/mg_{FILM})
PCL + PEG 10% CASTING FILMS	
2 h	0.25
6 h	0.40
15 h	0.67
24 h	0.82
PCL + PEG 10% ELECTROSPUN FILMS	
15 min	0.17
1 h	0.31
2 h	0.64

The morphological aspect of the fibers after the supercritical impregnation can be observed in Figure 7, where FESEM images taken at the same enlargement of samples processed at different impregnation times are reported. In particular, from Figure 7, it is possible to note that the plasticizing effect of the presence of alpha-tocopherol [47,48] contributes to lowering the T_m of the polymer, even more than CO_2 alone can. In fact, the fibers obtained after impregnation with alpha-tocopherol for 2 h have a less regular (and slightly melted) appearance than those obtained by processing the polymer with CO_2 in the absence of an active ingredient. The fibers obtained considering impregnation times equal to 1 h and 15 min have a stable structure. In particular, from the FESEM image related to 15 min impregnation time, it is possible to observe the presence of alpha-tocopherol in the form of nanoparticles on and in the fibers. Indeed, during the depressurization step, the carbon dioxide with the active ingredient solubilized in it returns to the gaseous form; therefore, the solute precipitates.

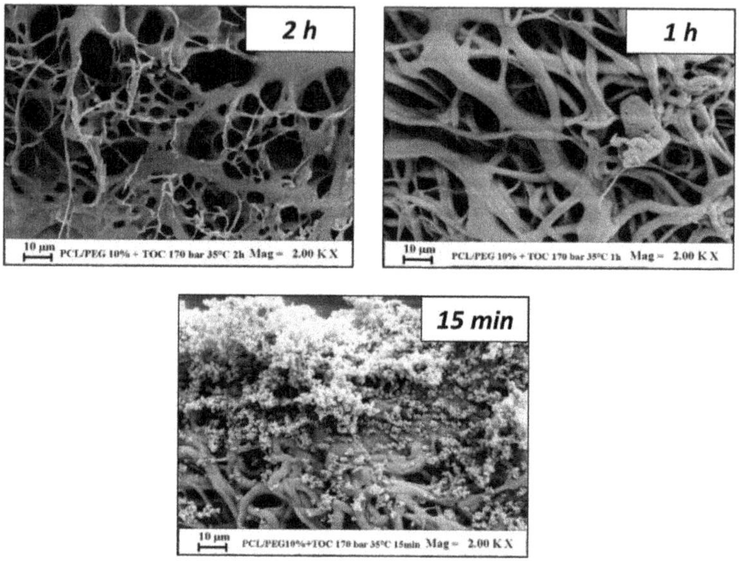

Figure 7. FESEM images of the electrospun fibers impregnated with alpha-tocopherol at 35 °C and 170 bar and different impregnation times: 2 h,1 h, and 15 min.

Electrospun films are more suitable for impregnation than those obtained by casting when the proper impregnation conditions are selected to avoid the denaturation of the structure because of the depression of the melting temperature caused by the interaction with $scCO_2$ and with alpha-tocopherol. Therefore, a good balance between impregnation time and film characteristics should be studied each time.

3.3. Migration Test of Alpha-Tocopherol into Fluid Simulant

Migration tests for the evaluation of alpha-tocopherol release from polymeric supports were carried out on the optimized samples considering, for each, the maximum impregnation time: PCL films produced by electrospinning impregnated for 2 h and PCL + PEG 10% films produced by casting impregnated for 24 h. The tests were carried out in static at 40 °C, using a 50% hydroalcoholic solution of ethanol as a simulant fluid for food products with a lipidic content. The released fluid was analyzed for one week, at predetermined time intervals, by UV-vis analysis at 295 nm through which the release percentage expressed as mg of alpha-tocopherol released/mg of theoretically impregnated alpha-tocopherol was estimated. The quantity of sample placed in contact with the simulant was evaluated to have the same initial theoretical load of impregnated antioxidant, about 9.5 mg per sample, for both techniques.

Figure 8 shows the graph comparing the alpha-tocopherol release kinetics from the impregnated samples produced by electrospinning and solvent casting. The samples produced by casting have, in the first hours, higher release kinetics than those produced by electrospinning, even though the fibers have a greater surface area exposed to the stimulant. After the first two days, the migration kinetics from electrospun samples becomes greater than those produced by casting. This trend is due to the fact that tocopherol, as observed in the SEM images of Figures 6 and 7, penetrates much better into the fibers, for contact times of 2 h, compared to casting films impregnated for 24 h, where instead the impregnation is more superficial, as the pores for PCL + PEG 10% samples are small and offer a smaller surface area exposed to the supercritical process. For this reason, the release from the fibers is more gradual and constant over time. For both techniques, however, about 50% of the impregnated tocopherol was released in 48 h.

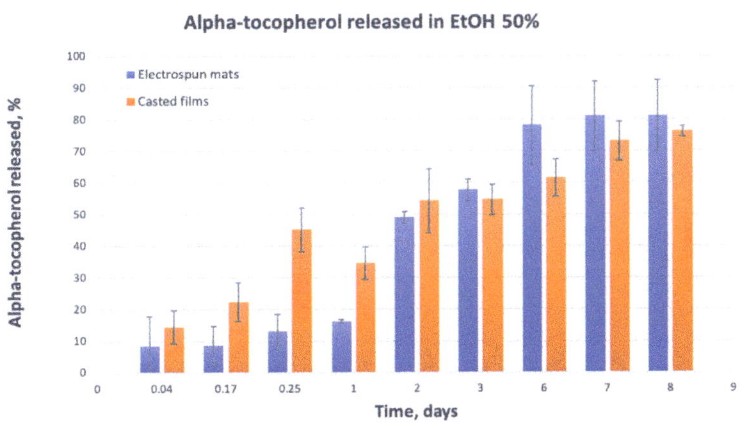

Figure 8. Release of alpha-tocopherol from electrospun sample impregnated for 2 h and from casted sample impregnated for 24 h, in EtOH 50% as a simulant, for a week at 40 °C.

4. Conclusions

This paper focused on the use of alternative green techniques for the preparation of active packaging films. The polymeric support was produced both by electrospinning and by solvent casting, and then it was loaded with alpha-tocopherol by impregnation with $scCO_2$. We demonstrated that the operating conditions have to be suitably selected as PCL has a low melting temperature, which is significantly reduced due to the presence of CO_2 in supercritical conditions. Furthermore, alpha-tocopherol has a plasticizing effect, which contributes to the negative modifying of the polymer structure. Therefore, for both tested green techniques combinations (casting with supercritical fluid impregnation and electrospinning with supercritical fluid impregnation), the optimal operating conditions, summarized in Table 5, were determined to obtain an active packaging.

Table 5. Summary of the optimized set condition for both electrospun and casted PCL films to be impregnated with $scCO_2$.

Conditions	Electrospun Films	Casted Films
Thickness (µm)	>100	>100
PEG content %	10	10
Impregnation time [h]	<2	<24

Author Contributions: Conceptualization, I.D.M.; Data curation, E.D.; Formal analysis, E.D.; Funding acquisition, R.C. and I.D.M.; Investigation, E.D.; Project administration, P.P.; Resources, P.P.; Supervision, R.C.; Writing—original draft, E.D.; Writing—review & editing, E.D., R.C., I.D.M. and P.P. All authors have read and agreed to the published version of the manuscript.

Funding: This research received no external funding.

Institutional Review Board Statement: Not applicable.

Informed Consent Statement: Not applicable.

Data Availability Statement: Not applicable.

Conflicts of Interest: The authors declare no conflict of interest.

References

1. Gowthaman, N.S.K.; Lim, H.N.; Sreeraj, T.R.; Amalraj, A.; Gopi, S. Advantages of biopolymers over synthetic polymers: Social, economic and environmental aspects. In *Biopolymers and Their Industrial Applications*; Elsevier: Amsterdam, The Netherlands, 2021; Volume 15, pp. 351–372. [CrossRef]
2. Dey, A.; Dhumal, C.V.; Sengupta, P.; Kumar, A.; Pramanik, N.K.; Alam, T. Challenges and possible solutions to mitigate the problems of single-use plastics used for packaging food items: A review. *J. Food Sci. Technol.* **2021**, *58*, 3251–3269. [CrossRef] [PubMed]
3. Chen, Y.; Awasthi, A.K.; Wei, F.; Tan, Q.; Li, J. Single-use plastics: Production, usage, disposal, and adverse impacts. *Sci. Total Environ.* **2021**, *752*, 141772. [CrossRef] [PubMed]
4. George, A.; Sanjay, M.R.; Srisuk, R.; Parameswaranpillai, J.; Siengchin, S. A comprehensive review on chemical properties and applications of biopolymers and their composites. *Int. J. Biol. Macromol.* **2020**, *154*, 329–338. [CrossRef]
5. Adeyeye, O.A.; Sadiku, E.R.; Reddy, A.B.; Ndamase, A.S.; Makgatho, G.; Sellamuthu, P.S.; Perumal, A.B.; Nambiar, R.B.; Fasikuldowu, V.O.; Ibrahim, I.D.; et al. The Use of Biopolymers in Food Packaging. In *Green Biopolymers and Their Nanocomposites*; Springer: Berlin/Heidelberg, Germany, 2019; pp. 137–158. [CrossRef]
6. Porta, R.; Sabbah, M.; Di Pierro, P. Biopolymers as Food Packaging Materials. *Int. J. Mol. Sci.* **2020**, *21*, 4942. [CrossRef]
7. Torres-Giner, S.; Figueroa-Lopez, K.J.; Melendez-Rodriguez, B.; Prieto, C.; Pardo-Figuerez, M.; Lagaron, J.M. Emerging Trends in Biopolymers for Food Packaging. *Sustain. Food Packag. Technol.* **2021**, 1–33. [CrossRef]
8. Lionetto, F.; Corcione, C.E. Recent Applications of Biopolymers Derived from Fish Industry Waste in Food Packaging. *Polymers* **2021**, *13*, 2337. [CrossRef]
9. Calva-Estrada, S.J.; Jimènez-Fernàndez, M.; Lugo-Cervantes, E. Protein-Based Films: Advances in the Development of Biomaterials Applicable to Food Packaging. *Food Eng. Rev.* **2019**, *11*, 78–92. [CrossRef]
10. Bilal, M.; Gul, I.; Basharat, A.; Qamar, S.A. Polysaccharides-based bio-nanostructures and their potential food applications. *Int. J. Biol. Macromol.* **2021**, *176*, 540–557. [CrossRef] [PubMed]
11. Mantecón-Oria, M.; Diban, N.; Berciano, M.T.; Rivero, M.J.; David, O.; Lafarga, M.; Tapia, O.; Urtiaga, A. Hollow Fiber Membranes of PCL and PCL/Graphene as Scaffolds with Potential to Develop In Vitro Blood-Brain Barrier Models. *Membranes* **2020**, *10*, 161. [CrossRef]
12. Fawal, G.E.; Hong, H.; Mo, X.; Wang, H. Fabrication of scaffold based on gelatin and polycaprolactone (PCL) for wound dressing application. *J. Drug Deliv. Sci. Technol.* **2021**, *63*, 102501. [CrossRef]
13. Malikmammadov, E.; Tanir, T.E.; Kiziltay, A.; Hasirci, V.; Hasirci, N. PCL and PCL-based materials in biomedical applications. *J. Biomater. Sci. Polym. Ed.* **2018**, *29*, 863–893. [CrossRef]
14. Dash, T.K.; Konkimalla, V.B. Polymeric Modification and Its Implication in Drug Delivery: Poly-ε-caprolactone (PCL) as a Model Polymer. *Mol. Pharm.* **2012**, *9*, 2365–2379. [CrossRef]
15. Hadi, A.; Mahsa, J.O.; Ayda, S. A review on techniques utilized for design of controlled release food active packaging. *Crit. Rev. Food Sci. Nutr.* **2020**, 1–21. [CrossRef]
16. Firouz, M.S.; Mohi-Alden, K.; Omid, M. A critical review on intelligent and active packaging in the food industry: Research and development. *Food Res. Int.* **2021**, *141*, 110113. [CrossRef]
17. Commission Regulation (EU). No. 10/2011 of 14 January 2011 on Plastic Materials and Articles Intended to Come into Contact with Food. *Off. J. Eur. Union L* **2011**, *12*, 1.
18. Kanatt, S.R. Development of active/intelligent food packaging film containing Amaranthus leaf extract for shelf-life extension of chicken/fish during chilled storage. *Food Packag. Shelf Life* **2021**, *24*, 100506. [CrossRef]
19. Moreirinha, C.; Vilela, C.; Silva, H.C.S.; Pinto, R.J.B.; Almeida, A.; Rocha, M.A.M.; Coelho, E.; Coimbra, M.A.; Silvestre, A.J.D.; Freire, C.S.R. Antioxidant and antimicrobial films based on brewers spent grain arabinoxylans, nanocellulose and feruloylated compounds for active packaging. *Food Hydrocoll.* **2020**, *108*, 105836. [CrossRef]
20. Ibrahim, S.; Elsayed, H.; Hasanin, M. Biodegradable, Antimicrobial and Antioxidant Biofilm for Active Packaging Based on Extracted Gelatin and Lignocelluloses Biowastes. *J. Polym. Environ.* **2020**, *29*, 472–482. [CrossRef]
21. Piri, H.; Moradi, S.; Amiri, R. The fabrication of a novel film based on polycaprolactone incorporated with chitosan and rutin: Potential as an antibacterial carrier for rainbow trout packaging. *Food Sci. Biotechnol.* **2021**, *30*, 683–690. [CrossRef] [PubMed]
22. Uzunlu, S.; Niranjan, K. Laboratory antimicrobial activity of cinnamaldehyde and pomegranate-based polycaprolactone films. *J. Appl. Polym. Sci.* **2017**, *134*, 45347. [CrossRef]
23. Sikora, J.; Majewski, L.; Puszka, A. Modern Biodegradable Plastics-Processing and Properties: Part I. *Materials* **2020**, *13*, 1986. [CrossRef] [PubMed]
24. Bastarrachea, L.J.; Wong, D.E.; Roman, M.J.; Lin, Z.; Goddard, J.M. Active Packaging Coatings. *Coatings* **2015**, *5*, 771–791. [CrossRef]
25. Zhang, C.; Li, Y.; Wang, P.; Zhang, H. Electrospinning of nanofibers: Potentials and perspectives for active food packaging. *Compr. Rev. Food Sci. Food Saf.* **2020**, *19*, 479–502. [CrossRef] [PubMed]
26. Sameen, D.E.; Ahmed, S.; Lu, R.; Li, R.; Dai, J.; Qin, W.; Zhang, Q.; Li, S.; Liu, Y. Electrospun nanofibers food packaging: Trends and applications in food systems. *Crit. Rev. Food Sci. Nutr.* **2021**, 1–14. [CrossRef] [PubMed]
27. Luraghi, A.; Peri, F.; Moroni, L. Electrospinning for drug delivery applications: A review. *J. Control Release* **2021**, *334*, 463–484. [CrossRef] [PubMed]

28. Pasha, H.Y.; Mohtasebi, S.S.; Tabatabaeekoloor, R.; Taherimehr, M.; Javadi, A.; Firous, M.S. Preparation and characterization of the plasticized polylactic acid films produced by the solvent-castin method for food packaging applications. *J. Food Process. Preserv.* **2021**, e16089. [CrossRef]
29. Peng, W.; Li, Y.; Zhang, C.; Que, F.; Weiss, J.; Zhang, H. Characterization and antioxidant activity of trilayer gelatin/dextran-propyl gallate/gelatin films: Electrospinning versus solvent casting. *LWT* **2020**, *128*, 109536. [CrossRef]
30. Tampau, A.; Gonzalez-Martinez, C.; Chiralt, A. Polylactic acid-based materials encapsulating carvacrol obtained by solvent casting and electrospinning. *J. Food Sci.* **2020**, *85*, 1177–1185. [CrossRef] [PubMed]
31. Drago, E.; Franco, P.; Campardelli, R.; De Marco, I. Zein electrospun fibers purification and vanillin impregnation in a one-step supercritical process to produce safe active packaging. *Food Hydrocoll.* **2022**, *122*, 107082. [CrossRef]
32. Franco, P.; Incarnato, L.; De Marco, I. Supercritical CO_2 impregnation of α-tocopherol into PET/PP films for active packaging applications. *J. CO_2 Util.* **2019**, *34*, 266–273. [CrossRef]
33. Mosquera, J.E.; Goñi, M.L.; Martini, R.E.; Gañàn, N.A. Mass transfer kinetics of CO_2 and eugenol in the supercritical impregnation of polyamide fibers: Experimental data and modeling. *J. Supercrit. Fluid.* **2020**, *166*, 105030. [CrossRef]
34. Bastante, C.C.; Silva, H.C.S.; Cardoso, L.C.; Serrano, C.M.; Martinez de la Ossa, E.J.; Freire, C.S.R.; Vilela, C. Biobased films of nanocellulose and mango leaf extract for active food packaging: Supercritical impregnation versus solvent casting. *Food Hydrocoll.* **2021**, *117*, 106709. [CrossRef]
35. Champeau, M.; Thomassin, J.-M.; Tassaing, T.; Jérome, C. Drug loading of polymer implants by supercritical CO_2 assisted impregnation: A review. *J. Control Release* **2015**, *209*, 248–259. [CrossRef]
36. Yoganathan, R.; Mammucari, R.; Foster, N.R. Impregnation of Ibuprofen into Polycaprolactone Using Supercritical Carbon Dioxide. *J. Phys. Conf. Ser.* **2010**, *215*, 012087. [CrossRef]
37. Campardelli, R.; Franco, P.; Reverchon, E.; De Marco, I. Polycaprolactone/nimesulide patches obtained by a one-step supercritical foaming + impregnation process. *J. Supercrit. Fluid.* **2019**, *146*, 47–54. [CrossRef]
38. Liparoti, S.; Franco, P.; Pantani, R.; De Marco, I. Supercritical CO_2 impregnation of caffeine in biopolymer films to produce anti-cellulite devices. *J. Supercrit. Fluids* **2022**, *179*, 105411. [CrossRef]
39. Yañez, F.; Martikainen, L.; Braga, M.E.M.; Alvarez-Lorenzo, C.; Concheiro, A.; Duarte, C.M.M.; Gil, M.H.; De Sousa, H.C. Supercritical fluid-assisted preparation of imprinted contact lenses for drug delivery. *Acta Biomater.* **2011**, *7*, 1019–1030. [CrossRef] [PubMed]
40. Shen, Z.; Huvard, G.S.; Warriner, C.; Mc Hugh, M.; Banyasz, J.L.; Mishra, M.K. CO_2-assisted fiber impregnation. *Polymers* **2008**, *49*, 1579–1586. [CrossRef]
41. Hussain, Y.; Grant, C.S. Ibuprofen impregnation into submicron polymeric films in supercritical carbon dioxide. *J. Supercrit. Fluids* **2012**, *71*, 127–135. [CrossRef]
42. Natu, M.V.; Gil, M.H.; De Sousa, H.C. Supercritical solvent impregnation of poly(ε-caprolactone)/poly(oxyethylene-b-oxypropylene-b-oxyethylene) and poly(ε-caprolactone)/poly(ethylene-vinyl acetate) blends for controlled release applications. *J. Supercrit. Fluids* **2008**, *47*, 93–102. [CrossRef]
43. Garcia-Casas, I.; Montes, A.; Valor, D.; Pereyra, C.; Martinez de la Ossa, E.J. Foaming of Polycaprolactone and Its Impregnation with Quercetin Using Supercritical CO_2. *Polymers* **2019**, *11*, 1390. [CrossRef] [PubMed]
44. Gaikwad, K.K.; Singh, S.; Ajji, A. Moisture absorbers for food packaging applications. *Environ. Chem. Lett.* **2019**, *17*, 609–628. [CrossRef]
45. Dumitriu, R.P.; Stoleru, E.; Mitchell, G.R.; Vasile, C.; Brebu, M. Bioactive Electrospun Fibers of Poly(ε-Caprolactone) Incorporating α-Tocopherol for Food Packaging Applications. *Molecules* **2021**, *26*, 5498. [CrossRef] [PubMed]
46. Raza, Z.A.; Abid, S.; Azam, A.; Rehman, A. Synthesis of alpha-tocopherol encapsulated chitosan nano-assemblies and their impregnation on cellulosic fabric for potential antibacterial and antioxidant cosmetotextiles. *Cellulose* **2020**, *27*, 1717–1731. [CrossRef]
47. Hwang, S.W.; Shim, J.K.; Selke, S.E.M.; Soto-Valdez, H.; Matuana, L.; Rubino, M.; Auras, R. Poly(L-lactic acid) with added α-tocopherol and resveratrol: Optical, physical, thermal and mechanical properties. *Polym. Int.* **2012**, *61*, 418–425. [CrossRef]
48. Jamshidian, M.; Tehrany, E.A.; Imran, M.; Akhtar, M.J.; Cleymand, F.; Desobry, S. Structural, mechanical and barrier properties of active PLA-antioxidant films. *J. Food Eng.* **2012**, *110*, 380–389. [CrossRef]

processes

Article

Graphite Classification Based on Improved Convolution Neural Network

Guangjun Liu [1,*], Xiaoping Xu [1], Xiangjia Yu [1] and Feng Wang [2]

[1] School of Sciences, Xi'an University of Technology, Xi'an 710054, China; xpxuxaut@sina.com (X.X.); xjyuxaut@sina.com (X.Y.)
[2] School of Mathematics and Statistics, Xi'an Jiaotong University, Xi'an 710049, China; wangf@mail.xjtu.edu.cn
* Correspondence: liuguangjun202106@126.com

Abstract: In the development of high-tech industries, graphite has become increasingly more important. The world has gradually entered the graphite era from the silicon era. In order to make good use of high-quality graphite resources, a graphite classification and recognition algorithm based on an improved convolution neural network is proposed in this paper. Based on the self-built initial data set, the offline expansion and online enhancement of the data set can effectively expand the data set and reduce the risk of deep convolution neural network overfitting. Based on the visual geometry group 16 (VGG16), residual net 34 (ResNet34), and mobile net Vision 2 (MobileNet V2), a new output module is redesigned and loaded into the full connection layer. The improved migration network enhances the generalization ability and robustness of the model; moreover, combined with the focal loss function, the superparameters of the model are modified and trained on the basis of the graphite data set. The simulation results illustrate that the recognition accuracy of the proposed method is significantly improved, the convergence speed is accelerated, and the model is more stable, which proves the feasibility and effectiveness of the proposed method.

Keywords: graphite; classification; transfer learning; focal loss; convolution neural network

Citation: Liu, G.; Xu, X.; Yu, X.; Wang, F. Graphite Classification Based on Improved Convolution Neural Network. *Processes* **2021**, *9*, 1995. https://doi.org/10.3390/pr9111995

Academic Editor: Roberta Campardelli

Received: 11 October 2021
Accepted: 4 November 2021
Published: 9 November 2021

1. Introduction

At present, nonmetallic mineral resources are widely used in emerging materials, energy, national defense, aerospace industry, and other fields [1–7]. Its strategic significance is gradually improved and its application field is expanding. Therefore, the research on graphite is of great significance [8–14]. In general, graphite can be divided into two categories: natural graphite and artificial graphite. They are similar in structure and physicochemical properties, but their use paths are quite different. In metallurgy, machinery industry, chemical industry, and many other application fields, it is particularly important to identify the type of graphite. If the two are confused, decision-making errors will result, leading to much resource waste and economic losses. Therefore, from the perspective of strategic demand, research on the intelligent classification of graphite has great far-reaching significance for the classification management of graphite resources.

Most of the traditional methods to identify graphite are microscope identification or chemical identification. These methods have a large human factor and high time cost, which bring great challenges to the industry [15]. Currently, artificial intelligence research based on deep learning provides the most advanced solution for computer vision. As a deep learning method, convolution neural network (CNN) is popular in different disciplines of image recognition application. Small details that people do not notice can be easily distinguished by CNN. This algorithm can directly extract visual patterns from pixel images with minimal preprocessing. CNN structure introduced by LeCun net (LeNet) architecture and Alex Krizhevsky net (AlexNet) makes CNN popular. Since then, CNN's popularity has grown exponentially with a variety of designs and applications [16–20].

Due to the high similarity of the four types of graphite images shown in Figure 1, the recognition rate of shallow CNN (such as AlexNet) is low, and the model is relatively simple, which may limit its classification accuracy. Considering the recognition accuracy, recognition time, and model parameters, VGG16, ResNet34, and MobileNet V2, which have more complex network structure and deeper layers, are selected to improve classification accuracy. These networks have less hardware requirements and lower computational complexity, and can effectively extract deep features of images while shortening training time. Furthermore, these networks avoid the complex method of manual feature extraction and improve the recognition accuracy. In view of the characteristics of graphite images and the small data set, the data set is preprocessed to achieve the purpose of data enhancement. At the same time, preprocessing resets the loading module of the migration network, debugs the output parameters of each layer connection, improves the convergence speed of the network, and adds a dropout layer optimization model to effectively integrate the migration network and the model.

(**a**) Flake graphite (**b**) Expanded graphite

(**c**) Artificial graphite (**d**) Carbon black

Figure 1. Images showing four types of graphite.

Traditional CNN models mostly use cross entropy loss function (CEL), but CEL only considers the accuracy of correct tags, ignoring the difference of incorrect tags, so most researchers have studied the loss function. Triplet loss was proposed by Schroff in 2015 [21]. It is mainly used to train samples with small differences. However, the training process of this function is unstable, the convergence speed is slow, and it may fall into a local optimum. In 2017, for target detection of dense objects, heproposed focal loss function (FL), which not only solved the imbalance problem of samples, but also solved the imbalance problem of difficult and easy samples, making the network give more attention to difficult samples [22]. Recognizing that the graphite data set image difference is small, not easy to distinguish, and a difficult sample is far more challenging than an easy sample, this paper selects FL as the loss function to improve the stability of the network. The simulation results show that the improved migration network combined with FL has stronger robustness and generalization ability, higher recognition accuracy, and achieves an ideal recognition effect.

2. Experimental Data

2.1. Data Set Construction

In this paper, the experimental samples were collected in Neixiang County, Shaanxi Province, China, from the flake graphite, artificial graphite, expanded graphite, and carbon black samples of Ruixin high-temperature composite products. The samples were photographed from different directions for 3 min by using an iPhone rear camera at the same position, and the resolution of the video was 1080 p, the captured graphite video was extracted by frame, 1 frame per second. After screening, 715 sample images were obtained and classified. During the experiment, all the images were obtained from the saved data. The four types of graphite images are shown in Figure 1. However, it is difficult to obtain a variety of graphite data sets. The images collected sometimes have the characteristics of uneven distribution and few images, so it was necessary to enhance the data set. The data enhancement method is given in Section 2.2.

2.2. Offline Expansion and Online Enhancement of Data Sets

The data enhancement method can be divided into offline enhancement and online enhancement. Offline enhancement directly processes the data set, which is often used when the data set is small. Online enhancement is suitable for large data sets. After the batch data are obtained in the process of model training, the batch data are enhanced, such as rotation, translation, folding, and other corresponding changes. In this paper, when the initial data set was small, offline expansion was adopted to directly process the data set and increase the sample size of the data set, and the collected images were flipped vertically, rotated by 90 degrees, flipped around an axis, and underwent affine transformation; finally, 3575 images were obtained. Moreover, in order to reduce the overfitting problem of deep CNN, online enhancement was used to process the data. Based on the characteristics of the graphite image, the training set images were randomly scrambled before input to reduce the influence of image order on the model. Each image in the training set was randomly cropped, the image pixels unified, and the image flipped horizontally according to the probability (here, setting $p = 0.5$). Finally, the processed data set was normalized, and the data of each channel were regularized with the mean value 0.5 and standard deviation 0.5, so as to achieve the purpose of data expansion and data enhancement.

According to the proportion, 2860 data sets (80%) were randomly selected as the training set to train the network, and 715 data sets (20%) were selected as the test set to verify the performance of the network. The training set and the test set were processed separately. The classification of data sets is shown in Table 1.

Table 1. Graphite data set.

	Flake Graphite	Expanded Graphite	Artificial Graphite	Carbon Black
Data set	1020	965	715	875
Training set	816	772	572	700
Test set	204	193	143	175

3. Experimental Principle and Method

3.1. Transfer Learning

In recent years, successful case studies regard deep CNN as the mainstream method to solve challenging computer vision tasks [23]. However, training a network from scratch requires many training data, much time, and a large graphic processing unit (GPU). We only have 3575 training and test images, and a small data set cannot train a deep CNN. Therefore, this paper combines the idea of transfer learning to train the convolution neural network.

Transfer learning refers to the process of using models that have been trained for different tasks, hoping that the model can have sufficient generalized information to solve new specific tasks [24]. This method uses CNN, a pretraining model on a huge database, to help learning target tasks [25]. The classified image in this paper is graphite, which has different visual performance, which has a visual performance different from cats and dogs. The visual performance learned from a large image may not be able to represent the graphite image very well, so it is necessary to modify the pretrained CNN structure to adapt to our task. Zeiler and Fergus provide evidence that the general image representation learned from the pretrained CNN is superior to the most advanced handmade features [26].

3.2. Convolution Neural Network Model

At present, common CNN include AlexNet, visual geometry group (VGG), residual net (ResNet), mobile net (MobileNet), etc.; the ready-made image representation learned from these deep networks is powerful and universal, and has been used to solve many visual recognition problems [27]. Considering the good performance of these ready-made CNN features, they have become the mainstream image features to solve most computer vision problems [28].

Deep convolution neural network has the characteristics of deep depth and good recognition effect. VGG16, ResNet34, and MobileNet V2 are representative CNNs in recent years, which have achieved excellent results in the field of image classification. VGG uses several smaller convolution kernels instead of larger ones to ensure that the network depth can continue to increase under the condition of the same receptive field. VGG16 has 16 hidden layers (13 convolution layers and 3 fully connected layers), which is a classic CNN model for image classification in the deep learning framework. ResNet puts forward residual structure and uses convolution to reduce dimension, which makes ResNet have advantages in parameters, depth, width, and calculation cost without introducing additional parameters and increasing network calculation complexity. There are less than 4000 data sets in this paper, and the depth of ResNet34 model can be used effectively to train the samples. As a lightweight network proposed in 2018, MobileNet V2 always uses depthwise (DW) convolution to extract features, and proposes inverted residual, which can reduce the use of memory in implementation [29]. Based on these advantages, this paper establishes graphite classification and recognition model based on VGG16, ResNet34, and MobileNet V2 networks; the comparative analysis is shown in Table 2.

Table 2. Comparative analysis of models.

Network	VGG16	ResNet34	MobileNet V2
Year	2014	2015	2018
Top-1 accuracy	71.5%	-	71.7%
Number of parameters	138,357,544	63,470,656	4,253,864
Number of layers	16	34	-
BN	No	Yes	Yes
Residual structure	No	Yes	Yes

3.3. Focus Loss Function

In image classification, the loss function is generally cross entropy loss function. Cross entropy is the distance between the real output (probability) and the predicted output (probability); therefore, the smaller the value of cross entropy, the closer the two probability distribution. For the multiclassification of graphite, there are a large number of simple negative samples in the data set. The losses caused by these samples in the iterative process occupy the majority of the total losses, which may lead us to deviate from the correct optimization direction. To improve this situation, the focus loss function is used in this paper; the formula is shown in Equation (1).

$$L_{\mathrm{FL}} = \begin{cases} -(1-y')^{\gamma}\log y',\ y=1 \\ -y^{\gamma}\log(1-y'),\ y=0 \end{cases} \tag{1}$$

Equation (1) gives the contribution of classified samples to the loss. Due to the effect of modulation factor γ, the model weakens the error contribution of easy-to-identify samples, so that the model can focus on the hard-to-classify samples more effectively in training. After many experiments, $\gamma = 0.25$ is set. On this basis, FL also introduces a balance factor α, as shown in Equation (2). This factor can be used to balance the imbalance between the number of positive and negative samples, that is, class imbalance. Because the class imbalance problem of the self-built graphite data set is not very prominent, we set γ= NAN.

$$L_{FL} = \begin{cases} -\alpha(1-y')^{\gamma}\log y', \ y=1 \\ -(1-\alpha)y^{\gamma}\log(1-y'), \ y=0 \end{cases} \tag{2}$$

3.4. Improved Migration Network Model

Combined with transfer learning, the model training process designed in this paper includes the following three stages: data expansion, transfer learning, and superparameter adjustment. The model training process is shown in Figure 2.

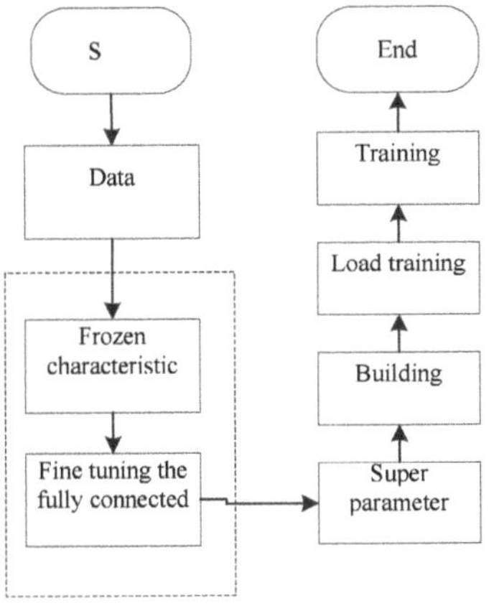

Figure 2. Flow chart of algorithm implementation.

Before loading new output modules, a global average pooling layer is added. Next, the SoftMax nonlinear final classification layer is used to set the final output to 4. Finally, the Adam optimizer is used to fine tune 100 epochs of all models. The initial learning rate is 0.0002, Nesterov momentum is 0.9, and batch size is 32.

In the above algorithm flow, this paper improves the three networks used, freezes the feature layer of each model architecture, redesigns new output modules and loads them into the full connection layer of their respective networks. The new full connection layer is shown in Figure 3.

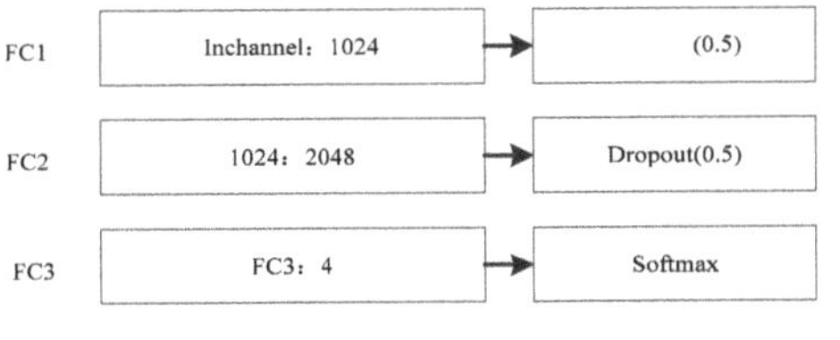

Figure 3. New output module.

In the new output module, three connection layers are set up to debug the output parameters of connection layers full connection 1 (FC1) and full connection 2 (FC2) to optimize the stability and generalization ability of the model. Because there are a large number of parameters in the full connection layer of the network, dropout is added in connection layers FC1 and FC2 to set the random loss probability $p = 0.5$. This technology can reduce the complex co-adaptation of neurons and make the model more effective for fusion. In this way, the structure of the migration network is improved, as shown in Figure 4.

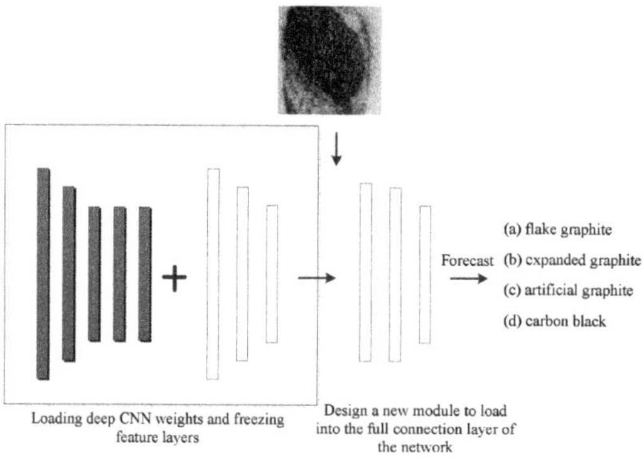

Figure 4. The structure of the improved migration network.

Herein, this model is called the improved convolutional neural network model based on transfer learning, which is applied to the graphite classification problem. The modified and optimized networks are renamed the improved visual geometry group 16 (I-VGG16), improved residual net 34 (I-ResNet34), and improved mobile net Vision 2 (I-MobileNet V2). On the basis of these improvements, the networks combined with FL loss function are recorded as I-VGG16 + FL, I-ResNet34 + FL, and I-MobileNet V2 + FL.

4. Result Analysis

4.1. Evaluation Index and Environmental Configuration

In this paper, the training effect of a network is evaluated by three indexes: accuracy (*acc*), loss, and running time. Test *acc* refers to the ratio of the model's output of correct results on the test set, and its definition formula is shown in Equation (3).

$$acc = \frac{n_{correct}}{n} \tag{3}$$

where $n_{correct}$ denotes the number of correct network identifications in the test set and n denotes the number of samples in the test set.

The training process of transfer learning is to minimize the loss function, and loss is the value of the loss function. In fact, the loss function calculates the mean square error (MSE, E) of the model on the test set.

$$E = \frac{1}{n}\sum_{i}(\hat{y} - y^{test})_i^2 \tag{4}$$

In this paper, accordingly, the epoch time is specified by the end of the whole network model. An epoch is a process in which all training samples are propagated forward and backward in the neural network model; that is, all training samples are trained once.

We used the 1.8.0 version of Python framework to complete the experimental simulation under the compiling environment of Python 3.8, and realized the image data preprocessing through transform.

4.2. Test Results

In order to verify the feasibility and effectiveness of the method presented, the graphite data set after data enhancement was imported into the method proposed in this paper. Through the experiment on the test set, the (i) unmodified model, (ii) improved model, and (ii) network model with the modified loss function, are all compared. Finally, the comparison curves of model loss and accuracy are obtained, as shown in Figures 5–7.

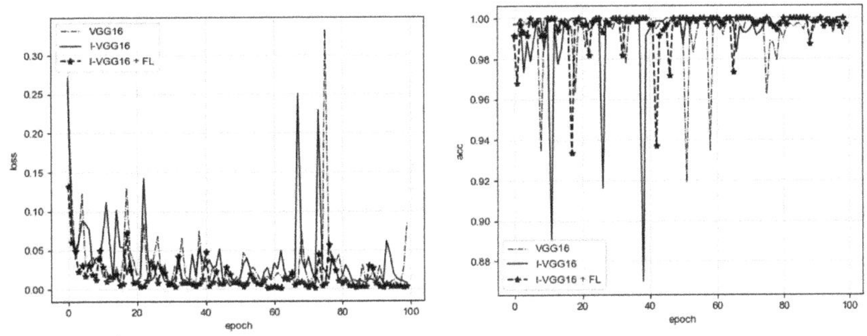

(a) Loss (b) Accuracy

Figure 5. VGG16 model loss and accuracy comparison chart.

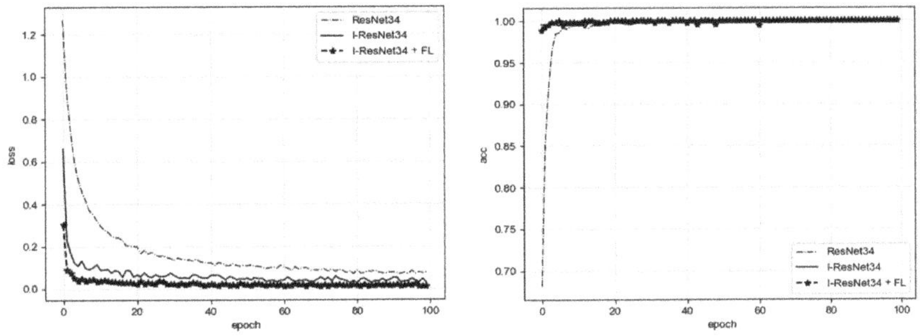

(a) Loss (b) Accuracy

Figure 6. ResNet34 model loss and accuracy comparison chart.

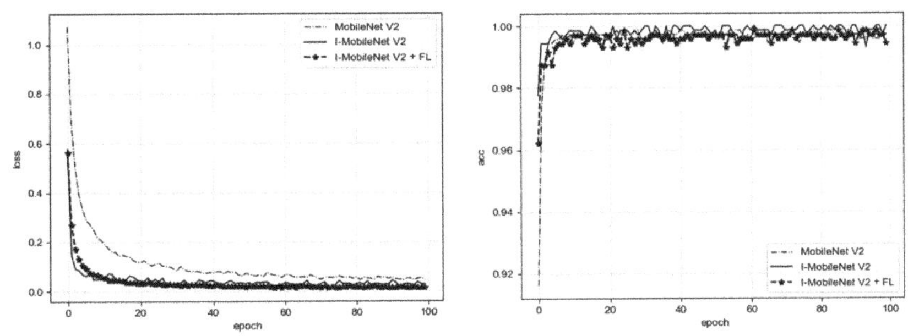

(a) Loss (b) Accuracy

Figure 7. MobileNet V2 model loss and accuracy comparison chart.

As shown in Figure 5, VGG16 has large loss value vibration, and the stability of the model is poor. After improvement of the model, the network performance of I-VGG16 and I-VGG16 + FL is obviously more stable and superior. Compared with VGG16, ResNet34 and MobileNet V2 have a deeper network and more stable model. It can be seen that the improved network effectively alleviates the problem of poor feature extraction ability caused by the secondary application of transfer learning. After using FL, the model is further optimized, the loss value is gradually reduced, and the accuracy is effectively improved, as shown in Figures 6 and 7.

In order to better understand the improved CNN model, the loss value, accuracy, and training time of the final experimental results of nine models in this paper are shown in Table 3. Table 3 shows that although the accuracy and training time of VGG16 are relatively poor, the loss is relatively minimal; overall, the training time of ResNet34 is relatively short and the accuracy of the optimized model is also improved. MobileNet V2, as a lightweight network, can reach more than 99% recognition accuracy, and with the improvement of the loss function, the loss value gradually decreases, which proves that the combination of FL and the improved model can effectively focus on the difficult sample image of the graphite data set. This shows the feasibility and effectiveness of a convolutional neural network in graphite classification. The comparison chart of the loss value and accuracy of the improved CNN model of FL is shown in Figure 8, which intuitively shows that the improved model is more stable, can converge quickly in several epochs, and the accuracy is floating within 93%, which proves the effectiveness of the improved algorithm.

Table 3. Comparison of loss value, accuracy, and training time of different algorithm models.

Model	Loss Value ($\times 10^{-2}$)	Accuracy (%)	Training Time (min)
VGG16	0.36	92.93	74.73
I-VGG16	0.16	93.71	76.13
I-VGG16 + FL	0.10	95.69	75.88
ResNet34	6.75	93.29	40.13
I-ResNet34	2.61	99.18	40.72
I-ResNet34 + FL	0.97	99.84	41.27
MobileNet V2	4.37	98.57	39.10
I-MobileNet V2	1.68	99.81	45.18
I-MobileNet V2 + FL	1.23	99.57	46.82

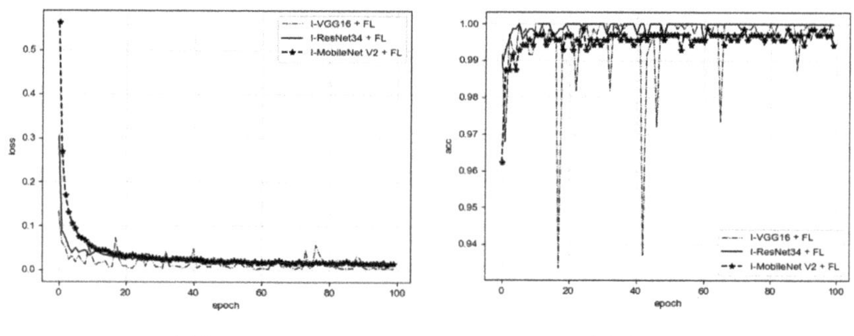

(**a**) Loss (**b**) Accuracy

Figure 8. Comparison of the improved CNN model combined with FL.

As can be seen from these experimental results, a graphite classification and recognition method using an improved convolution neural network is effective and achieves the purpose of the paper.

5. Conclusions

In this paper, an improved CNN combined with focal loss is applied to graphite image classification and recognition. The recognition model based on CNN does not need complicated image preprocessing steps in the early stage, and the data enhancement can also be completed online in the training process, which makes the model building process more concise. The graphite data set is expanded and enhanced to solve the problem of insufficient training samples. Through the simulation experiment on the processed data set, it is concluded that the redesigned full connection layer is helpful to improve the classification performance of the network. After optimizing the loss function, the performance of the network model is further improved by debugging the superparameters of the model, and the relatively high accuracy and low loss value of graphite classification are realized. In the future work, we will try deep networks other than the three kinds of CNN that were implemented. Therefore, it is our key research content in the future to explore the influence of different freezing ratios and freezing layers on the classification effect of the graphite data set.

Author Contributions: Conceptualization, X.X. and X.Y.; methodology, F.W.; software, X.X.; validation, X.X., X.Y. and F.W.; formal analysis, G.L.; investigation, G.L.; resources, X.X.; data curation, G.L.; writing—original draft preparation, X.X.; writing—review and editing, X.Y.; visualization, G.L.; supervision, X.X.; project administration, G.L.; funding acquisition, X.X. All authors have read and agreed to the published version of the manuscript.

Funding: This work is supported by the Innovation Capability Support Program of Shaanxi Province of China (Grants No. 2020PT-023), National Natural Science Foundation of China (Grants No. 11801438), and Natural Science Basic Research Plan in Shaanxi Province of China (Grants No. 2018JQ1089).

Institutional Review Board Statement: Not applicable.

Informed Consent Statement: Not applicable.

Data Availability Statement: The data sets used and/or analyzed during the current study are available from the corresponding author on reasonable request.

Conflicts of Interest: It is declared by the authors that this article is free of conflict of interest.

References

1. Wang, C.Y.; Zhao, J.L.; Du, X.H. Hydrogen production from ammonia borane hydrolysis catalyzed by non-noble metal-based materials: A review. *J. Mater. Sci.* **2021**, *56*, 2856–2878. [CrossRef]
2. Li, H.R.; Zhang, X.L.; Qi, Y.; Sun, R. Study on process of high pressure water jet cleaning mold of matrix PDC bit. *Coal Mine Mach.* **2021**, *42*, 93–96.
3. Ma, Y.J.; Fang, M.; Li, G.X.; Huang, M.; Zhang, N. Study on wear resistance and thermal stability of graphite reinforced TPU materials. *Mod. Plast. Process. Appl.* **2021**, *33*, 1–3.
4. Lu, J.J.; Liu, J.H.; Qian, G.Y.; Wang, Z.; Ma, J. Preparation of anode material for lithium-ion battery from waste silicon powder by ball milling. *Min. Metall.* **2021**, *30*, 12–18.
5. Wu, E.H.; Li, J.; Hou, J.; Xu, Z.; Huang, P.; Jiang, Y.; Chen, F. Investigation on preparation and electrical properties of graphite/$VO_{2.4-x}$ composite powders. *Rare Met. Cem. Carbides* **2021**, *49*, 57–61.
6. Feng, L.L.; Chen, Y.; Li, J.G.; Tang, S.Y.; Du, J.Z.; Li, T.Y.; Li, X.G. Research progress in carbon-based composition molded bipolar plates. *Chin. J. Eng.* **2021**, *43*, 585–593.
7. Dong, Y.R.; Kong, G.L.; Zhang, Y.; Ma, L.; Wei, L.M. Preparation and application of micro silicon-graphite-carbon anodes for lithium-ion batteries. *Micronanoelectron, Technol.* **2021**, *58*, 379–385.
8. Liu, M.Q.; Jiao, Y.Y.; Qin, J.C.; Zhongjun, L.; Wang, J. Boron doped C_3N_4 nanodots/nonmetal element (S, P, F, Br) doped C_3N_4 nanosheets heterojunction with synergistic effect to boost the photocatalytic hydrogen production performance. *Appl. Surf. Sci.* **2021**, *541*, 148558. [CrossRef]
9. Yang, S.Z.; Liang, J.; Feng, B.; Liu, P.; Yang, X.F.; Liu, Q.C. Study on carbon coating modification of nana-silicon/graphite anode materials. *Funct. Mater.* **2021**, *52*, 03130–03134.
10. Zhou, S.Z.; Nie, T.T.; Zhang, J.Y. Research and analysis of purification technology for graphite. *Carbon Tech.* **2021**, *40*, 60–62.
11. Cui, A.L.; Feng, G.X.; Zhao, Y.F. Synthesis and separation of mellitic acid and graphite oxide colloid through electrochemical oxidation of graphite in deionized water. *Electrochem. Commun.* **2009**, *11*, 409–412. [CrossRef]
12. Sheng, Z.H.; Lin, S.; Chen, J.J. Catalyst-free synthesis of nitrogen-doped graphene via thermal annealing graphite oxide with melamine and its excellent electrocatalysis. *ACS Nano* **2011**, *5*, 4350–4357. [CrossRef] [PubMed]
13. Laffont, L.; Pugliara, A.; Hungria, T.; Lacaze, J. Stem observation of a multiphase nucleus of spheroidal graphite. *J. Mater. Res. Technol.* **2020**, *9*, 4665–4671. [CrossRef]
14. Kumar, V.S.P.; Deshpande, P.A. Synergistic effect of metal-nonmetal substitution on oxygen activation in Pd/C- and Pd/N-substituted TiO2. *Comput. Mater. Sci.* **2019**, *162*, 349–358. [CrossRef]
15. Li, Y.; Zhang, S.Y. Present situation and prospect of high purity graphite production process. *Sci. Technol. Innov.* **2018**, *22*, 166–167.
16. Lecun, Y.; Bottou, L. Gradient-based learning applied to document recognition. *Proc. IEEE* **1998**, *86*, 2278–2324. [CrossRef]
17. Kim, S.; Song, W.J.; Kim, S.H. Double weight-based SAR and infrared sensor fusion for automatic ground target recognition with deep learning. *Remote Sens.* **2018**, *10*, 72. [CrossRef]

18. Bouti, A.; Mahraz, M.A.; Riffi, J. A robust system for road sign detection and classification using LeNet architecture based on convolutional neural network. *Soft Comput.* **2020**, *24*, 121–128. [CrossRef]
19. Krizhevsky, A.; Sutskever, I.; Hinton, G.E. ImageNet classification with deep convolution Neural Networks. *Adv. Neural Inf. Process. Syst.* **2012**, *25*, 1097–1105.
20. Han, X.; Zhong, Y.; Cao, L. Pre-trained AlexNet architecture with pyramid pooling and supervision for high spatial resolution remote sensing image scene classification. *Remote Sens.* **2017**, *9*, 848. [CrossRef]
21. Schroff, F.; Kalenichenko, D.; Philbin, J. Facenet: A unified embedding for face recognition and clustering. In Proceedings of the IEEE Conference on Computer Vision and Pattern Recognition, Boston, MA, USA, 7–12 June 2015; pp. 815–823.
22. He, K.M.; Zhang, X.Y.; Ren, S.Q.; Jian, S. Deep residual learning for image recognition [EB/OL]. *arXiv* **2021**, arXiv:1512.03385.
23. Simonyan, K.; Zisserman, A. Very deep convolutional networks for large-scale image recognition [EB/OL]. *arXiv* **2021**, arXiv:1409.1556.
24. Lu, J.; Behbood, V.; Hao, P.; Zuo, H.; Xue, S.; Zhang, G. Transfer learning using computational intelligence: A survey. *Knowledge-Based Syst.* **2015**, *80*, 14–23. [CrossRef]
25. Tajbakhsh, N.; Shin, J.Y.; Gurudu, S.R.; Todd Hurst, R.; Kendall, C.B.; Gotway, M.B.; Liang, J. Convolutional neural networks for medical image analysis: Full training or fine tuning? *IEEE Trans. Med. Imaging* **2016**, *35*, 1299–1312. [CrossRef]
26. Zeiler, M.D.; Fergus, R. Visualizing and understanding convolutional net-works [EB/OL]. *arXiv* **2021**, arXiv:1311.2901.
27. Razavian, A.S.; Azizpour, H.; Sullivan, J.; Carlsson, S. CNN features off-the-shelf: An astounding baseline for recognition. In Proceedings of the 2014 IEEE Conference on Computer Vision and Pattern Recognition Workshops, Columbus, OH, USA, 23–28 June 2014; pp. 512–519.
28. Azizpour, H.; Razavian, A.S.; Sullivan, J.; Maki, A.; Carlsson, S. From generic to specific deep representations for visual recognition [EB/OL]. *arXiv* **2021**, arXiv:1406.5774.
29. Sandler, M.; Howard, A.; Zhu, M.L.; Zhmoginov, A.; Chen, L.-C. MobileNetV2: Inverted residuals and linear bottlenecks [EB/OL]. *arXiv* **2021**, arXiv:1801.04381.

processes MDPI

Article

Environmental and Sustainability Analysis of a Supercritical Carbon Dioxide-Assisted Process for Pharmaceutical Applications

Paolo Trucillo [1,2,*], Roberta Campardelli [1,3] and Iolanda De Marco [1,4]

[1] Department of Industrial Engineering (DIIn), University of Salerno, Via Giovanni Paolo II 132, 84084 Fisciano, Italy; roberta.campardelli@unige.it (R.C.); idemarco@unisa.it (I.D.M.)
[2] Department of Chemical, Material and Industrial Production Engineering, University of Naples Federico II, Piazzale V. Tecchio 80, 80125 Napoli, Italy
[3] Department of Civil, Chemical and Environmental Engineering (DICCA), University of Genoa, Via Opera Pia 15, 16145 Genova, Italy
[4] Research Centre for Biomaterials BIONAM, University of Salerno, Via Giovanni Paolo II 132, 84084 Fisciano, Italy
* Correspondence: paolo.trucillo@unina.it

Abstract: Drug delivery systems (DDS) are artificial devices employed to enhance drug bioavailability during administration to a human body. Among DDS, liposomes are spherical vesicles made of an aqueous core surrounded by phospholipids. Conventional production methods are characterized by several drawbacks; therefore, Supercritical assisted Liposome formation (SuperLip) has been developed to overcome these problems. Considering that the use of high pressures involves high energy cost, in this paper, sustainability indicators were calculated to quantitatively evaluate the emissions related to the attainment of liposomes containing daunorubicin (a model antibiotic drug) using the SuperLip process. The indicators were depicted using a spider diagram to raise the actual weaknesses of this technique; some variations were proposed in the process layout to solve the critical issues. According to the literature, many studies related to the pharmaceutical industry are expressed in terms of solid, liquid waste, and toxic emissions; however, liposomes have never explicitly been considered for an analysis of environmental sustainability.

Keywords: supercritical fluids; liposomes; pharmaceutical applications; biomedical

Citation: Trucillo, P.; Campardelli, R.; De Marco, I. Environmental and Sustainability Analysis of a Supercritical Carbon Dioxide-Assisted Process for Pharmaceutical Applications. *Processes* **2021**, *9*, 1788. https://doi.org/10.3390/pr9101788

Academic Editor: Irena Zizovic

Received: 15 September 2021
Accepted: 5 October 2021
Published: 8 October 2021

1. Introduction

Drug delivery systems (DDS) are artificial devices employed to enhance drug bioavailability during topical delivery [1]. Several systems and complexes have been developed at micro and nano levels to achieve high entrapment efficiency of therapeutic agents [2], targeted delivery to specific human tissues, and improved protection of the entrapped drug from degradation phenomena [3].

Among DDS, liposomes are spherical vesicles made of an aqueous core surrounded by one or more layers of phospholipids [4], generally employed for pharmaceutical [5], cosmetic [6], and nutraceutical purposes [7]. Currently, the main liposomes producing countries are the United States of America, Republic of China, Japan, and the western countries of Europe [8], with the following market share: pharmaceutic industries (61.7%), cosmetics (22.8%), and nutraceutical industries (15.6%) [9–12].

The well-known conventional methods for liposome production are generally characterized by low entrapment efficiencies of active principles and difficult replicability of Particle Size Distribution (PSD), due to discontinuous process layouts [13]. The Supercritical assisted Liposome formation (SuperLip) technique has been recently developed to overcome these problems, consisting of the inversion of the traditional production steps of production [14] through an atomization step directly into a supercritical medium containing the phospholipids. This process has been successfully tested for the entrapment of proteins, antioxidants, dietary supplements, dyes, and antibiotics [14,15].

The SuperLip process has been developed primarily at a lab scale; however, its configuration layout is continuous, which guarantees its replicability on a larger scale [16,17]. Comparing to other processes proposed in the literature, SuperLip has several advantages, as described in the sketches reported in Figure 1, where SR represents the Solvent Residue and EE the Encapsulation Efficiency (in particular, see Figure 1b).

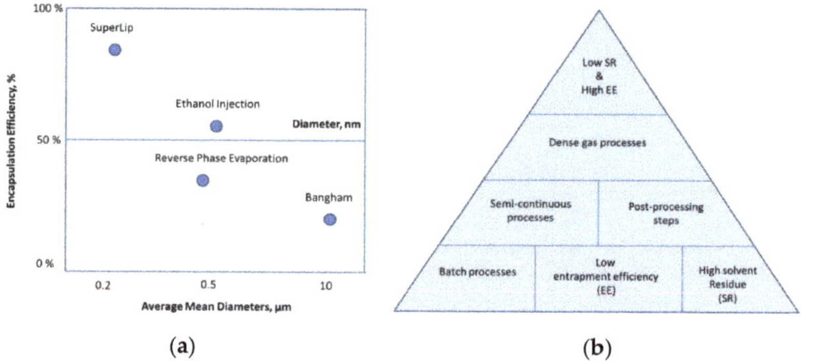

Figure 1. (**a**) A features map of encapsulation efficiencies and mean size of liposomes produced using different techniques. (**b**) Pyramid sketch of the main advantages and disadvantages of liposomes production (SR: Solvent Residue, EE: Encapsulation Efficiency).

According to the literature published on SuperLip, it is possible to affirm that other well-known processes, such as ethanol injection [18–20] or reverse-phase evaporation [21,22], resulted in the production of larger liposomes (around 500 nm) and encapsulation efficiencies between 40% and 60%. In particular, the conventional Bangham method [23,24] results in vesicles' mean dimensions highly variable, from 1 to 100 μm, and encapsulation efficiencies are generally lower than 30%. Concerning Figure 1b, the bottom of the pyramid is characterized by the worst operating conditions. These processes are characterized by low entrapment efficiencies, and amounts of solvent residues above the Food and Drug Administration (FDA) imposed limits [25]. Therefore, a high solvent residue also means that these processes create liposomes formulations with a high level of toxicity [26]. In the second level of the pyramid, semi-continuous processes and post-processing steps are reported, such as Reverse Phase Evaporation and Microfluidic channel techniques [27], that result in the production of quasi-homogeneous samples. Due to this not optimal homogeneity, vesicles mean dimensions are reduced after extrusion or sonication [28]. At the third level of this pyramid, dense gas processes find a good location, also in terms of reduced solvent residue, thanks to the use of carbon dioxide in supercritical conditions. These processes, such as Supercritical AntiSolvent (SAS) [29], Depressurization of an Expanded Solution into Aqueous Media (DESAM) [30], Depressurization of a CO_2-expanded liquid organic solution (DELOS) [31], and Supercritical Reverse Phase Evaporation (SRPE) [32], were developed to avoid the high cost of post-processing steps, avoiding loss of expensive molecules. These methods were obtained after great technical efforts and encountered a success after proposal in the academic community. Some improvements were still needed to produce liposomes available to be sold in the market with a good balance among profitability, environmental impact, and energy consumption [21,33]. SuperLip process was demonstrated to provide all these advantages.

For the reasons listed above, it was considered attractive to focus on the main advantage of SuperLip: the low solvent residue, as indicated in previous work [17]. To better explain this advantage, a working map has been proposed in Figure 2, creating a strict correspondence among two important operating parameters (Gas to Liquid Ratio, i.e., feeding ratio, calculated as carbon dioxide over ethanol flow rates on a mass basis, and mean diameter of the liposomes produced). In this diagram, the surface of each circled area represents the concentration of solvent residue obtained in different operating conditions; whereas, the center of each circle is related to a specific Encapsulation Efficiency and a specific Gas To Liquid Ratio. As indicated in Figure 2, vesicles produced with the conventional technique are characterized by a high level of solvent residue (20,000 ppm), measured after evaporation; whereas, small circles are related to liposomes produced with an ethanol residue lower than 150 ppm.

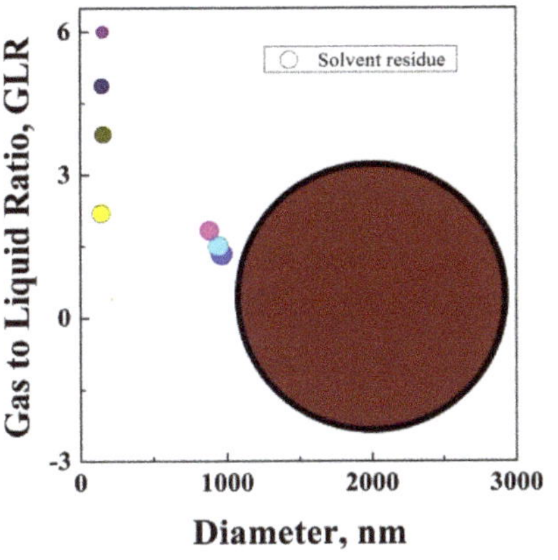

Figure 2. Bubble diagrams: comparison of solvent residue amount among SuperLip process (small circles) and conventional method (large circle). The surface of circles express the concentration of ethanol in ppm, in the final aqueous suspension.

After these considerations, the elimination of solvent residue becomes fundamental in pharmaceutical processes [34]; in particular, SuperLip eliminates large parts of its solvent from the top of the main process unit. The remaining amount can be eliminated using rotary evaporation performed on the recovered liquid suspension. This step can avoid the pharmaceutical formulations to be toxic for cells [18,34]. Moreover, the commercial profitability of SuperLip has been already demonstrated, in terms of economic and financial analysis [35].

Solvent residue causes a significant environmental impact during the production of drug carriers. Therefore, the most common way to calculate the environmental impact is represented by the analysis of sustainability indicators, or the Life Cycle Assessment, largely used in many fields, such as energy [36], beverages and foods [37–41], pharmaceutical delivery [42–44] systems, cosmetics [45], and wastewater treatments [46]. Concerning pharmaceutical industry, a few papers [43,47–49] are related to the management of solid waste and solvent treatment; moreover, liposomes have never been considered for a sustainability evaluation.

Therefore, the aim of this work is the assessment of the environmental impact of the SuperLip process. The eco-balance of this technique will be evaluated to study the effects of liposomes production using this supercritical assisted technique, from the acquiring of raw materials and reagents to the manufacture of the final produced vesicles. An inventory of materials employed in this process and energy consumption will be provided, evaluating the inputs and the outputs of the process, and making a final analysis on the results, according to market profitability reference. A model drug such as daunorubicin, which is generally employed against leukemia [50], will be considered for this analysis. The results of this study will also improve the proposed technique and certify its quality, with the final aim to assess the profitability of a scale-up for this process, to achieve high volumes of commercialization of this liposome-based products.

2. Process Description

2.1. Apparatus

SuperLip process consists of three feeding lines: carbon dioxide is pumped at the flow rate of 6.5 g/min using an Ecoflow pump (mod LDC-M-2, Lewa, Germany), until reaching the pressure of 100 bar; an ethanol/phospholipids solution is fed at the flow rate of 3.5 mL/min, using a high-pressure precision pump (Model 305, Gilson, France). Ethanol and carbon dioxide are first mixed and then heated up to 40 °C, using thin Band Heaters (3 × 120 W, Watlow Italy, Milano, Italy). The carbon dioxide over ethanol feeding ratio is called Gas to Liquid Ratio of the Expanded Liquid (GLR-EL), and it has been set at 2.4. The ethanol + lipids + carbon dioxide mixture is sent to a stainless-steel vessel (500 cm^3) that works at the pressure of 100 bar and temperature of 40 °C, heated using Band Heaters (2 × 400 W, Watlow Italy, Milan, Italy).

A third feeding line sends water (plus a dissolved hydrophilic drug) to the system; another high-pressure precision pump supports this feeding line at the flow rate of 10 mL/min (Model 305, Gilson, France). The water flow rate is atomized in droplets in the formation vessel, using an 80 μm nozzle.

The production of liposomes occurs in the vessel of SuperLip, by first creating water droplets and then the lipid layer around. Liposomes are collected from the bottom of the vessel using an on/off valve. The separation of the ethanol/carbon dioxide expanded liquid occurs from the top of the vessel, where an exit line has been designed. This line is heated at 30 °C using a tubular resistance (275 W, Watlow Italy, Milan, Italy). A stainless-steel separator (300 cm^3) is employed to separate ethanol and carbon dioxide at the pressure of 10 bar. A rotameter (mod. N.5–2500, Serval 115022, ASA, Italy) is used to measure carbon dioxide flow rate.

Liposomes are produced from SuperLip in aqueous suspension. However, a reduced amount of ethanol is still present in the final solution; therefore, liposomes suspensions are sent to rotary evaporation, operating at 30 °C under vacuum at a stirring rate of 120 rpm (for 30 min), in order to eliminate solvent residue without damaging vesicles produced.

2.2. Materials and Procedures

The raw materials for the production of liposomes production are essentially phospholipids, that are provided by several companies such as Sigma Aldrich (Milan, Italy) or Lipoid (Ludwigshafen, Germany). Daunorubicin has been purchased from Sigma Aldrich, Milan, Italy; whereas, distilled water was self-produced using a lab-scale distillation column, separated from the SuperLip plant. Carbon dioxide is provided by Morlando Group, Naples, Italy, and it is stocked into an external tank with a volume of 800 L. The carbon dioxide needs to be cooled using a cooling bath at the temperature of −10 °C; once that carbon dioxide is in liquid state, it is pumped to the system, where it is again heated up to 40 °C. The pumps guarantee the pressure of 100 bar constant to achieve supercritical conditions for carbon dioxide. Ethanol and water are pumped into the system as well. The heart of the production is characterized by ethanol and carbon dioxide pre-mixing and heating, followed by the mixture in the formation vessel, together with the atomized droplets of water + drug. The final product is the liposomes suspension, which is subjected to solvent elimination post-treatment. Ethanol and carbon dioxide are separated from the formation vessel and sent to depressurization and splitting. In Table 1, the process details and main activities are described.

Table 1. Process details and assumptions.

Process	**Characteristics and Details**
Energy supply to facility	Italian energy mix medium voltage
Production	
Pressurization	t_1 = 0.25 h; carbon dioxide and ethanol supply; energy supply
Operating conditions stabilization	T = 40 °C; P = 100 bar; t_2 = 0.2 h; carbon dioxide and ethanol supply; energy supply

Table 1. *Cont.*

Process	Characteristics and Details
Injection	T = 40 °C; P = 100 bar; t_3 = 1 h; carbon dioxide and ethanol supply; water solution; energy supply
Separation	T = 30 °C; P = 10 bar; t_4 = 1 h
Stocking	T = 4 °C; P = 1 bar; t_5 = 0.5 h
Carbon dioxide supply to facility	Transport by truck, 28 t from Sant'Antimo (Italy) To the University of Salerno (Italy), distance = 67 km
Reagents supply to facility	Transport by truck, 28 t from Milan to the University of Salerno (Italy), distance = 800 km

3. Methodology

As indicated in similar studies [51], this work aims to evaluate the emissions related to the use of the SuperLip technique to produce a liposomal formulation containing an active principle (daunorubicin). The system boundaries are characterized by the operating parameters described in this section and are highlighted in Figure 3. Equipment impacts were not included.

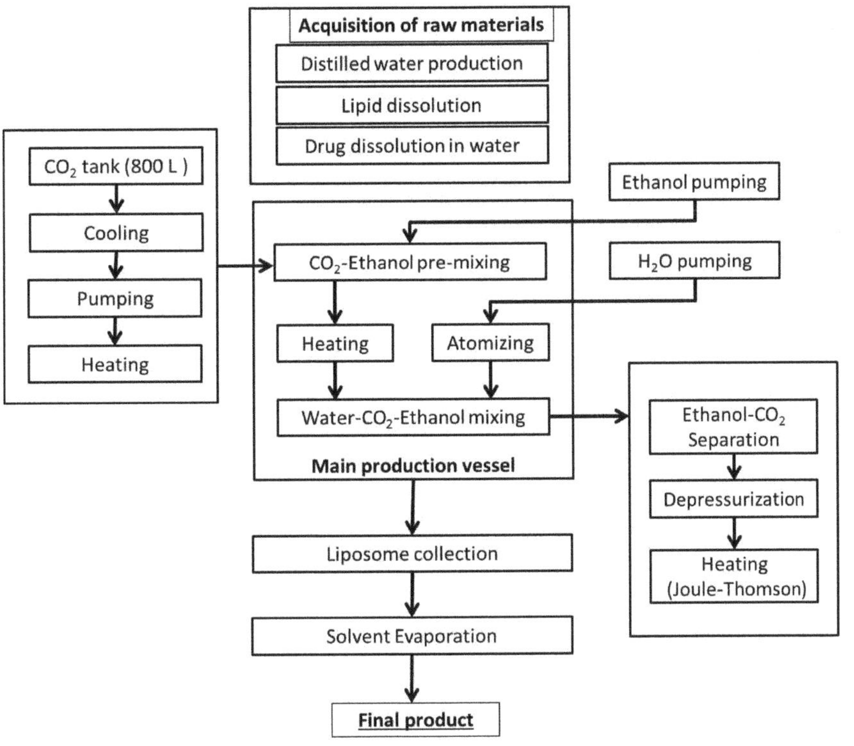

Figure 3. SuperLip system boundaries.

An indicator is an index used to define the sustainability conditions of a working process, giving a practical sense and perception of the system. It generally does not work as a preliminary index, but it contains information about an already developed phenomenon or process. It is a way to give a precise meaning to the raw data of the process.

Due to the large number of collected values related to a process plant and several existing indexes, the application of this methodology to different processes could be difficult. For this reason, a sustainability scale can be defined by enclosing two scenarios representing the best case (100% sustainable process) and the worst case (0% sustainable). The final score related to each indicator is represented by a combination of worst, best case, and actual value, i.e., the real value of the parameters measured in the process. The general formula is the following:

$$\text{Percent Score} = \frac{\text{Actual} - \text{Worst}}{\text{Best} - \text{Worst}} \times 100\%$$

Sustainability indicators were studied according to the calculation of the following percent scores [52,53] defined in Table 2.

Table 2. Sustainability indicators calculation formula.

Formula	Best	Worst
$\text{Global warming potential} = \frac{\text{Total mass of } CO_2 \text{ released}}{\text{Mass of product}}$	No CO_2 released	All CO_2 released
$\text{Global warming intensity} = \frac{\text{Total mass of } CO_2 \text{ released}}{\text{Sales revenue}}$	No CO_2 released	All CO_2 released
$\text{Specific energy intensity} = \frac{\text{Total energy of the process}}{\text{Mass of product}}$	Min. theoretical energy (Gibbs)	5.85×10^{11} KJ/Kg [54]
$\text{Energy intensity} = \frac{\text{Total energy of the process}}{\text{Sales revenue}}$	0	2.294×10^{9} KJ/EUR [55]
$\text{Specific liquid waste volume} = \frac{\text{Total liquid volume rated as waste}}{\text{Mass of product}}$	0	100%
$\text{Reaction mass efficiency} = \frac{\text{Mass of product}}{\text{Total mass of reagents}}$	100%	0%
Total material consumption = Total mass input *	2.5×10^{-2} Kg	1 Kg
$\text{Mass intensity} = \frac{\text{Total mass input}}{\text{Mass of product}}$	1	40 Kg/Kg [56]
$\text{Value mass intensity} = \frac{\text{Total mass input}}{\text{Sales revenue}}$	0	52 Kg/EUR [57]
$\text{Fractional water consumption} = \frac{\text{Volume of fresh water consumed}}{\text{Mass of product}}$	0	2.95 m^3/kg [58]
$\text{Water intensity} = \frac{\text{Volume of fresh water consumed}}{\text{Sales revenue}}$	0	1.55 m^3/EUR [58]
$\text{Recycled material fraction} = \frac{\text{Recycled mass input}}{\text{Total mass input}}$	1	0 Kg/Kg

* Total material consumption was calculated considering the mass of a unit of product, equal to 0.025 Kg (in this study) in the best conditions and 40 times that value in the worst condition [56].

4. Results and Discussion

The sustainability indicators, whose formulas were reported in the Methods Section, were calculated taking into account the actual values of SuperLip working conditions, considering the best and worst values indicated for each specific situation. Environmental and economic indicators such as Global Warming Potential and Global Warming Intensity were considered. These two indicators need to be shown together; indeed, the first one correlates the emissions of carbon dioxide (on mass basis) to the mass of product obtained. The second one correlates the emissions of carbon dioxide to the economic value of the products sold. In other words, these two indicators compare the environmental impact and the profitability of the process, in order to understand if the process is lacking in both areas, or lacking in just one of the two. This comparison has the potential to indicate the points of strength of the process and the main weaknesses.

More indicators about energy consumption were also evaluated, providing correlations among power consumption and the mass of products or the sales revenue generated by those products. Another environmental indicator was determined by liquid waste volume and recycled material fraction. The system's productivity was also evaluated in terms of reaction efficiency, i.e., the transformation of raw materials into products through the process. Total mass consumption was put in correlation with the mass of product and also to the sales revenue. Actual values shown in Table 3 represent the real situation of SuperLip working in standard conditions.

Actual values were inserted in the score calculation formula, and the scores for each indicator were obtained in terms of percentage (see Table 4 under the column "before optimization"). In this context, 0% represents "totally not sustainable" and 100% "totally sustainable".

Table 3. List of actual values calculated for each sustainability indicator.

Indicator	Description of the Parameter	Value	Unit
Global warming potential	Total mass of CO_2 released	1.06×10^{-2}	Kg
	Mass of product	2.50×10^{-2}	Kg
	Ratio	43%	Kg/Kg
Global warming intensity	total mass of CO_2 released	1.06×10^{-2}	Kg
	sales revenue	27.5	EUR
	Ratio	3.9×10^{-4}	Kg/EUR
Specific energy intensity	total energy consumed in the process	34.7	KJ
	mass of product	2.50×10^{-2}	Kg
	Ratio	1389.20	KJ/Kg
Energy intensity	total energy consumed in the process	34.7	KJ
	sales revenue	27.5	EUR
	Ratio	1.2629	KJ/EUR
Specific liquid waste volume	total volume of liquid rated as waste (ethanol)	6.58×10^{-3}	Kg
	Mass of product	2.50×10^{-2}	Kg
	Ratio	26%	
Reaction mass efficiency	Mass of product	2.50×10^{-2}	Kg
	Total mass of reagents	3.42×10^{-2}	Kg
	Ratio	73%	
Total material consumption	total mass input	3.424×10^{-2}	Kg
Mass intensity	total mass input	3.42×10^{-2}	Kg
	mass of product	2.50×10^{-2}	Kg
	Ratio	1.370	Kg/Kg
Value mass intensity	Total mass input	3.42×10^{-2}	Kg
	sales venue	27.5	EUR
	Ratio	1.25×10^{-3}	Kg/EUR
Fractional water consumption	volume of fresh water consumed	2.50×10^{-2}	m^3
	mass of product	2.50×10^{-2}	Kg
	Ratio	1.0000	m^3/Kg
water intensity	volume of fresh water consumed	2.50×10^{-2}	m^3
	sales venue	27.5	EUR
	Ratio	9.09×10^{-4}	m^3/EUR
Recycled material fraction	Recycled mass input	0	Kg
	total mass input	3.42×10^{-2}	Kg
	Ratio	0	Kg/Kg

Table 4. Scores calculated before and after optimization of the process.

	Before Optimization	After Optimization
Global warming potential	57%	96%
Global warming intensity	100%	100%
Specific energy intensity	100%	100%
Energy intensity	100%	100%
Specific liquid waste volume	74%	97%
Reaction mass efficiency	73%	85%
Total material consumption	99%	99%
Mass intensity	99%	99%
Value mass intensity	100%	100%
Fractional water consumption	66%	83%
Water intensity	100%	100%
Recycled material fraction	0%	45%

To achieve better control of parameters and increase the previously calculated scores, some modifications could be proposed to the layout of the process. In particular, energy does not require specific intervention. Therefore, no problems were registered in terms of the operating cost of the process.

The weakest points of the process SuperLip emerged as the feeding of CO_2, ethanol, and water. To improve these pumping steps and increase the related sustainability scores, 90% recirculation of ethanol has been proposed via rotary evaporation followed by condensation. The additional instrument energy required is negligible, according to the volumes of production. In this manner, the specific liquid waste volume will be 10% of the previously calculated one.

Another possible modification is the 90% recirculation of carbon dioxide employed in the process. In this manner, global warming potential will be calculated considering only 10% of the carbon dioxide releasing mass.

Consequently, reaction mass efficiency will be positively increased to 85%, and recycled material fraction will become the sum of 90% of ethanol recirculated plus 90% of recirculated carbon dioxide. Moreover, a 50% water recirculation has been proposed to the process after recovering the processing water. The effect of these new calculations and process layout brought to the definition of a new scenario, as expressed in Table 4.

The sustainability analysis calculated and reported in Table 4 was translated into a spider diagram (shown in Figure 4), i.e., a visual tool used to organize scores and compare them logically and quickly.

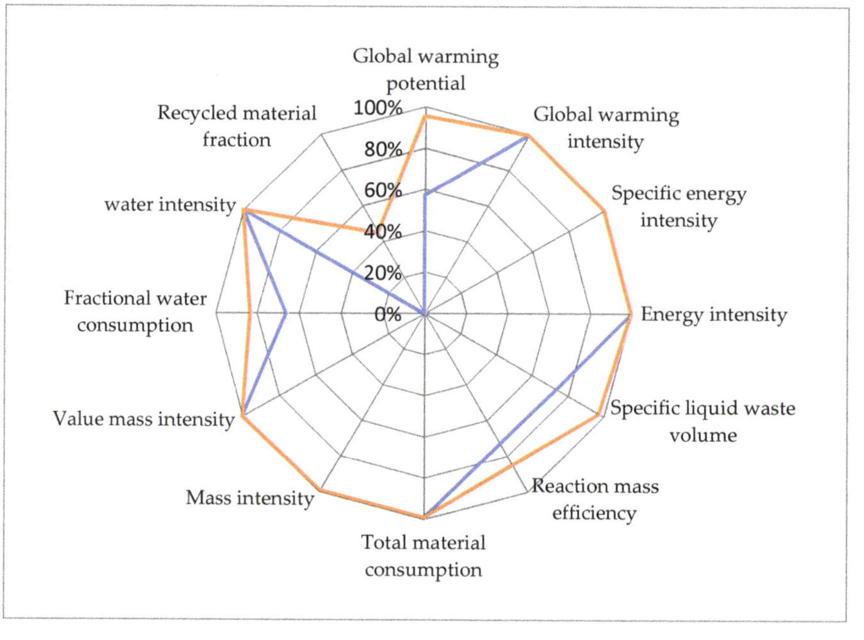

Figure 4. Spider diagram representing the sustainability analysis of the SuperLip process before (blue) and after (orange) optimization.

In this diagram, the blue line represents the previous situation, while the orange line is related to the case in the new process configuration; it is possible to say that sustainability indicators significantly increased after process modification (see Figure 4 and Table 4).

According to results shown in Table 4, the differentiation among Global Warming potential and Global Warming Intensity results to be particularly important. Sustainability analysis generally embraces all three spheres of a process: environmental impact, economic convenience, and social impact. In this case, GWP and GWI represent the intersection between environment and economic impacts. In fact, after a careful analysis of this process, it appears very clear that the GWP, that is the environmental impact related to the product mass, has a score of only 57%; on the other hand, the GWI, that is the environmental impact correlated to the sales of the product, has a top score of 100%. In other words, the economic value of SuperLip products is so high that almost justifies the process carbon dioxide emissions; however, the other indicator shows that the environmental impact is not negligible. Therefore, the idea of modifying the process layout adding a recirculation step, resolves the environmental problem while maintaining a high economic value of the products.

This positive effect was also registered, after the introduction for recirculation, in terms of liquid waste and the fraction of water consumed in the process. The overall variation of the SuperLip layout resulted in an overall increase in the efficiency of the process.

5. Conclusions

In this work, we started by analyzing a process employed to produce drug carriers at high working pressures. After establishing the process economic profitability, we realized that further analysis on sustainability was needed. Considering the concept of sustainable development, we studied the possibility of using resources without compromising their future availability; in our case, this resulted in a proper recirculation of the process input materials.

The SuperLip process was studied in terms of power consumption, sales revenue, and global warming to balance them simultaneously. The key points were characterized by waste management in terms of recycling, and energy recovery through process efficiency enhancement, reducing the impact of CO_2 emissions on the mass of products obtained from the process. Process indicators were calculated and analyzed in the SuperLip working process, from cradle to grave, not just considering it as a black box.

After proposing recirculation, several indicators improved significantly, such as global warming potential, from 57% to 96%; specific liquid waste volume, from 74% to 97%; reaction mass efficiency, from 73% to 85%; fractional water consumption, from 66% to 83%, and recycled material fraction, from 0% to 45%. The analysis resulted in being successful in demonstrating the sustainability potential of the SuperLip process. Further studies will regard the possibility of scaling up this analysis to other industrial processes to produce polymeric drug carriers.

Author Contributions: Conceptualization, P.T., R.C. and I.D.M.; methodology, P.T., R.C. and I.D.M.; validation, R.C. and I.D.M.; formal analysis, P.T.; data curation, P.T.; writing—original draft preparation, P.T.; writing—review and editing, R.C. and I.D.M. All authors have read and agreed to the published version of the manuscript.

Funding: This research received no external funding.

Institutional Review Board Statement: Not applicable.

Informed Consent Statement: Not applicable.

Conflicts of Interest: The authors declare no conflict of interest.

References

1. Tiwari, G.; Tiwari, R.; Sriwastawa, B.; Bhati, L.; Pandey, S.; Pandey, P.; Bannerjee, S.K. Drug delivery systems: An updated review. *Int. J. Pharm. Investig.* **2012**, *2*, 2.
2. Patra, J.K.; Das, G.; Fraceto, L.F.; Campos, E.V.R.; del Pilar Rodriguez-Torres, M.; Acosta-Torres, L.S.; Diaz-Torres, L.A.; Grillo, R.; Swamy, M.K.; Sharma, S. Nano based drug delivery systems: Recent developments and future prospects. *J. Nanobiotechnol.* **2018**, *16*, 71. [CrossRef] [PubMed]
3. Ito, Y. Drug delivery systems. In *Photochemistry for Biomedical Applications*; Springer: Berlin/Heidelberg, Germany, 2018; pp. 231–275.
4. Pattni, B.S.; Chupin, V.V.; Torchilin, V.P. New developments in liposomal drug delivery. *Chem. Rev.* **2015**, *115*, 10938–10966. [CrossRef] [PubMed]
5. Ozer, A.Y. Alternative applications for drug delivery: Nasal and pulmonary routes. In *Nanomaterials and Nanosystems for Biomedical Applications*; Springer: Berlin/Heidelberg, Germany, 2007; pp. 99–112.

6. Van Tran, V.; Moon, J.-Y.; Lee, Y.-C. Liposomes for delivery of antioxidants in cosmeceuticals: Challenges and development strategies. *J. Control. Release* **2019**, *300*, 114–140. [CrossRef] [PubMed]
7. Taylor, T.M.; Weiss, J.; Davidson, P.M.; Bruce, B.D. Liposomal nanocapsules in food science and agriculture. *Crit. Rev. Food Sci. Nutr.* **2005**, *45*, 587–605. [CrossRef]
8. Zhou, X.; Zhao, G. Global liposome research in the period of 1995–2014: A bibliometric analysis. *Scientometrics* **2015**, *105*, 231–248. [CrossRef]
9. Radhika, P.R.; Singh, R.B.M.; Sivakumar, T. Nutraceuticals: An area of tremendous scope. *Int. J. Res.* **2011**, *2*, 410–415.
10. Aditya, N.; Espinosa, Y.G.; Norton, I.T. Encapsulation systems for the delivery of hydrophilic nutraceuticals: Food application. *Biotechnol. Adv.* **2017**, *35*, 450–457. [CrossRef]
11. Crommelin, D.J.; Storm, G. Liposomes: From the bench to the bed. *J. Liposome Res.* **2003**, *13*, 33–36. [CrossRef]
12. Singh, H.; Thompson, A.; Liu, W.; Corredig, M. Liposomes as food ingredients and nutraceutical delivery systems. In *Encapsulation Technologies and Delivery Systems for Food Ingredients and Nutraceuticals*; Elsevier: Amsterdam, The Netherlands, 2012; pp. 287–318.
13. Mozafari, M.R. Liposomes: An overview of manufacturing techniques. *Cell. Mol. Biol. Lett.* **2005**, *10*, 711.
14. Trucillo, P.; Campardelli, R.; Reverchon, E. Supercritical CO_2 assisted liposomes formation: Optimization of the lipidic layer for an efficient hydrophilic drug loading. *J. CO_2 Util.* **2017**, *18*, 181–188. [CrossRef]
15. Trucillo, P.; Campardelli, R.; Aliakbarian, B.; Perego, P.; Reverchon, E. Supercritical assisted process for the encapsulation of olive pomace extract into liposomes. *J. Supercrit. Fluids* **2018**, *135*, 152–159. [CrossRef]
16. Trucillo, P.; Campardelli, R.; Reverchon, E. A versatile supercritical assisted process for the one-shot production of liposomes. *J. Supercrit. Fluids* **2019**, *146*, 136–143. [CrossRef]
17. Trucillo, P.; Campardelli, R.; Scognamiglio, M.; Reverchon, E. Control of liposomes diameter at micrometric and nanometric level using a supercritical assisted technique. *J. CO_2 Util.* **2019**, *32*, 119–127. [CrossRef]
18. Dua, J.; Rana, A.; Bhandari, A. Liposome: Methods of preparation and applications. *Int. J. Pharm. Stud. Res.* **2012**, *3*, 14–20.
19. Pons, M.; Foradada, M.; Estelrich, J. Liposomes obtained by the ethanol injection method. *Int. J. Pharm.* **1993**, *95*, 51–56. [CrossRef]
20. Laouini, A.; Jaafar-Maalej, C.; Limayem-Blouza, I.; Sfar, S.; Charcosset, C.; Fessi, H. Preparation, characterization and applications of liposomes: State of the art. *J. Colloid Sci. Biotechnol.* **2012**, *1*, 147–168. [CrossRef]
21. Meure, L.A.; Foster, N.R.; Dehghani, F. Conventional and dense gas techniques for the production of liposomes: A review. *AAPS PharmSciTech* **2008**, *9*, 798. [CrossRef]
22. Otake, K.; Shimomura, T.; Goto, T.; Imura, T.; Furuya, T.; Yoda, S.; Takebayashi, Y.; Sakai, H.; Abe, M. Preparation of liposomes using an improved supercritical reverse phase evaporation method. *Langmuir* **2006**, *22*, 2543–2550. [CrossRef]
23. Deamer, D.W. From "banghasomes" to liposomes: A memoir of Alec Bangham, 1921–2010. *FASEB J.* **2010**, *24*, 1308–1310. [CrossRef]
24. Liu, W.; Ye, A.; Liu, W.; Liu, C.; Singh, H. Stability during in vitro digestion of lactoferrin-loaded liposomes prepared from milk fat globule membrane-derived phospholipids. *J. Dairy Sci.* **2013**, *96*, 2061–2070. [CrossRef]
25. He, T.; Liang, Q.; Wang, J.; Luo, G. Microfluidic Fabrication of Liposomes as Drug Carriers. *Prog. Chem.* **2018**, *30*, 1734–1748.
26. Grimaldi, N.; Andrade, F.; Segovia, N.; Ferrer-Tasies, L.; Sala, S.; Veciana, J.; Ventosa, N. Lipid-based nanovesicles for nanomedicine. *Chem. Soc. Rev.* **2016**, *45*, 6520–6545. [CrossRef] [PubMed]
27. Mufamadi, M.S.; Pillay, V.; Choonara, Y.E.; Du Toit, L.C.; Modi, G.; Naidoo, D.; Ndesendo, V.M. A review on composite liposomal technologies for specialized drug delivery. *J. Drug Deliv.* **2011**, 1–19. [CrossRef] [PubMed]
28. Pradhan, P.; Guan, J.; Lu, D.; Wang, P.G.; Lee, L.J.; Lee, R.J. A facile microfluidic method for production of liposomes. *Anticancer Res.* **2008**, *28*, 943–947.
29. William, B.; Noémie, P.; Brigitte, E.; Géraldine, P. Supercritical fluid methods: An alternative to conventional methods to prepare liposomes. *Chem. Eng. J.* **2020**, *383*, 123106. [CrossRef]
30. Meure, L.A.; Knott, R.; Foster, N.R.; Dehghani, F. The depressurization of an expanded solution into aqueous media for the bulk production of liposomes. *Langmuir* **2009**, *25*, 326–337. [CrossRef]
31. Zhao, L.; Temelli, F. Preparation of liposomes using a modified supercritical process via depressurization of liquid phase. *J. Supercrit. Fluids* **2015**, *100*, 110–120. [CrossRef]
32. Huang, Z.; Li, X.; Zhang, T.; Song, Y.; She, Z.; Li, J.; Deng, Y. Progress involving new techniques for liposome preparation. *Asian J. Pharm. Sci.* **2014**, *9*, 176–182. [CrossRef]
33. Beh, C.C.; Mammucari, R.; Foster, N.R. Lipids-based drug carrier systems by dense gas technology: A review. *Chem. Eng. J.* **2012**, *188*, 1–14. [CrossRef]
34. Cortesi, R.; Esposito, E.; Gambarin, S.; Telloli, P.; Menegatti, E.; Nastruzzi, C. Preparation of liposomes by reverse-phase evaporation using alternative organic solvents. *J. Microencapsul.* **1999**, *16*, 251–256. [CrossRef]
35. Trucillo, P.; Campardelli, R.; Iuorio, S.; De Stefanis, P.; Reverchon, E. Economic analysis of a new business for liposome manufacturing using a high-pressure system. *Processes* **2020**, *8*, 1604. [CrossRef]
36. González-García, S.; Dias, A.C.; Clermidy, S.; Benoist, A.; Bellon Maurel, V.; Gasol, C.M.; Gabarrell, X.; Arroja, L. Comparative environmental and energy profiles of potential bioenergy production chains in Southern Europe. *J. Clean. Prod.* **2014**, *76*, 42–54. [CrossRef]
37. De Marco, I.; Riemma, S.; Iannone, R. Life cycle assessment of supercritical CO_2 extraction of caffeine from coffee beans. *J. Supercrit. Fluids* **2018**, *133*, 393–400. [CrossRef]
38. Berlin, J. Environmental life cycle assessment (LCA) of Swedish semi-hard cheese. *Int. Dairy J.* **2002**, *12*, 939–953. [CrossRef]

39. Biswas, W.K.; Naude, G. A life cycle assessment of processed meat products supplied to Barrow Island: A Western Australian case study. *J. Food Eng.* **2016**, *180*, 48–59. [CrossRef]
40. De Marco, I.; Riemma, S.; Iannone, R. Uncertainty of input parameters and sensitivity analysis in life cycle assessment: An Italian processed tomato product. *J. Clean. Prod.* **2018**, *177*, 315–325. [CrossRef]
41. Gazulla, C.; Raugei, M.; Fullana-I-Palmer, P. Taking a life cycle look at crianza wine production in Spain: Where are the bottlenecks? *Int. J. Life Cycle Assess.* **2010**, *15*, 330–337. [CrossRef]
42. De Marco, I.; Riemma, S.; Iannone, R. Life cycle assessment of supercritical impregnation: Starch aerogel + a-tocopherol tablets. *J. Supercrit. Fluids* **2019**, *143*, 305–312. [CrossRef]
43. Emara, Y.; Lehmann, A.; Siegert, M.W.; Finkbeiner, M. Modeling pharmaceutical emissions and their toxicity-related effects in life cycle assessment (LCA): A review. *Integr. Environ. Assess. Manag.* **2019**, *15*, 6–18. [CrossRef]
44. Wernet, G.; Conradt, S.; Isenring, H.P.; Jiménez-González, C.; Hungerbühler, K. Life cycle assessment of fine chemical production: A case study of pharmaceutical synthesis. *Int. J. Life Cycle Assess.* **2010**, *15*, 294–303. [CrossRef]
45. Guilbot, J.; Kerverdo, S.; Milius, A.; Pomrehn, F. Life cycle assessment of surfactants: The case of an alkyl polyglucoside used as a self emulsifier in cosmetics. *Green Chem.* **2013**, *15*, 3337–3354. [CrossRef]
46. Lassaux, S.; Renzoni, R.; Germain, A. Life cycle assessment of water from the pumping station to the wastewater treatment plant. *Int. J. Life Cycle Assess.* **2007**, *12*, 118–126.
47. Mata, T.M.; Martins, A.A.; Neto, B.; Martins, M.L.; Salcedo, R.L.R.; Costa, C.A.V. Lca tool for sustainability evaluations in the pharmaceutical industry. *Chem. Eng. Trans.* **2012**, *26*, 261–266.
48. Emara, Y.; Siegert, M.-W.; Lehmann, A.; Finkbeiner, M. Life cycle management in the pharmaceutical industry using an applicable and robust LCA-based environmental sustainability assessment approach. In *Designing Sustainable Technologies, Products and Policies*; Springer: Cham, Switzerland; Berlin/Heidelberg, Germany, 2018; pp. 79–88.
49. Raymond, M.J.; Slater, C.S.; Savelski, M.J. LCA approach to the analysis of solvent waste issues in the pharmaceutical industry. *Green Chem.* **2010**, *12*, 1826–1834. [CrossRef]
50. Zhang, Y.; Zhai, M.; Chen, Z.; Han, X.; Yu, F.; Li, Z.; Xie, X.; Han, C.; Yu, L.; Yang, Y. Dual-modified liposome codelivery of doxorubicin and vincristine improve targeting and therapeutic efficacy of glioma. *Drug Deliv.* **2017**, *24*, 1045–1055. [CrossRef] [PubMed]
51. Goedkoop, M.; Heijungs, R.; Huijbregts, M.; De Schryver, A.; Struijs, J.; van Zelm, R. ReCiPe 2008, A Life Cycle Impact Assessment Method Which Comprises Harmonised Category Indicators at the Midpoint and the Endpoint Level. First Edition. Report I: Characterization. 2009. Available online: http://www.lcia-recipe.net (accessed on 10 September 2021).
52. Ruiz-Mercado, G.J.; Smith, R.L.; Gonzalez, M.A. Sustainability indicators for chemical processes: I. Taxonomy. *Ind. Eng. Chem. Res.* **2012**, *51*, 2309–2328. [CrossRef]
53. Ruiz-Mercado, G.J.; Smith, R.L.; Gonzalez, M.A. Sustainability indicators for chemical processes: II. Data needs. *Ind. Eng. Chem. Res.* **2012**, *51*, 2329–2353. [CrossRef]
54. GlaxoSmithKline plc. Annual Report 2019. Available online: www.gsk.com/annualreport (accessed on 1 September 2021).
55. Air Products. 2010 Sustainability Report. Available online: http://www.airproducts.com/responsibility/2010AnnualReport.htm (accessed on 1 September 2021).
56. Constable, D.J.; Curzons, A.D.; Cunningham, V.L. Metrics to 'green'chemistry—Which are the best? *Green Chem.* **2002**, *4*, 521–527. [CrossRef]
57. ICIS Chemical Business. ICIS Pricing Glycerine Report. Available online: http://www.icispricing.com/il_shared/Samples/SubPage170.asp (accessed on 1 September 2021).
58. Sustainability Reporting 2009. Available online: http://www.bp.com/subsection.do?categoryID=9032624&contentID=7061085 (accessed on 1 September 2021).

processes

MDPI

Article

Gas to Liquids Techno-Economics of Associated Natural Gas, Bio Gas, and Landfill Gas

Federico Galli [1,*], Jun-Jie Lai [2], Jacopo De Tommaso [1], Gianluca Pauletto [3] and Gregory S. Patience[1]

1 Polytechnique Montreal, Department Chemical Engineering, CP 6079, Succ CV, Montreal, QC H3C 3A7, Canada; jacopo.de-tommaso@polymtl.ca (J.D.T.); gregory-s.patience@polymtl.ca (G.S.P.)
2 Department of Chemical Engineering, National Taiwan University, No. 1, Section 4, Roosevelt Rd, Da'an District, Taipei City 106319, Taiwan; e1252081@gmail.com
3 SYPOX, Department of Chemistry, Technical University of Munich, 85747 Garching, Germany; gianlucapauletto@hotmail.it
* Correspondence: federico.galli@polymtl.ca; Tel.: +1-438-836-7612

Abstract: Methane is the second highest contributor to the greenhouse effect. Its global warming potential is 37 times that of CO_2. Flaring-associated natural gas from remote oil reservoirs is currently the only economical alternative. Gas-to-liquid (GtL) technologies first convert natural gas into syngas, then it into liquids such as methanol, Fischer–Tropsch fuels or dimethyl ether. However, studies on the influence of feedstock composition are sparse, which also poses technical design challenges. Here, we examine the techno-economic analysis of a micro-refinery unit (MRU) that partially oxidizes methane-rich feedstocks and polymerizes the syngas formed via Fischer–Tropsch reaction. We consider three methane-containing waste gases: natural gas, biogas, and landfill gas. The FT fuel selling price is critical for the economy of the unit. A Monte Carlo simulation assesses the influence of the composition on the final product quantity as well as on the capital and operative expenses. The Aspen Plus simulation and Python calculate the net present value and payback time of the MRU for different price scenarios. The CO_2 content in biogas and landfill gas limit the CO/H_2 ratio to 1.3 and 0.9, respectively, which increases the olefins content of the final product. Compressors are the main source of capital cost while the labor cost represents 20–25% of the variable cost. An analysis of the impact of the plant dimension demonstrated that the higher number represents a favorable business model for this unit. A minimal production of 7,300,000 kg y^{-1} is required for MRU to have a positive net present value after 10 years when natural gas is the feedstock.

Keywords: techno-economic analysis; GtL; Fischer–Tropsch; ASPEN–Python; Monte Carlo simulation

Citation: Galli, F.; Lai, J.-J.; De Tommaso, J.; Pauletto, G.; Patience, G.S., Gas to Liquids Techno-Economics of Associated Natural Gas, Bio Gas, and Landfill Gas. *Processes* **2021**, *9*, 1568. https://doi.org/10.3390/pr9091568

Academic Editors: Roberta Campardelli, Paolo Trucillo and Adam Smoliński

Received: 26 July 2021
Accepted: 26 August 2021
Published: 1 September 2021

1. Introduction

Since 2018, USA has produced more than 10 million bbl d^{-1} of crude oil [1], while Canada produces half of that. At a price of 75 USD bbl^{-1} (July 2021), crude oil remains an important source of revenue for these countries. At extraction sites, regardless of the technology used to recover the oil, pumps extract natural gas with the oil. The prohibiting costs of infrastructure (installing gas purification stations as pipelines and a compressor) make venting or flaring the preferred alternative for remote wells. Methane is the second highest contributor to greenhouse gases, accounting for 16% of global emissions after carbon dioxide (65%) [2], and its global warming potential is 37 ± 10 times more than that of carbon dioxide over a 100-year period [3]. Moreover, according to the U.S. Environmental Protection Agency (EPA), methane is also the second largest greenhouse gas emitted in North America.

Methane emissions largely come from fermentation (biogas), associated natural gas, and landfill gas [4,5]. Flaring from oil batteries is an associated emission of CO_2 directly correlated with solution gas extraction. According to the Alberta Energy Regulator (AER) report "Upstream Petroleum Industry Flaring and Venting Report", 382×10^3 m^3 solution gas was flared from crude bitumen and crude oil batteries in 2019, while only 144×10^3 m^3 was vented in Alberta [6] (Figure 1).

Figure 1. Satellite-detected natural gas flared in June 2020 over a 30-day span—obtain from SkyTruth [7].

To limit greenhouse gas emissions, governments have applied carbon taxes that are proportional to the quantity of gas flared or vented [8]. In Alberta, for example, this tax is 0.04 CAD kg^{-1} CO_{2eq} (2021), which will rise to 0.05 CAD kg^{-1} CO_{2eq} in 2022, and is expected to reach 0.16 CAD kg^{-1} CO_{2eq} in 2030, which represents an important stimulus to find alternative means to convert natural gas into useful products.

Many companies offer solutions to transform solution gas into methanol, DME, or fuels [9–12]. Most of these technologies reform methane into syngas via an endothermic or exothermic reaction and then react the syngas to produce the target product, often adopting Fischer–Tropsch synthesis (FT). FT converts syngas (CO, H_2) into in hydrocarbons, olefins and to a lesser extent, alcohols. This reaction occurs with metal catalysts (Fe, Co [13,14], Rh) and within a temperature range of 150 °C to 300 °C. With Fe as the catalyst, the water–gas shift reaction (WGS) occurs as well [15]. WGS converts the CO and H_2O into CO_2 and H_2.

Scaling down issues, the energy required to reform natural gas with water or CO_2 (endothermic reforming), and the quality of the feedstock limit the application of these new technologies to remote locations.

A case study conducted in Nigeria demonstrated that a GtL based on FT synthesis has capital costs of USD 100,000 per daily barrel capacity. The authors applied the cost-to-capacity methodology with a scaling exponent of 0.66 [16], and estimated that production costs amounted to USD 900,000. However, applying a single exponent to scale-up (or scale-down) the cost of a technology is a gross approximation because different unit operations and equipment possess a scale exponents range from 0.3 to 1 [17,18].

Mohajerani et al. instead employed different exponents and studied the economy of GtL for the Canadian context [19]. The base case production was 50,000 bbl d^{-1} and the flowsheet included an air separation unit (ASU) upstream from the reforming reactor. ASU accounts for 30% of the costs and therefore this design is impractical for smaller units. In addition to safety issues, at the scale considered in this manuscript, ASU is uneconomical [20,21].

Dong et al. compared GtL to LNG technology with sales volumes of 5 Mt to 6 Mt [22]. They assumed a market price for GtL diesel from 120 USD bbl^{-1} to 160 USD bbl^{-1} and a process that includes an ASU for catalytic partial oxidation. They concluded that GtL is an economical alternative to LNG. However, in the actual economic scenario, the oil price is at approximately 70 USD bbl^{-1} and there is the necessity to scale down the GtL unit to face the needs of smaller and more remote producers. Our study seeks to address this lacuna. Since the waste gas is a small part of their main product, we reasonably assume that the scale of this process would not be excessively large. We conceived a micro-refinery unit

(MRU) that couples a CPOX and an FT reactor on a battery oil unit and recycles the heat generated by the reactions and produces paraffins that can be mixed with oil, increasing production yield, while at the same time reducing the blended gas flaring [23,24]. This unit is economical because it employs the catalytic partial oxidation of natural gas at high pressure (2 MPa), and air as an oxidant, avoiding excessive air separation costs [21,25,26]. Collodi et al. [20] demonstrated that ASUs are uneconomic for production rates below 1000 $t\,d^{-1}$ for GtL technologies that convert natural gas into methanol. Furthermore, pure oxygen poses safety issues since the gas mixture and products are in the flammability range. On the other hand, nitrogen is inert and increases reactor volumes and compression costs.

A few studies have referred to mobile units that treat up to 20 $m^3\,min^{-1}$ of natural gas, with a technology that sacrifices the efficiency to renounce to high ASU costs.

Natural gas is the main interest of North American economies, as it is the main fuel in the transition towards renewable energy [27]. However, in the short and long term, different sources of methane will feed units such as the MRU, including biogas [28,29] and landfill gas [30,31]. China and the USA produced 23.5 $Gm^3\,y^{-1}$ of biogas in 2014 [32] and the International Economic Agency estimates that its demand will double by 2040 [33]. On the other hand, 29% of the world's waste is landfilled, and this quantity will double by 2050 [34]. The MRU treats methane emissions and turns this waste into a more economic product. Few studies have considered the influence of feedstock composition or the technical challenges of designing a plant [35,36].

Here, we simulate the MRU process with ASPEN Plus and developed a Python code to conduct an economic analysis [37].We study the techno-economic analysis of the application of the MRU for treating biogas, natural gas, and landfill gas.

2. Materials and Methods

2.1. Process Simulation

The following description applies to the landfill gas and the biogas scenarios as well. A Python 3.9.4 code selects the composition of the sweet feedstock stream by a Monte Carlo analysis (Table 1) while its mole flow is set to 50 $kmol\,h^{-1}$ (approximately 20 $m^3\,min^{-1}$, see Supporting Materials). The MRU scale is larger compared to the Canadian production, but falls in the 30th percentile of the permitted flaring of Texas oil producers [6,38]. The Soave–Redlich–Kwong (RKS) equation of state calculates the fugacities of the components [39]. Aspen v.11 calculated the mass and energy balances for the two compressors (methane-rich gas and air) and the CPOX reactor, which we simulated as a Gibbs reactor. The compression ratio of the reciprocal compressors should be lower than 5 [40], with a maximum outlet temperature below 433 K, therefore, we simulated three sequential one-stage intercooled compressors for each stream to meet these specifications. The outlet pressure of both compressors is 2 MPa. A design specification (Design/Spec) fixes the molar ratio of carbon and oxygen in the stream entering in the CPOX reactor equal to 2 by adjusting the flowrate of air (simulated as a mixture of 21% of O_2 and 79% of N_2). A thermodynamic analysis individuated the best ratio to be 1.7 [24,25] to minimize the coke formation. However, here we assumed the stoichiometric ratio because the objective was not to optimize the catalyst formulation to minimize coke formation.

The CPOX reactor operates at 2 MPa and we assumed a negligible pressure drop across the catalytic bed. Some of the authors optimized this configuration in a previous work [25]. We simulated an adiabatic Gibbs reactor. CPOX converts the gas into CO and H_2 (plus H_2O and CO_2). The outlet temperature of the CPOX reactor ranges from 933 K to 1033 K. A heat exchanger (shell and tube), with an exchange area of 1.834 m^2, reduces the temperature of this stream to 548 K. We fixed its value to $S + 2\Delta S$, where S and ΔS are the average heat exchange area and its standard deviation obtained by 50 simulations changing the gas composition, respectively. Water at 298 K is the utility stream for the heat exchanger.

We coded the kinetics of the FT reaction in Python 3.9.4. We employed a model that considers an iron catalyst and calculates the paraffin, olefin and alcohol distributions based on 7 micro-kinetics rate steps [41]. The model also considers the water–gas shift reaction (see Supporting Information). We assumed a residence time of 1.6 min and calculated CO conversion, product distribution (α) and the ratio between hydrocarbons and olefins and alcohols.

The Python 3.9.4 program iteratively solves the Aspen flowsheet and the FT reactor, changing the initial feed gas composition 3000 times (Figure 2). It then stores the results of each cycle in an Excel file, that calculates the distribution of the output variables. These values, with their uncertainties, passed to the techno-economic analysis.

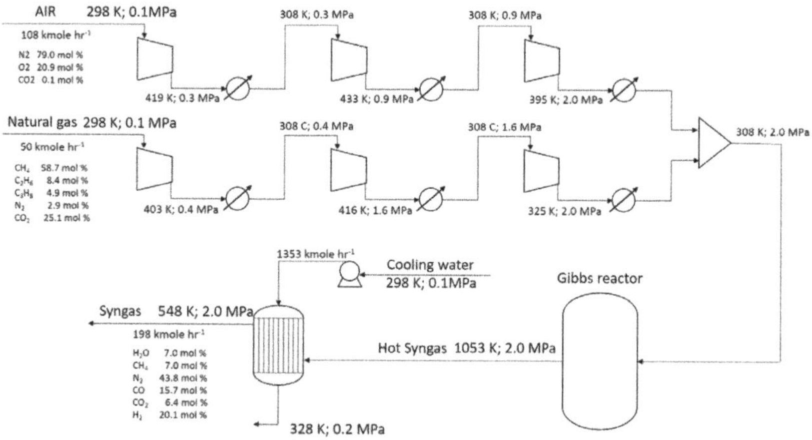

Figure 2. Schematic flowsheet of the simulation. The results reported represent one of the 3000 Monte Carlo cycles performed by the Python code. A complete list of all the simulated results for the three scenarios is available in the Supporting Materials.

Table 1. Composition distribution in the mole % for natural gas, landfill gas, and biogas. We assumed a triangular distribution for natural gas and a normal distribution for landfill gas. For each Monte Carlo cycle, Python 3.9.4 initializes the composition of the feedstock according to the probability distribution of each component and then normalizes the values to 100%.

	Natural Gas [42]			Landfill Gas [43]		Biogas [44,45]	
	Min	Most Likely Value	Max	μ	σ	μ	σ
CH_4	19.2	9.0	99.5	52.5	2.5	48.0	5.8
C_2H_6	0.0	9.8	93.5	0.0	0.0	0.0	0.0
C_3H_8	0.0	5.8	41.0	0.0	0.0	0.0	0.0
N_2	0.0	3.4	80.2	3.5	0.5	16.4	6.7
CO_2	0.0	2.9	39.7	50.0	3.3	31.9	4.1
O_2	0.0	0.0	0.0	0.5	0.2	3.8	2.1
CO	0.0	0.0	0.0	0.1	0.03	0.0	0.0

2.2. Techno-Economic Analysis

First, we calculated the capital cost for each equipment, and then the total capital investment (*TCI*) as the sum of direct capital cost, indirect capital, cost and working capital. Then, we estimated the profitability of each scenario via the Net Present Value (NPV) [46]. We selected reciprocal piston stainless steel compressors for both air and methane-rich feedstock. The pump for cooling water is a reciprocal stainless steel pump. For CPOX and FT reactors, we employed stainless steel vessels with an inner diameter of 0.5 m and 1 m, respectively. The height of the CPOX and FT reactor is 1 m and 6 m, respectively.

Empirical correlations estimated the cost of unit operations ($C_{BM,i}$), the purchase price ($C_{P,i}$) and the bare module factor ($F_{BM,i}$) [47] ($CEPCI_{2004}$ = 400). We actualized the cost of the equipment using the 2020 CEPCI index (Equation (1), 618.7 for heat exchangers and reactors, and 1080.2 for compressors and pumps) [48,49]

$$C_{\text{BM,i}} = C_{\text{P,i}} \cdot F_{\text{BM,i}} \cdot \frac{\text{CEPCI}_{2020}}{400} \tag{1}$$

A second Python 3.9.4 algorithm generates the averages and standard deviations of compressors' duties, pump duty, cooling water flow rate, and FT product flowrate and composition. From these, it calculates the cost of the equipment [47] and the total capital investment (TCI). TCI includes the direct capital costs (C_{Direct}) as well as the indirect capital costs (C_{Indirect}), and working capital costs (WC) (Equation (2)):

$$\begin{cases} TCI = FCI + WC \\ WC = (C_{\text{Pur}} + C_{\text{Del}}) \cdot 0.05 \\ C_{\text{Pur}} = \sum_{i=1}^{N} C_{\text{BM,i}} \\ C_{\text{Del}} = C_{\text{Pur}} \cdot 0.1 \\ FCI = C_{\text{Direct}} + C_{\text{Indirect}} \\ C_{\text{Direct}} = (C_{\text{Pur}} + C_{\text{Del}}) \cdot (1 + f_{\text{Inst}} + f_{\text{Instr}} + f_{\text{Pipe}} + f_{\text{Elec}} + f_{\text{Buil}} + f_{\text{Fac}} + f_{\text{Impr}}) \\ C_{\text{Indirect}} = (C_{\text{Pur}} + C_{\text{Del}}) \cdot (1 + f_{\text{Eng}} + f_{\text{Constr}} + f_{\text{Leg}} + f_{\text{Fee}} + f_{\text{Cont}}) \end{cases} \tag{2}$$

The direct capital costs include purchased equipment installation (f_{Inst}), instrumentation and controls (f_{Instr}), piping (f_{Pipe}), electrical systems (f_{Elec}), building (f_{Buil}), yard improvement (f_{Impr}), and service facilities (f_{Fac}). The indirect capital costs include engineering and supervision (f_{Eng}), construction expenses (f_{Constr}), legal expenses (f_{Leg}), the contractor's fee (f_{Fee}), and contingency (f_{Cont}). We assigned a factor to calculate these costs (Table 2, Equation (2)).

Table 2. Capital cost factors.

Item	Fraction of Delivered Equipment, $(C_{\text{Pur}} + C_{\text{Del}})$
Purchased Equipment Installation, f_{Inst}	0.15
Instrumentation and Controls, f_{Instr}	0.36
Piping, f_{Pipe}	0.16
Electrical Systems, f_{Elec}	0.10
Building, f_{Buil}	0.00
Service Facilities, f_{Fac}	0.30
Yard Improvement f_{Impr}	0.00
Engineering and Supervision, f_{Eng}	0.01
Construction Expenses, f_{Constr}	0.34
Legal Expenses, f_{Leg}	0.04
Contractor's Fee, f_{Fee}	0.17
Contingency, f_{Cont}	0.32

The plant benefits from an accelerated depreciation over 5 y: 20% the first year, then in an sum-of-the-years' digits method for the remaining 4 y. We considered electricity, waste disposal, and cooling water as utilities.

Variable costs (C_{VAR}) include labor costs (C_{Lab}), supervision (C_{Sup}), maintenance (C_{Main}), supplies (C_{Supp}), laboratory and research (C_{Res}), royalties (C_{Roy}), catalyst (C_{Cat}), and utilities (C_{Ut}), as shown in Equation (3). For labor costs, we assumed 1 operator per shift over four shifts per day. Since the MRU integrates an already existing plant, one additional operator is enough to operate the unit. The operator cost is 34 USD h^{-1}. The annual operating labor cost is 0.295 MUSD y^{-1}. Concerning utilities, the electricity cost is 0.012 USD kW^{-1} h [50]. The waste disposal cost includes the hazardous (145 USD t^{-1}) and non-hazardous materials (36 USD t^{-1}) [51,52]. We assumed that both the hazardous component and the non-hazardous component account for 1% of the whole product [46].

The cooling water cost is 0.08 USD m^{-3} and we only accounted for the make up water. In particular, we calculated the make up water flowrate as 0.2%/K [46,53] of the cooling water mass flowrate considering the temperature difference between the inlet and outlet of the heat exchanger. We also accounted for the carbon dioxide emission cost (50 USD t^{-1}):

$$\begin{cases} C_{VAR} = C_{Lab} + C_{Sup} + C_{Main} + C_{Supp} + C_{Res} + C_{Roy} + C_{Cat} + C_{Ut} \\ C_{Sup} = C_{Lab} \cdot 0.05 \\ C_{Main} = FCI \cdot 0.06 \\ C_{Supp} = C_{Main} \cdot 0.05 \\ C_{Res} = C_{Lab} \cdot 0.05 \\ C_{Roy} = 0 \\ C_{Cat} = \text{Yearly Catalyst} \cdot \text{Product price} \cdot 0.005 \end{cases} \quad (3)$$

We set royalties to zero because the unit is internal proprietary technology [54].

Regarding raw material and product prices, we devised nine different cases that represent the optimal (Case 3), average, and worst case (Case 1) scenarios (Table 3).

Table 3. Our model evaluates the payback period and the net return after 10 y, considering feedstock and product prices variation.

Value	Case 1	Case 2	Case 3
Feedstock Price	0.23 USD kg^{-1} [55]	0	$-50 \cdot tCO_{2eq}$ In Feed, USD
Product Price (USD kg^{-1})	0.2	0.3	0.4 [56]

To the variable costs, we added fixed charges as ($C_{Charges}$): taxes (C_{Tax}), financing (C_{Fin}), insurance (C_{Insu}), and renting material (C_{Rent}) —and plant overheads($C_{Overhead}$), as illustrated in Equation (4):

$$\begin{cases} C_{Charges} = C_{Tax} + C_{Fin} + C_{Insu} + C_{Rent} + C_{Overhead} \\ C_{Tax} = FCI \cdot 0.02 \\ C_{Fin} = 0 \\ C_{Insu} = FCI \cdot 0.01 \\ C_{Rent} = 0 \\ C_{Overhead} = (C_{Lab} + C_{Sup} + C_{Main}) * 0.2 \end{cases} \quad (4)$$

We assume zero financing and renting costs because we did not consider lending money for the construction of the MRU and there is no need to rent further equipment. Eventually, we considered general expenses ($C_{General}$), constituted by administrative expenses (C_{Admin}), distribution and selling (C_{Distr}), and development (C_{Devel})—as illustrated Equation (5):

$$\begin{cases} C_{General} = C_{Admin} + C_{Distr} + C_{Devel} \\ C_{Admin} = (C_{Lab} + C_{Sup} + C_{Main}) \cdot 0.1 \\ C_{Distr} = (C_{Admin} + C_{Overhead} + C_{Charges} + C_{VAR} - C_{Roy} - C_{Cat}) \cdot 0.05 \\ C_{Devel} = (C_{Admin} + C_{Overhead} + C_{Charges} + C_{VAR} - C_{Roy} - C_{Cat}) \cdot 0.05 \end{cases} \quad (5)$$

For the calculation of the net return and the payback time, we assumed an inflation rate of 2% and an income tax of 28%, which is conservative in Canada, whose taxation is from 11.5% to 16% depending on the province [57], but it is in line with the rest of the Western nations. Eventually, we considered that the MRU operates at 0% capacity the first year (−1) (we considered construction and installation), then at 50%, 90%, 100% each progressive successive year. We applied a yearly present worth factor (PWF) of 13% for the calculation of the NPV (Table 4).

Table 4. Present worth factor applied (value of USD 1 at year y). Year 0 corresponds to the start of the operation.

Year	−1	0	1	2	3	4	5	6	7	8	9	10	11	12
PWF	1.2	1.1	0.9	0.8	0.7	0.6	0.5	0.5	0.4	0.4	0.3	0.3	0.2	0.2

3. Results and Discussion

The MRU produces a syngas whose H_2/CO ratio is understoichiometric, due to the presence of CO_2 and higher hydrocarbons in the feedstock (from 0.7 to 1.7, as shown in Figure 3).

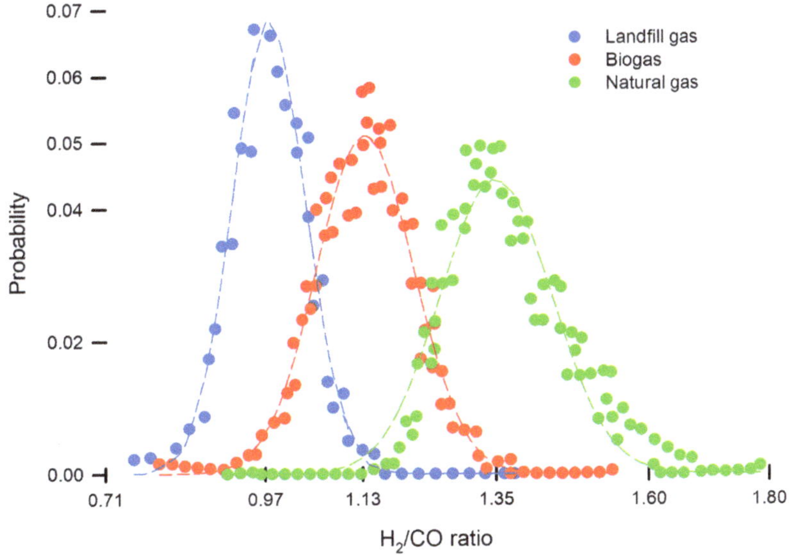

Figure 3. Three thousand Monte Carlo iterations calculating the distribution of the hydrogen/carbon monoxide molar ratio after the CPOX unit for the three feedstocks considered.

The high concentration of CO_2 in landfill gas limitS hydrogen production for the co-presence of the dry reforming reaction that is endothermic and produces an equimolar mixture of CO and H_2 [58]. In this scenario, the CPOX reactor of the MRU operates more as an autothermal dry reformer with an excess of oxygen. Indeed, the equilibrium temperature reached in the Gibbs reactor depends on the content of CO_2, ranging from 973 K to 1123 K (Figure S1b in Supporting Information). Similarly, the most likely value of the CO conversion in the Fischer–Tropsch reactor follows the same trend, ranging from 50% to 65%, which are values that agree with single-pass FT reactors [59].

For NG, we calculated a hydrocarbon production of 1,956,100 kg y^{-1}. However, the MRU is only economical when we discount the carbon tax for the avoided flaring, with a payback time ranging from (0.71 ± 0.11) y to (0.66 ± 0.09) y depending on the product price assumed. It is difficult to compare our results to those in the literature because, as explained in the introduction, there are not so many techno-economic models and most of them focus on large-scale units.

A sensitivity analysis (Table 5) on the production of hydrocarbons revealed that an iron catalyst is unsuitable to achieve an economical payback time because of its higher selectivity towards C_1–C_3 products. The production to have a payback time lower than 10 y is 7.3 kt y^{-1}. A cobalt-based catalyst is therefore preferred for GtL units because of its higher intrinsic activity [60,61]. We continued our analysis accordingly, assuming a production of 1×10^3 kt y^{-1}.

Table 5. Payback time (PT, y) correlates with the hydrocarbon production (P, kg y^{-1}), $\ln(PT) = \frac{28\,790\,000 \pm 2\,036\,000}{P} - 1.64 \pm 0.21$, $R^2 = 0.97$. With a production of 1.05×10^7 kg y^{-1}, the MRU breaks—even 3 y after its installation.

Case	Production, kg y^{-1}	Average Payback Time, y	1σ
1	1.96×10^6	-	-
2	3.91×10^6	-	-
3	5.87×10^6	33.54	14.96
4	7.82×10^6	5.68	0.47
5	9.78×10^6	3.13	0.19
6	1.17×10^7	2.16	0.12
7	1.37×10^7	1.65	0.08
8	1.56×10^7	1.33	0.06
9	1.76×10^7	1.12	0.05

Under the most probable operating conditions (2–2, shown in Table 3), the natural gas scenario is economically the most interesting. When the natural gas feedstock is available at a price of zero, and when the liquid product sells similarly to crude oil, the cumulative NPV goes towards zero after 3.5 y, even with an accelerated depreciation of 5 years. This demonstrates that a portable, modular MRU plant, is economically self-sufficient without any direct (investment) or indirect (on the avoided emission) subsidy. After all the expenses, the MRU produces enough liquid to pay for carbon taxes for the CO_2 emissions related to the combustion of the remaining flue gases. Oil and gas companies as well as local governments must at least aim to achieve this chemistry (CO conversion, product distribution), and these economics to operate their wells. Even more interesting is the case where the natural gas feedstock comes with a negative price, equivalent to the avoided carbon emissions. With the carbon tax increasing in the near future, this case will be even more beneficial.

The nature of the feedstock does not heavily affect the economics. The MRU performs better for feedstocks richer in methane, but at the same time, it operates quite well over wide range of natural gas compositions (Figure 3).

Emissions from flared stranded gas pose the main threat (in terms of volumes) to the environment. We optimized our operating conditions (H_2/CO ratio) for this feedstock source. By injecting water, or by simply changing the air intake, the biogas and the landfill cases could be more viable. Moreover, steam condensation requires additional CAPEXs and OPEXs due to heat tracing. However, our ultimate objective was to benchmark the three cases in the easiest manner, to demonstrate that the process is attractive even when the best possible outcome is not achieved.

We studied the influence of the scale of the MRU, changing the natural gas flowrate to 25 kmol h^{-1} (approximately 9 m^3 min^{-1}, lower scale) and 100 kmol h^{-1} (approximately 40 m^3 min^{-1}, higher scale). In the actual economic scenario, the reduced production due to a lower gas flow is insufficient to have a positive net present value, even if CAPEXs are 60% compared to the base scale analyzed (natural gas flared flowrate of 19 m^3 min^{-1}, as shown in Figure 4).

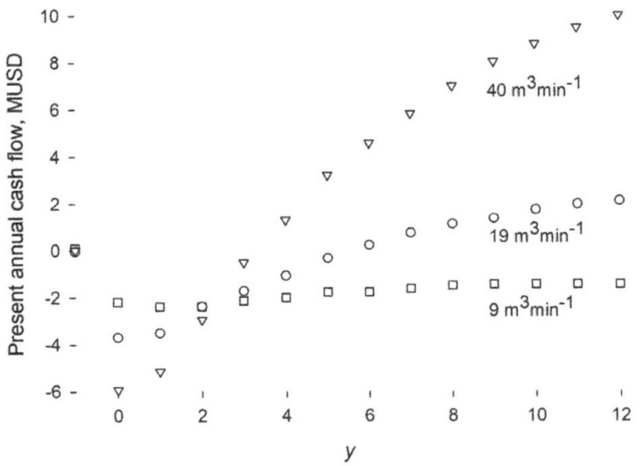

Figure 4. The MRU is uneconomical for battery units that flare less than $10\,m^3\,min^{-1}$ of natural gas. Here, the FT fuel is sold at the same price as raw oil and the natural gas cost is $0\,USD\,m^{-3}$ (scenario 2–2).

MRU benefits to some extent from the economies of the scale. Even if the CAPEXs and OPEXs increase accordingly, higher production reduces the payback time if we assume a natural gas flowrate of $40\,m^3\,min^{-1}$ (Table 6). This flowrate represents approximately 25 % of the Texas permitted flares in 2010 [38].

Table 6. If associated gas costs the same as natural gas, the MRU is uneconomical at all scales studied. In other cases, the higher the quantity of natural gas treated, the lower the payback time. Results are expressed as the average of 3000 simulations $\pm 1\sigma$.

Raw Material Price, USD kg^{-1}			Product Price, USD kg^{-1}	
	Flowrate, kmol h^{-1}	**0.20**	**0.30**	**0.40**
0.2322	25	-	-	-
	50	-	-	-
	100	-	-	-
0	25	-	13 ± 1	3.7 ± 0.13
	50	13 ± 2	3.1 ± 0.19	1.8 ± 0.09
	100	4.0 ± 0.3	1.8 ± 0.11	1.2 ± 0.07
Negative	25	1.8 ± 0.11	1.3 ± 0.15	1.1 ± 0.10
	50	0.56 ± 0.1	0.49 ± 0.06	0.44 ± 0.05
	100	0.22 ± 0.03	0.21 ± 0.03	0.20 ± 0.02

Similarly, MRU benefits from numbering up [62,63]. The unit itself needs minimum maintenance, control and supervision, and runs lean in terms of personnel and direct and indirect capital cost expenses (Equation (2)). This is because of its small scale and its modularity [18]. A bigger plant is not portable and needs at least one extra operator, plus higher direct and indirect capital costs (5.9 MUSD for the case where we considered $40\,m^3\,min^{-1}$ of natural gas flared instead of a total of 3.7 MUSD for the base case, Equation (2)). Furthermore, the attractiveness goes with a quick payback time and a relatively cheap plant. This way, oil companies can approach the increasing carbon tax and public concern in a timely manner, whilst having a delocalized unit that they build over one year, depreciate in five, and dismantle and move between wells at low cost. The same applies in the case of landfills and biogas in remote places, where local authorities need a simple, affordable, and modular plant to address the environmental threats of uncontrolled emissions.

We foresee the MRU as a lean plant, where the investment itself plays the biggest impact on annual expenses (Figure 5). Compressors represent the main source of fixed costs for the MRU, with 70 ± 8%, 71 ± 8%, and 57 ± 3% for natural gas, biogas, and landfill gas, respectively. For natural gas, the bare module model overestimates the cost of the air compressor (USD 562,000 ± USD 42,000) because the empirical regressions only account for the unit duty (Figure 5). In the other two scenarios, both the compressors were estimated to cost between USD 200,000 to USD 210,000 (including installation, control, etc.), which is more reasonable (Figure 5). The model estimates the reactors' cost at USD 170,000, which represent between 11% and 14% of the purchase costs. The total fixed investment resulted in 2.7 MUSD, 3.3 MUSD, and 4.7 MUSD for landfill gas, biogas, and natural gas (Figure 5), respectively (uncertainty below 1%). Utilities account for between USD 200,000 y^{-1} and USD 270,000 y^{-1} with waste disposal as the main source of cost (waste disposal is proportional to annual production).

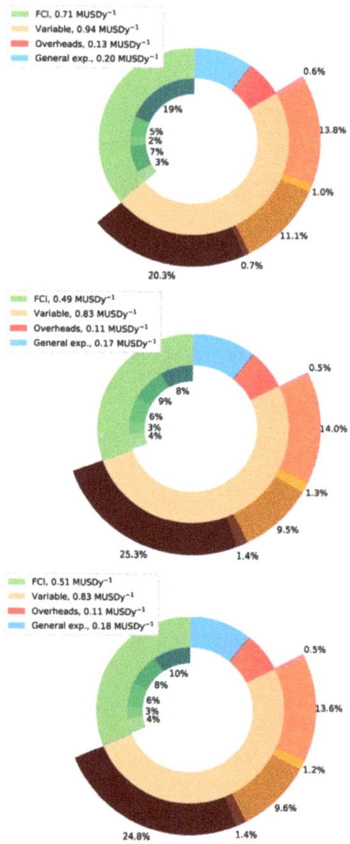

Figure 5. Depreciation of fixed capital investments (FCIs) per year (linear depreciation over 5 y), Variable costs, plant overheads and general expenses (administration distribution, selling, and R&D) for the MRU in the base case scenario with the three feedstocks (price = 0 USD y^{-1} and FT fuel price = 0.3 USD kg^{-1}). We outlined the contribution percentage of the main units to the FCIs (air compressor, natural gas compressor, water pump, CPOX reactor, FT reactor) and to variable costs (labor, catalyst, maintenance, supervision, utilities, and supplies). Top: natural gas; middle: biogas; bottom: landfill gas.

According to the rule of thumb which says that FT-based GtL units cost (FCIs) approximately 100,000 USD bbl^{-1} d of production [16], we calculated that our unit respects this estimation for the biogas and landfill scenario (USD 95,000 bbl^{-1} d to USD 105,000 bbl^{-1} d), while for the natural gas, the overestimation of the FCI costs for compressor is a bias.

We designed and simulated a CPOX reactor with a stoichiometric hydrocarbon/oxygen ratio. With this configuration, the biogas and the landfill gas in the current economic scenario (case 2–2 of Table 3) resulted with a negative actual cash flow (Figure 6). The presence of CO_2 and higher hydrocarbon reduces the total FT-fuel production. It is out of the scope of this paper to optimize the condition for both these methane-rich feedstocks, which, however, resulted in a slightly negative annualized actual cash flow: −3.8 MUSD and −4.4 MUSD in year $y = 12$ for biogas and landfill gas, respectively, with an initial investment of −2.5 MUSD and −2.6 MUSD. We speculate that the optimized operating conditions and oxidant feed (steam or using green hydrogen to increase the CO/H_2 ratio in the FT reactor) also make the MRU economical for biogas and landfill gas, feedstocks that are suitable for GtL technologies [64–66].

A combined techno-economic assessment and LCA may also find that optimal operating conditions reduce emissions [67] .

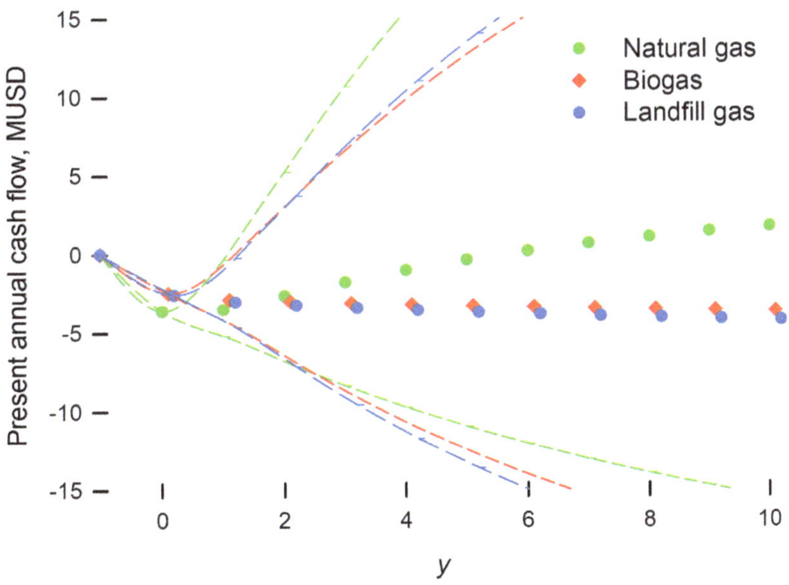

Figure 6. Present annual cash flow of the MRU for the three feedstocks studied. Lines represent the best possible scenario (cost of feedstock discounted by the carbon tax (−50 USD t^{-1} g CO_{2eq}) and an FT fuel price of 0.2 USD kg^{-1}); and the worst possible scenario (cost of feedstock of 0.23 USD kg^{-1} and an FT fuel price of 0.4 USD kg^{-1}). Symbols represent the base-case scenario (cost of feedstock of 0 USD kg^{-1} and an FT fuel price of 0.3 USD kg^{-1}).

4. Conclusions

Flaring natural gas is the only solution for battery units in remote locations. However, environmental concerns and new GtL processes create new avenues to solve this issue. The MRU integrates the existing flaring sites on the well and converts 40% to 75% of the stranded gas into liquid hydrocarbons, which are an extra source of revenue for producers. The MRU has a payback time of less than 10 years regardless of the assumed fuel price. On the other hand, if we consider the stranded gas to cost the same as natural gas, or the MRU to treat less than 10 m^3 min^{-1}, the technology we propose is not viable. The oil industry is mercurial, and the stability of production, safety, technology maturity, and oil price are the major sources of uncertainty. Moreover, the chemical industry should strive towards plant automatization, as a part of the accelerated development of the 4.0 technologies [68]. In the North American context, regional or provincial bodies can either decide to grant incentives up-front on the CAPEXs, or alternatively, focus on the final product. We did not account for governmental economic incentives that reduce CAPEXs from 30% to 50% (and their relative depreciation) in our calculation. We also neglected a scenario where local bodies decide to subsidize the FT liquid itself. This last scenario is particularly appreciated by the local authorities, because it can decide to bet only on winning technologies, is mature enough to ensure steady operation and production in the long term. A third kind of incentive is the carbon tax, planned to increase to 135 USD by 2030, becoming the biggest contributor to the NPV of the plant. Concerning the Fischer–Tropsch synthesis, Co or Rh are more indicative compared to iron-based catalysts due to their higher α and CO per-pass conversion. MRU is also economical with feedstock such as biogas or landfill gas, which will become the main source of carbon when the energetic transition towards renewables will be achieved.

Supplementary Materials: The following are available online at https://www.mdpi.com/article/10.3390/pr9091568/s1, Figure S1: Comparison between the calculated kinetic constants from our Python model and the experimental data at 573 K. Paraffin formation (a), olefins absorption (b), and alcohols desorption (d) show good agreement with the experimental data. The olefins desorption step (c) resulted as underestimated because the value of the parameter $\alpha_{P,HC6re}$ was not reported in the manuscript. We assumed $\alpha_{P,HC6re} = 1 - \alpha_{P,HC6}$. Moreover, the value of E0HC6re is not present in Table 6. We assumed $E_{0,HC6re} = E_{0,HC6}$ (lines 30–31 of kinetics.py), Figure S2: Distributions of the feedstock compressor duty (a), CPOX temperature (b), H_2/CO ratio in syngas (c), CO conversion in the FT reactor (d), and asf chain growth probability factor (e) for the three scenarios, Code—Python.rar: Source code of the Python algorithm.

Author Contributions: Conceptualization, F.G. and J.D.T.; methodology, F.G., J.-J.L. and J.D.T.; software, F.G. and J.-J.L.; validation, F.G., J.D.T., G.P. and G.S.P.; formal analysis, F.G.; resources, G.S.P.; data curation, F.G. and G.S.P.; writing—original draft preparation, F.G., J.-J.L. and J.D.T.; writing—review and editing, F.G., G.P. and G.S.P.; visualization, F.G. and G.S.P.; supervision, G.S.P. and F.G.; funding acquisition, G.S.P. All authors have read and agreed to the published version of the manuscript.

Funding: This research was undertaken, in part, thanks to funding from the Canada Research Chair program and Canada Foundation for Innovation. J.-J.L. received the "2021 Winter Research Internship Program" scholarship funded by Polytechnique Montreal— project number WRIP21, Project 17: GtL micro-refinery unit: conversion of methane into liquid hydrocarbons. Supervised by Pr. Gregory Patience.

Institutional Review Board Statement: Not applicable.

Informed Consent Statement: Not applicable.

Data Availability Statement: The data presented in this study are available in the article and the Supplementary Materials.

Conflicts of Interest: The authors declare no conflict of interest.

Abbreviations

The following abbreviations are used in this manuscript:

AER	Alberta Energy Regulator
ASU	Air separation Unit
CAPEXs	Capital expenses
CEPCI	Chemical engineering plant cost index
CPOX	Catalytic partial oxidation
EPA	Environmental Protection Agency
FCIs	Fixed capital investments

FT	Fischer–Tropsch
GtL	Gas to liquid
LNG	Liquefied natural gas
MRU	Micro refinery unit
NG	Natural gas
NPV	Net present value
OPEXs	Operating expenses
PWF	Present worth factor
TCI	Total capital investment
WGS	Water–gas shift

References

1. US Energy Information Administration. Upstream Petroleum Industry Flaring and Venting Report Industry Performance for Year Ending 31 December 2019. 2020. Available online: https://static.aer.ca/prd/documents/sts/ST60B-2020.pdf (accessed on 11 April 2021)
2. Edenhofer, O. *Climate Change 2014: Mitigation of Climate Change*; Cambridge University Press: Cambridge, MA, USA, 2015; Volume 3.
3. Derwent, R.G. Global Warming Potential (GWP) for Methane: Monte Carlo Analysis of the Uncertainties in Global Tropospheric Model Predictions. *Atmosphere* **2020**, *11*, 486. [CrossRef]
4. European Commission. Reducing Greenhouse Gas Emissions: Commission Adopts EU Methane Strategy as Part of European Green Deal. 2020. Available online: https://ec.europa.eu/commission/presscorner/detail/en/ip_20_1833 (accessed on 21 July 2021)
5. European Commission. Reducing Methane Emissions: Opportunities and Barriers in Waste and Agriculture through Biogas Production. 2020. Available online: https://ec.europa.eu/info/events/reducing-methane-emissions-opportunities-and-barriers-waste-and-agriculture-through-biogas-production-2020-jul-17_en (accessed on 21 July 2021)
6. US Energy and Information Administration. PETROLEUM & OTHER LIQUIDS. 2021. Available online: https://www.eia.gov/dnav/pet/hist/LeafHandler.ashx?n=pet&s=mcrfpus2&f=m (accessed on 11 April 2021)
7. Skytruth. Satellite-Detected Natural Gas Flaring. 2020. Available online: https://skytruth.org/flaring/ (accessed on 7 June 2021)
8. Government of Canada. How Carbon Pricing Works. 2021. Available online: https://www.canada.ca/en/environment-climate-change/services/climate-change/pricing-pollution-how-it-will-work/putting-price-on-carbon-pollution.html (accessed on 21 July 2021)
9. CompactGTL. CompactGTL's Project in Kazakhstan. 2018. Available online: http://www.compactgtl.com/about/projects/ (accessed on 17 February 2021)
10. Greyrock. Technology Greyrock Chemical and Fuel Production. 2008. Available online: http://www.greyrock.com/technology/ (accessed on 17 February 2021)
11. Hargreaves, N. Roll out of Smaller Scale GTL Technology at ENVIA Energy's Plant in Oklahoma City, USA. 2017. Available online: https://www.velocys.com/press/ppt/Gastech2017_Velocys_FINAL_4.3_web.pdf (accessed on 2 March 2021)
12. Infra. 2020. Available online: https://en.infratechnology.com/technology/RR# (accessed on 3 March 2021)
13. Pirola, C.; Scavini, M.; Galli, F.; Vitali, S.; Comazzi, A.; Manenti, F.; Ghigna, P. Fischer–Tropsch Synthesis: EXAFS Study of Ru and Pt bimetallic Co Based Catalysts. *Fuel* **2014**, *132*, 62–70. [CrossRef]
14. Di Michele, A.; Sassi, P.; Comazzi, A.; Galli, F.; Pirola, C.; Bianchi, C.L. Co- and Co(Ru)-Based Catalysts for Fischer-Tropsch Synthesis Prepared by High Power Ultrasound. *Mater. Focus* **2015**, *4*, 295–301. [CrossRef]
15. Louyot, P.; Neagoe, C.; Galli, F.; Pirola, C.; Patience, G.S.; Boffito, D.C. Ultrasound-assisted impregnation for high temperature Fischer-Tropsch catalysts. *Ultrason. Sonochemistry* **2018**, *48*, 523–531. [CrossRef] [PubMed]
16. Kanshio, S.; Agogo, H.O.; Chior, T.J. Techno-Economic Assessment of Mini-GTL Technologies for Flare Gas Monetization in Nigeria. Paper Presented at the SPE Nigeria Annual International Conference and Exhibition, Lagos, Nigeria, 31 July–2 August 2017; doi:10.2118/189120-MS [CrossRef]
17. Remer, D.S.; Chai, L.H. Design cost factors for scaling-up engineering equipment. *Chem. Eng. Prog.* **1990**, *86*, 77–82.
18. Patience, G.S.; Boffito, D.C. Distributed production: Scale-up vs. experience. *J. Adv. Manuf. Process.* **2020**, *2*, e10039. [CrossRef]
19. Mohajerani, S.; Kumar, A.; Oni, A.O. A techno-economic assessment of gas-to-liquid and coal-to-liquid plants through the development of scale factors. *Energy* **2018**, *150*, 681–693. [CrossRef]
20. Collodi, G.; Azzaro, G.; Ferrari, N.; Santos, S. Demonstrating Large Scale Industrial CCS through CCU—A Case Study for Methanol Production. *Energy Procedia* **2017**, *114*, 122–138. [CrossRef]
21. Banaszkiewicz, T.; Chorowski, M.; Gizicki, W. Comparative analysis of cryogenic and PTSA technologies for systems of oxygen production. *AIP Conf. Proc.* **2014**, *1573*, 1373–1378. [CrossRef]
22. Dong, L.; Wei, S.; Tan, S.; Zhang, H. DGTL or LNG: Which is the best way to monetize 'stranded' natural gas? *Pet. Sci.* **2008**, *5*, 388–394. [CrossRef]
23. Trevisanut, C.; Jazayeri, S.M.; Bonkane, S.; Neagoe, C.; Mohamadalizadeh, A.; Boffito, D.C.; Bianchi, C.; Pirola, C.; Visconti, C.G.; Lietti, L.; et al. Micro-syngas technology options for GtL. *Can. J. Chem. Eng.* **2016**, *94*, 613–622. [CrossRef]

24. Pauletto, G.; Mendil, M.; Libretto, N.; Mocellin, P.; Miller, J.T.; Patience, G.S. Short contact time CH4 partial oxidation over Ni based catalyst at 1.5MPa. *Chem. Eng. J.* **2021**, *414*, 128831. [CrossRef]
25. Pauletto, G.; Galli, F.; Gaillardet, A.; Mocellin, P.; Patience, G.S. Techno economic analysis of a micro Gas-to-Liquid unit for associated natural gas conversion. *Renew. Sustain. Energy Rev.* **2021**. [CrossRef]
26. Zichittella, G.; Pérez-Ramírez, J. Status and prospects of the decentralised valorisation of natural gas into energy and energy carriers. *Chem. Soc. Rev.* **2021**, *50*, 2984–3012. [CrossRef] [PubMed]
27. Delborne, J.A.; Hasala, D.; Wigner, A.; Kinchy, A. Dueling metaphors, fueling futures: "Bridge fuel" visions of coal and natural gas in the United States. *Energy Res. Soc. Sci.* **2020**, *61*, 101350. [CrossRef]
28. Yuan, X.C.; Lyu, Y.J.; Wang, B.; Liu, Q.H.; Wu, Q. China's energy transition strategy at the city level: The role of renewable energy. *J. Clean. Prod.* **2018**, *205*, 980–986. [CrossRef]
29. Baena-Moreno, F.M.; Pastor-Pérez, L.; Wang, Q.; Reina, T. Bio-methane and bio-methanol co-production from biogas: A profitability analysis to explore new sustainable chemical processes. *J. Clean. Prod.* **2020**, *265*, 121909. [CrossRef]
30. Heo, J.; Lee, B.; Lim, H. Techno-economic analysis for CO2 reforming of a medium-grade landfill gas in a membrane reactor for H2 production. *J. Clean. Prod.* **2018**, *172*, 2585–2593. [CrossRef]
31. Gao, R.; Zhang, C.; Lee, Y.J.; Kwak, G.; Jun, K.W.; Kim, S.K.; Park, H.G.; Guan, G. Sustainable production of methanol using landfill gas via carbon dioxide reforming and hydrogenation: Process development and techno-economic analysis. *J. Clean. Prod.* **2020**, *272*, 122552. [CrossRef]
32. Statista. Leading Biogas Producing Countries in 2014. 2021. Available online: https://www.statista.com/statistics/481840/biogas-production-worldwide-by-key-country/ (accessed on 16 August 2021)
33. International Energy Agency. The Outlook for Biogas and Biomethane to 2040. 2021. Available online: https://www.iea.org/reports/outlook-for-biogas-and-biomethane-prospects-for-organic-growth/the-outlook-for-biogas-and-biomethane-to-2040 (accessed on 16 August 2021)
34. The World Bank. What a Waste 2.0. A Global Snapshot of Solid Waste Management to 2050. 2021. Available online: hdl.handle.net/10986/30317 (accessed on 16 August 2021)
35. Mukherjee, R.; Reddy Asani, R.; Boppana, N.; El-Halwagi, M.M. Performance evaluation of shale gas processing and NGL recovery plant under uncertainty of the feed composition. *J. Nat. Gas Sci. Eng.* **2020**, *83*, 103517. [CrossRef]
36. Basini, L.E.; Busto, C.; Villani, M. Process for the Production of Methanol from Gaseous Hydrocarbons. WO 2020/058859 A1, 9 July 2020.
37. De Tommaso, J.; Rossi, F.; Moradi, N.; Pirola, C.; Patience, G.S.; Galli, F. Experimental methods in chemical engineering: Process simulation. *Can. J. Chem. Eng.* **2020**, *98*, 2301–2320. [CrossRef]
38. US Department of Energy, Office of Oil and Natural Gas. Natural Gas Flaring and Venting: State and Federal Regulatory Overview, Trends, and Impacts. 2019. Available online: https://www.energy.gov/sites/prod/files/2019/08/f65/Natural%20Gas%20Flaring%20and%20Venting%20Report.pdf (accessed on 28 June 2021)
39. Soave, G. Equilibrium constants from a modified Redlich-Kwong equation of state. *Chem. Eng. Sci.* **1972**, *27*, 1197–1203. [CrossRef]
40. Mokhatab, S.; Poe, W.A. *Handbook of Natural Gas Transmission and Processing*; Gulf Professional Publishing: Waltham, MA, USA, 2012.
41. Zhou, L.; Froment, G.F.; Yang, Y.; Li, Y. Advanced fundamental modeling of the kinetics of Fischer–Tropsch synthesis. *AIChE J.* **2016**, *62*, 1668–1682. [CrossRef]
42. Johnson, M.; Kostiuk, L.; Spangelo, J. A Characterization of Solution Gas Flaring in Alberta. *J. Air Waste Manag. Assoc.* **2001**, *51*, 1167–1177. [CrossRef]
43. Berger, S.; Dun, S.; Gavrelis, N.; Helmick, J.; Mann, J.; Nickle, R.; Perlman, G.; Warner, S.; Wilhelmi, J.; Williams, R.; Zarus, G. *Landfill Gas Primer: An Overview for Environmental Health Professionals*; ATSDR: Atlanta, GA, USA, 2001.
44. Li, Y.; Alaimo, C.P.; Kim, M.; Kado, N.Y.; Peppers, J.; Xue, J.; Wan, C.; Green, P.G.; Zhang, R.; Jenkins, B.M.; Vogel, C.F.A.; Wuertz, S.; Young, T.M.; Kleeman, M.J. Composition and Toxicity of Biogas Produced from Different Feedstocks in California. *Environ. Sci. Technol.* **2019**, *53*, 11569–11579. [CrossRef]
45. Ammenberg, J.; Bohn, I.; Feiz, R. Systematic Assessment of Feedstock for an Expanded Biogas Production: A Multi-Criteria Approach. 2017. Available online: http://www.diva-portal.org/smash/record.jsf?pid=diva2%3A1156008&dswid=4879 (accessed on 5 May 2021).
46. Green, D.W.; Southard, M.Z. *Perry's Chemical Engineers' Handbook*; McGraw-Hill Education: San Francisco, CA, USA, 2019
47. Ulrich, G.D.; Vasudevan, P.T. *Chemical Engineering Process Design and Economics: A Practical Guide*; Process Publishing: New York, NY, USA, 2003.
48. Vatavuk, W.M. Updating the CE Plant Index. 2002. Available online: https://www.chemengonline.com/Assets/File/CEPCI_2002.pdf (accessed on 5 May 2021)
49. Jenkins, S. 2021 CEPCI UPDATES: FEBRUARY (PRELIM.) AND JANUARY (FINAL). 2021. Available online: https://www.chemengonline.com/2021-cepci-updates-february-prelim-and-january-final/ (accessed on 5 May 2021)
50. Sönnichsen, N. Canada's Average Industrial Electricity Prices by major City 2019 by Select City (in Canadian Cents per Kilowatt Hour). 2020. Available online: https://www.statista.com/statistics/579159/average-industrial-electricity-prices-canada-by-major-city/ (accessed on 5 May 2021)

51. Town of Calgary. Landfill Rates. 2021. Available online: https://www.calgary.ca/uep/wrs/landfill-information/landfill-rates.html (accessed on 5 May 2021)
52. Town of Edmonton. Waste Management Fees & Rates. 2021. Available online: https://www.edmonton.ca/programs_services/garbage_waste/rates-fees.aspx (accessed on 5 May 2021)
53. Vengateson, U. Cooling Towers: Estimate Evaporation Loss and Makeup Water Requirements. 2020. Available online: https://www.chemengonline.com/cooling-towers-estimate-evaporation-loss-and-makeup-water-requirements/?printmode=1 (accessed on 5 May 2021)
54. Patience, G.; Boffito, D.C. Method and Apparatus for Producing Chemicals from a Methane-Containing Gas. U.S. Patent 9,243,190, 26 January 2016.
55. American Petroleum Institute. Gasoline, Diesel and Crude Oil Prices. 2021. Available online: https://gaspricesexplained.com/#/?section=gasoline-diesel-and-crude-oil-prices (accessed on 5 May 2021)
56. Index Mundi. Crude Oil, Average Spot Price of Brent, Dubai and West Texas Intermediate, Equally Weighed. 2021. Available online: https://www.indexmundi.com/commodities/?commodity=crude-oil (accessed on 5 May 2021)
57. TaxTips.ca. 2020 Corporate Income Tax Rates. 2021. Available online: https://www.taxtips.ca/smallbusiness/corporatetax/corporate-tax-rates-2020.htm (accessed on 6 May 2021)
58. Ponugoti, P.V.; Janardhanan, V.M. Mechanistic Kinetic Model for Biogas Dry Reforming. *Ind. Eng. Chem. Res.* **2020**, *59*, 14737–14746. [CrossRef]
59. Raje, A.; Inga, J.R.; Davis, B.H. Fischer-Tropsch synthesis: Process considerations based on performance of iron-based catalysts. *Fuel* **1997**, *76*, 273–280. [CrossRef]
60. Dancuart, L.; Steynberg, A. Fischer-Tropsch Based GTL Technology: A New Process? In *Fischer-Tropsch Synthesis, Catalyst and Catalysis*; Davis, B.; Occelli, M., Eds.; Studies in Surface Science and Catalysis; Elsevier: Amsterdam, The Netheralnds, 2007; Volume 163, pp. 379–399. [CrossRef]
61. Teimouri, Z.; Abatzoglou, N.; Dalai, A.K. Kinetics and Selectivity Study of Fischer–Tropsch Synthesis to C5+ Hydrocarbons: A Review. *Catalysts* **2021**, *11*. [CrossRef]
62. Bhosekar, A.; Ierapetritou, M. A framework for supply chain optimization for modular manufacturing with production feasibility analysis. *Comput. Chem. Eng.* **2021**, *145*, 107175. [CrossRef]
63. Weber, R.S.; Askander, J.A.; Barclay, J.A. The scaling economics of small unit operations. *J. Adv. Manuf. Process.* **2021**, *3*, e10074. [CrossRef]
64. Okeke, I.J.; Mani, S. Techno-economic assessment of biogas to liquid fuels conversion technology via Fischer-Tropsch synthesis. *Biofuels Bioprod. Biorefining* **2017**, *11*, 472–487. [CrossRef]
65. Hernandez, B.; Martin, M. Optimization for biogas to chemicals via tri-reforming. Analysis of Fischer-Tropsch fuels from biogas. *Energy Convers. Manag.* **2018**, *174*, 998–1013. [CrossRef]
66. Zhao, X.; Naqi, A.; Walker, D.M.; Roberge, T.; Kastelic, M.; Joseph, B.; Kuhn, J.N. Conversion of landfill gas to liquid fuels through a TriFTS (tri-reforming and Fischer–Tropsch synthesis) process: A feasibility study. *Sustain. Energy Fuels* **2019**, *3*, 539–549; Correction in **2019**, *3*, 2142–2142. [CrossRef]
67. Galli, F.; Pirola, C.; Previtali, D.; Manenti, F.; Bianchi, C.L. Eco design LCA of an innovative lab scale plant for the production of oxygen-enriched air. Comparison between economic and environmental assessment. *J. Clean. Prod.* **2018**, *171*, 147–152. [CrossRef]
68. Ramasamy, C.; Chun, T.H.; Kee, C.M.; Fong, G.T. Industrial Revolution 4.0 Contribution of chemical engineer. *Turk. J. Comput. Math. Educ. (TURCOMAT)* **2021**, *12*, 5037–5041.

processes MDPI

Article

Supercritical Assisted Production of Lutein-Loaded Liposomes and Modelling of Drug Release

Paolo Trucillo [1,2,*], Mathieu Martino [3] and Ernesto Reverchon [1]

1 Department of Industrial Engineering, University of Salerno, Via Giovanni Paolo II, 132, 84084 Fisciano, Italy; ereverchon@unisa.it
2 Department of Chemical, Material and Industrial Production Engineering, University of Naples Federico II, Piazzale V. Tecchio, 80, 80125 Napoli, Italy
3 Aix Marseille, CNRS, Centrale Marseille, M2P2, 13451 Marseille, France; mathieu.martino@univ-amu.fr
* Correspondence: paolo.trucillo@unina.it; Tel.: +39-(329)-656-6043

Abstract: In this work, a lipophilic ophthalmic drug, lutein, has been entrapped in liposomes, using a supercritical assisted process. Effects of pressure, temperature, and drug to lipid ratio variation were studied on mean diameters and lutein encapsulation efficiency. Liposomes with diameters between 153 ± 38 and 267 ± 56 nm were produced, and lutein encapsulation efficiencies between 86.5 ± 0.4% and 97.8 ± 1.2% were obtained. A Scanning Electron Microscope confirmed spherical shape and mean dimensions of vesicles. The variation of temperature for the production of liposomes showed a significant impact on lutein retention time in the double lipidic layer. Lutein drug release from liposomes produced at 35 °C ended in almost 4.5 days; whereas, liposomes produced at 40 °C showed a faster lutein release in 3 days; then, vesicles obtained at 45 °C released their lutein content in only 2 days. Drug release raw data were well-fitted using Weibull model (R^2 up to 99%).

Keywords: liposomes; supercritical fluids; carbon dioxide; high pressure systems; antibiotics; lutein; drug delivery; Weibull model

Citation: Trucillo, P.; Martino, M.; Reverchon, E. Supercritical Assisted Production of Lutein-Loaded Liposomes and Modelling of Drug Release. *Processes* **2021**, *9*, 1162. https://doi.org/10.3390/pr9071162

Academic Editor: Dimitrios I. Gerogiorgis

Received: 31 May 2021
Accepted: 1 July 2021
Published: 4 July 2021

1. Introduction

Drug delivery systems have been developed in different shapes and configurations to achieve one main goal: drug protection until reaching target cells [1]. This resulted in a significantly reduced drug loss, a most powerful targeted drug delivery, and a drastic reduction of the side effects [2].

Among drug carriers, liposomes are spherical particles characterized by an inner aqueous core and a lipidic external double layer [3]. These vesicles are particularly versatile, since they can entrap hydrophilic compounds in the inner core and lipophilic drugs in the double lipidic layer [4]. However, amphiphilic molecules can also be easily loaded, considering them as surfactants that cover the inner aqueous core [5]. Liposomes have been clearly recognized as the most used non-toxic drug carriers, since they are biocompatible with cell barriers and have the ability to fuse with them [6]. Moreover, due to their high biocompatibility and enhanced bioavailability of the loaded drugs, liposomes are considered the main key for the development and administration of a large number of drugs, such as antibiotics, proteins, dyes and markers, dietary supplements, and chromophores molecules; last but not least, liposomes are nowadays largely employed for the administration of vaccines, included the ones against COVID-19 disease [7].

Despite the importance of liposomes, which has been nowadays well-accepted by the majority of the scientific community [8], the actual most commonly employed methods of production are essentially simple post-processing developments of the conventional Bangham method [9,10]. This means that the most-used method for the production of these vesicles is thin-layer hydration [11,12], which was pioneered in 1970s, but after the very fast evolution of science, it became unacceptable to guarantee high-level quality and stability for human applications [13]. The main disadvantages were the low reproducibility of batch configurations, micrometric dimensions of vesicles produced, the necessity of post-processing steps, as well as the low encapsulation efficiency and the high solvent residue of ethanol or chloroform [14,15].

For all these reasons, CO_2-assisted high-pressure methods have been proposed to solve almost all these problems [16,17], since carbon dioxide can be employed as a non-toxic co-solvent during drug carrier formation. Among the proposed techniques, SuperLip is a CO_2-assisted process which inverts the traditional steps of hydration of a lipidic layer, by first creating the droplet of water and, then, a double lipidic layer around it [18]. This is guaranteed by a pressurized environment, where CO_2 is mixed with ethanol, forming an expanded liquid, as a support for lipid deposition around water droplets. This process is already employed in some pharmaceutical, cosmetic, textile, and nutraceutical applications [19,20], obtaining encapsulation efficiencies of hydrophilic, amphiphilic, and lipophilic compounds, up to 99% at nanometric level, with a low solvent residue and high cellular uptake. Additionally, the high time stability of liposomes produced in this manner has already been demonstrated using this process [21], as well as their capability to deliver entrapped molecules directly to the target cells [22].

This process has been employed several times for the encapsulation of hydrophilic and lipophilic active principles; in this paper, the encapsulation of a molecule with significant antioxidant properties [23,24], lutein, has been attempted. Lutein belongs to the family of carotenoids and has a high potential if used as a dietary supplement for the prevention of ocular illnesses involving retinal damages and cataracts [25], but it is also used to reduce cardiovascular disease risk [26], to replace yellow dyes in beverage coloration [27], and, last but not least, in anti-carcinogenic formulations. However, this molecule suffers for facile degradation, due to the exposition to light and heat; for this reason, the encapsulation in drug carriers such as liposomes enhances this molecule bioavailability, improving the effect on human cells [28–32].

Lutein has been successfully entrapped into liposomes using supercritical assisted techniques: in the work by Zhao et al., liposomes mean dimensions ranged between 148 nm and 195 nm, while encapsulation efficiencies ranged between 57% and 97.8% [33]. In the work by Tan et al. [34], lutein loaded liposomes with diameters ranging between 76 and 135 nm, and EE between 83% a 92%.

In this study, the effect of SuperLip operating parameters has been investigated in order to provide controlled and tunable drug release, according to the time required by prolonged circulation of molecules until reaching target tissues. Despite the fact that SuperLip process has already been employed for the encapsulation of drugs for ocular delivery [35], the effect of process conditions on the post-production drug release administration has never been investigated. Another goal of this study is to propose the modelling of the raw data obtained from drug release, which has not been yet investigated on liposomes produced using SuperLip process. In this work, Weibull equation has been applied to raw data obtained from drug release tests.

2. Materials and Methods

2.1. Materials

L-α-phosphatidylcholine from egg yolk (PC, about 60% purity) was purchased from Sigma-Aldrich, Milan, Italy and used as a source of phospholipids. Absolute ethanol (purity 99.8%) was obtained from Carlo Erba, Milan, Italy. Carbon dioxide (>99.7% purity) was purchased from Morlando Group, Naples, Italy. Distilled water was produced directly in our laboratories. Lutein was provided by Shaanxi Superior Bio Technology Co., Shanxi, China; this company is a Natural Herbal Extract Manufacturer, that extracted lutein from marigold flowers.

2.2. Preparation of Preliminary Solutions

In details, 500 mg of phosphatidylcholine was dissolved in 100 mL ethanol by being stirred at room temperature at 300 rpm for 20 min. The solution was then filtered using grade 600 cellulose filter paper (Fisher Scientific, Milan, Italy), under vacuum. Then, 2.5 mg or 5 mg of lutein (depending on the chosen lipid to drug ratio) was dissolved in the filtered solution, in order to obtain 0.5% *w*/*w* or 1% *w*/*w* lutein to lipid feeding ratio. The final solution was left stirring at room temperature at 300 rpm for 20 min more.

2.3. Description of SuperLip Process Layout

The description of the process is depicted in Figure 1 and is described as follows. This process is characterized by three feeding lines: carbon dioxide (P1), ethanol (P2), and water (P3), while the main elements are a High-Pressure Vessel (HPV) and a Separator (S), as reported in the sketch of Figure 1.

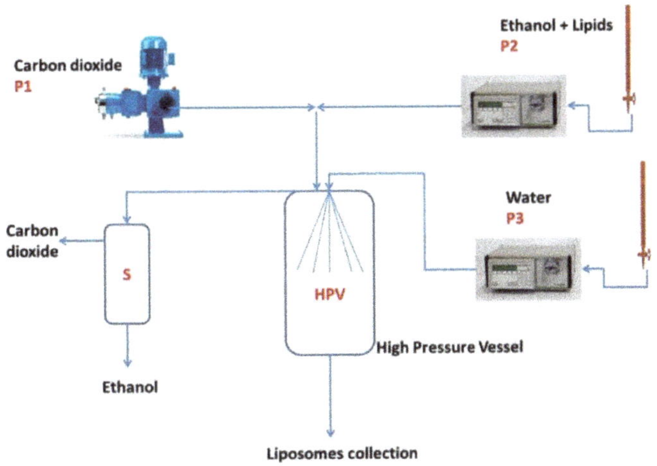

Figure 1. Experimental set-up of SuperLip process; P1 is the CO_2 pump, P2 is the ethanol pump, P3 the water pump, S stands for Separator, HPV is the High-Pressure Vessel.

Liquid carbon dioxide was pumped to the system from a 30 L internal tank, where CO_2 was stocked at a temperature of 25 °C and 60 bar.

Carbon dioxide was pumped using an Ecoflow pump, mod. LDC-M-2, Lewa, Germany. The ethanol, in which lipids were previously dissolved, was pre-mixed with carbon dioxide, at a temperature of 40 °C and pressures of 100 or 150 bar. The mixer was a stainless-steel tank of 300 cm^3, in which an expanded liquid was formed that was injected in the HPV (stainless-steel vessel of 500 cm^3 with a height of 30 cm and internal diameter of 7.5 cm). Water was directly pumped into the HPV, obtaining the droplets of water after high-pressure atomization using an 80 µm nozzle. Water and carbon dioxide were continuously pumped to the system until the end of the ethanol solution. Water (P3) and ethanol (P2) were pumped using two different high pressure precision pumps (Model 305, Gilson, France). Water flow rate was set at 10 mL/min, ethanol flow rate was set at 3.5 mL/min, corresponding to about 2.76 g/min at the indicated conditions; whereas, carbon dioxide flow rate was set at 6.5 g/min. Flow rate ratio of carbon dioxide to ethanol was named Gas to Liquid Ratio (of the expanded liquid) and set at about 2.4 on mass basis.

Liposomes were continuously collected from the bottom of the high-pressure vessel using an on-off valve. A stainless-steel vessel (internal volume of 300 cm^3) was used to separate ethanol and carbon dioxide from the liposomal solution, working at a slow decompression, down to 10 bar. At these conditions, carbon dioxide was separated from ethanol, and sent to a rotameter (mod. N.5-2500, Serval 115022, ASA, Italy) to measure CO_2 flow rate, before being released to the atmosphere.

Experiments were performed with varying operating pressures (100 and 150 bars), temperatures (35, 40, and 45 °C), and drug to lipid ratios (0.5 and 1%, *w/w*). Each test belonging to this experimental campaign has been reported in Table 1, together with its specific operating conditions of production.

Table 1. Experimental conditions (pressure, temperature, drug to lipid ratio) used in this process.

Test Name	Drug to Lipid Ratio (%, *w/w*)	Pressure (bar)	Temperature (°C)
Lut1	1	100	40
Lut1′	1	100	40
Lut1″	1	100	40
Lut2	1	100	35
Lut3	1	100	45
Lut4	1	150	35
Lut5	1	150	45
Lut6	0.5	100	35
Lut7	0.5	100	45
Lut8	0.5	150	35
Lut9	0.5	150	45

The liposomal suspensions obtained at these conditions were analyzed (liposome mean particle size, particle size distribution, and encapsulation efficiency) in order to determine the optimal operating conditions. Then, each sample was stocked in glass bottles and stored at 4 °C in the absence of light.

2.4. Characterization of the Produced Samples

Liposomal suspensions were characterized by a Dynamic Light Scattering (DLS) instrument (Zetasizer nano S, Alfatest, Cinisello Balsamo, Italy). These analyses allow the determination of the particle mean size and particle size distribution of liposomes in aqueous suspensions, considered as colloidal suspensions [36]. Liposome size was characterized in terms of Mean Diameter (MD), Standard Deviation (SD), and PolyDispersity Index (PDI). The light source of the Zetasizer instrument is a He-Ne laser (4 mW, 633 nm) and the backscatter angle is 173°. In this work, for each sample, 5 measurements were performed at 25 °C in a 10 mm quartz glass cell (Hellma Italia Srl, Milan, Italy).

Then, 5 tubes (volume of 15 mL) were used to prepare the sample for centrifugation. In each one, 13 mL of the liposomal suspension were added. The 5 tubes were balanced with a sixth one, of the same weight, loaded with the same weight of pure water. The samples were centrifuged for 60 min, at 6500 rpm and −9 °C. After this operation, 4 mL of supernatant was removed from the top of the falcon and lutein absorbance was measured using a Micro-Volume UV–Vis Spectrophotometer (Biospec-nano, Shimadzu Scientific Instruments, Columbia, MD, USA). According to a calibration line previously obtained for lutein dissolved in a 50/50 *v/v* ethanol/water solution, it was possible to determine the absorbance of lutein in the supernatant. Recalling the Lambert–Beer equation, absorbance was converted to concentration using the linear correlation. This gave the possibility to calculate the lutein entrapped percentage with respect to the initial lutein fed to the system, as the complement, to 100%. The resulting number represented effective encapsulation efficiency (EE) in the double lipidic layer of liposomes, according to Equation (1).

The maximum lutein absorbance wavelength [37] was obtained by a spectrum measurement on lutein from 200 nm to 800 nm; encapsulation efficiency tests and drug release tests were performed at the wavelength of 445 nm, after detecting the maximum absorbance peak on a lutein spectrum.

$$EE(\%) = \frac{[\text{Lutein Total}] - [\text{Lutein Free}]}{[\text{Lutein Total}]} \times 100 \quad (1)$$

Encapsulation efficiencies are reported in terms of mean values and standard deviations, after three runs and 5 measurements repetitions.

2.5. Scanning Electron Microscope Observations

The morphology of the vesicles produced using SuperLip process was observed using a Scanning Electron Microscope (SEM mod. LEO 1525; Carl Zeiss SMT AG, Oberkochen, Germany). A drop of liposome suspension was spread on a carbon tab placed on an aluminum stub and dried at air for 48 h. Then, dried samples were coated with a gold layer using a sputter coater (thickness 250 Å, model B7341; Agar Scientific, Stansted, UK). After becoming conducible, samples were observed using SEM.

2.6. Drug Release Test

Drug release tests were also performed for SuperLip-produced vesicles. The external bulk used for drug release in this work was Phosphate Buffer Saline (PBS). It was prepared in this manner: 8 g NaCl, 0.2 g KCl, 1.44 g Na_2HPO_4, 0.24 g KH_2PO_4 for 1 L distilled water. The 7-cm-long membranes (or dialysis sacks, 12 kDa cut-off) were washed in a bath prepared dissolving 146 g Ethylene Diamine Tetraacetic acid (EDTA, $C_{10}H_{16}N_2O_8$) and 10 mg Sodium Bicarbonate ($NaHCO_3$) in 500 mL distilled water. In each experiment, the dialysis sack was submerged in the washing bath at 50 °C for 30 min. Then, the membrane was clamped from one side, and 5 mL liposomal suspension was added inside, together with a magnet for internal stirring, and then it was clamped also from the other side. Then, the double-clamped membrane was added to a 300 mL PBS volume of external bulk, pre-heated at a temperature of 37 °C (considered the average human body temperature). The external bulk was also stirred with another larger magnet. The same probe of the Micro-Volume UV–Vis Spectrophotometer (Biospec-nano, Shimadzu Scientific Instruments, Columbia, MD, USA) was used to measure the absorbance of lutein released in the external bulk. Additionally, in this case, the absorbance of lutein was continuously measured at the same wavelength of 445 nm, using another lutein calibration line, obtained in PBS buffer. Drug release tests were observed until 110 h, corresponding to about 4.5 days.

2.7. Drug Release Raw Data Modelling

Mathematical modelling has been applied to model drug release raw data from liposomes [36,37]. In this work, raw data obtained from drug release tests were modelled according to Weibull equation [38–40], according to Equation (2).

$$y = A\left(1 - e^{-(k(t-t_c))^d}\right) \quad (2)$$

In this equation, A is the amplitude (100%) of the release percentage, whereas t is the time and t_c is the central value of time, while k and d are the two working variables.

Diagrams reported in the final section of results show the raw data and their fitting curve; model parameters and R^2 values are tabled at the end.

3. Results

3.1. Repeatability Experiments

The typical SuperLip experiment consisted in preparing the three feeding lines: ethanol containing 500 mg phospholipids and the defined amount of lutein, carbon dioxide, and pure water. These three species were pumped to the system until reaching steady state conditions. To demonstrate the advantages of the continuity of this process, experiments of repeatability were performed, in different days, using newly prepared ethanolic solutions, working at 100 bar and setting the temperature at 40 °C and the drug to lipid ratio at 1% on mass basis. The operator was always the same and the DLS and Encapsulation efficiency (EE) measurements were performed for each run. Results are reported in Table 2.

Table 2. Mean diameter (MD), PolyDispersity Index (PDI), and encapsulation efficiency (EE) obtained in reproducibility tests.

Test	MD ± SD (nm)	PDI	EE ± SD (%)
Lut1	153 ± 38	0.25	86.5 ± 0.4
Lut1′	176 ± 40	0.23	87.1 ± 1.0
Lut1″	169 ± 44	0.26	89.5 ± 0.6

As it is possible to see in Table 2 and Figure 2, the three experiments gave almost repeatable data (Lut1, Lut1′, and Lut1″). In particular, Lut1 had mean diameter of 153 ± 38 nm, Lut1′ was 176 ± 40 nm large, and Lut1″ had mean dimensions of 169 ± 44 nm. The three particle size distributions have a similar PDI. According to the direct use of liposomes, this process is able to produce one-shot nanometric liposomes that are directly available for drug release; mean dimensions lower than 200 nm are also available for biomedical and pharmaceutical applications [41,42]. These repeatability data were also confirmed by SEM images reported in the following part of this results section. Moreover, the encapsulation efficiencies of lutein were similar, with values between 86.5 ± 0.4% and 89.5 ± 0.6%.

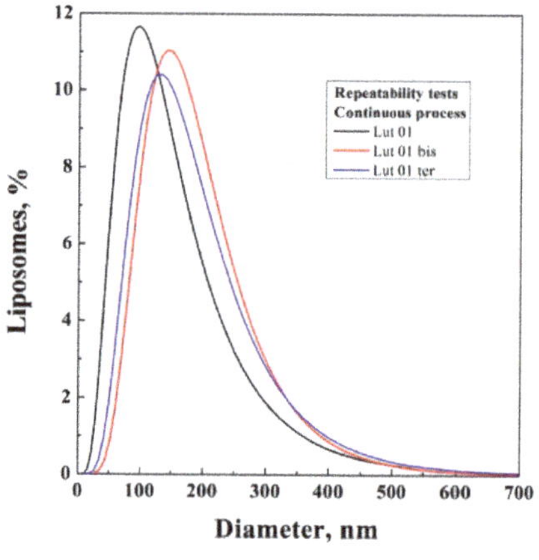

Figure 2. Particle size distribution obtained in repeatability tests.

3.2. *Study of the Effect of Operating Parameters*

The effect of operating temperature on mean particle size and lutein encapsulation efficiency on the liposomes produced using SuperLip, was studied at 35 °C, 40 °C, and 45 °C. These temperatures were chosen in order to inhibit chemical degradation of carotenoids [43,44], maintaining their physical properties. Therefore, lutein is subject to minimum degradation at temperatures lower than 60 °C [45]. Moreover, the critical temperature of carbon dioxide is reached at 31.1 °C. Setting the lower limit of 31.1 °C and the higher limit of 60 °C, we decided to work in a thin temperature operating window among 35 °C and 45 °C.

The variation of temperature did not have a significant effect on the particle size distribution, but it influenced the permeability of the double lipidic layer of the liposomes produced. In the literature, one of the most studied effects regards the drug release activated by an external stimulus such as the increase of temperature [46,47]; the increase of temperature could facilitate drug release, since the external lipidic barrier improves its permeability after receiving the stimulus. Therefore, in this study, we applied this theory of the fast diffusion to the liposomes produced at a higher temperature, which could be employed for a fast drug administration, while protecting lutein from degradation. Since in this work, we employed carbon dioxide as co-solvent medium, the low critical temperature of CO_2 (31.1 °C) allowed us to process lutein without any risk of thermal degradation.

Then, the effect of pressure was studied (100 and 150 bar), by leaving constant the temperature at 35 °C and drug to lipid ratio at 1% on mass basis. At the end, the effect of DLR was studied (0.5% and 1% on mass basis) by leaving the pressure constant at 100 bar and the temperature at 45 °C. Results are reported in Table 3.

Table 3. Size, size distribution, and encapsulation efficiency for the experimental design performed for the encapsulation of lutein.

Test	Pressure (bar)	Temperature (°C)	Drug to Lipid Ratio (% mass)	Liposome Mean Diameter nm (MD ± SD)	Encapsulation Efficiency % (EE ± SD)
Lut1	100	40	1	153 ± 38	86.5 ± 0.4
Lut2	100	35	1	206 ± 41	97.1 ± 0.5
Lut3	100	45	1	267 ± 56	96.0 ± 0.2
Lut4	150	35	1	192 ±38	88.0 ± 1.2
Lut5	150	45	1	196 ± 39	87.0 ± 0.1
Lut6	100	35	0.5	145 ± 54	97.8 ± 1.2
Lut7	100	45	0.5	163 ± 39	86.7 ± 1.6
Lut8	150	35	0.5	158 ± 55	91.3 ± 2.2
Lut9	150	45	0.5	178 ± 63	84.7 ± 2.5

According to the operating conditions in the first set of experiments, the sample named Lut1 produced liposomes of 153 ± 38 nm, working at 40 °C, 100 bar, and 1% DLR. Then, the sample named Lut2 was produced at 35 °C, with the same pressure and DLR, obtaining a mean value of 206 ± 41 nm. The reduction of temperatures produced an increase of mean diameter of about 50 nm. However, when the temperature was further increased to 45 °C, (experiment named Lut3), the mean size became 267 ± 56 nm (see Figure 3a). Even if the temperature was increased from 35 °C to 45 °C, there was not a dramatic increase of the mean size of liposomes, according to the mean dimensions needed for pharmaceutical issues [48,49]; moreover, the encapsulation of lutein was still successful and could be applied for specific cellular applications.

The effect of pressure was also studied (see Figure 3b). Liposomes produced at 35 °C (Lut2) gave a mean diameter of 206 ± 41 nm at 100 bar; however, by increasing the pressure at 150 bar and leaving the temperature at 35 °C (Lut4) and DLR at 1% w/w, the mean size of liposomes became 192 ± 38 nm, with a similar mean size and standard deviation (Figure 3b). Even if the mean size =variation did not follow the increase of temperature, the EE was 97.1 ± 0.5% for Lut2 and 88.0 ± 1.2% for Lut4. This was not necessarily linked to pressure variation. The Lut5 experiments was produced at 45 °C and 150 bar (leaving DLR constant at 1% w/w), showing a mean size of 178 ± 63 nm and an EE of 87.0 ± 0.1%, obtaining EE overlapping results with Lut4.

At the end of this campaign of experiments, the effect of the drug to lipid ratio was studied. The drug to lipid ratio was reduced to 0.5% w/w. At 100 bar and 35 °C (Lut6), 145 ± 54 nm liposomes were produced, with an EE of 97.8 ± 1.2%, which was the highest value obtained in terms of drug EE for this designed process for lutein encapsulation. Indeed, the reduced amount of lutein entrapped in the double layer of phospholipids contributed to reduce particle mean size, resulting in a reduced space occupied by drugs in the external region of vesicles. However, when the temperature was increased to 45 °C (Lut7), liposomes produced showed a mean size of 163 ± 39 nm and an EE of 86.7 ± 1.6%. The last case regards the reduced DLR to 0.5% w/w and the increased pressure to 150 bar: in the case of 35 °C (Lut8), mean size is 158 ± 55 nm and EE of 91.3 ± 2.2%, whereas at 45 °C (Lut9), mean dimensions are 178 ± 63 nm and EE 84.7 ± 2.5%. Again, a decreased temperature contributed to higher encapsulation efficiency, due to the fact that the external lipidic barrier becomes more compact than the one obtained during a higher temperature SuperLip process (45°C).

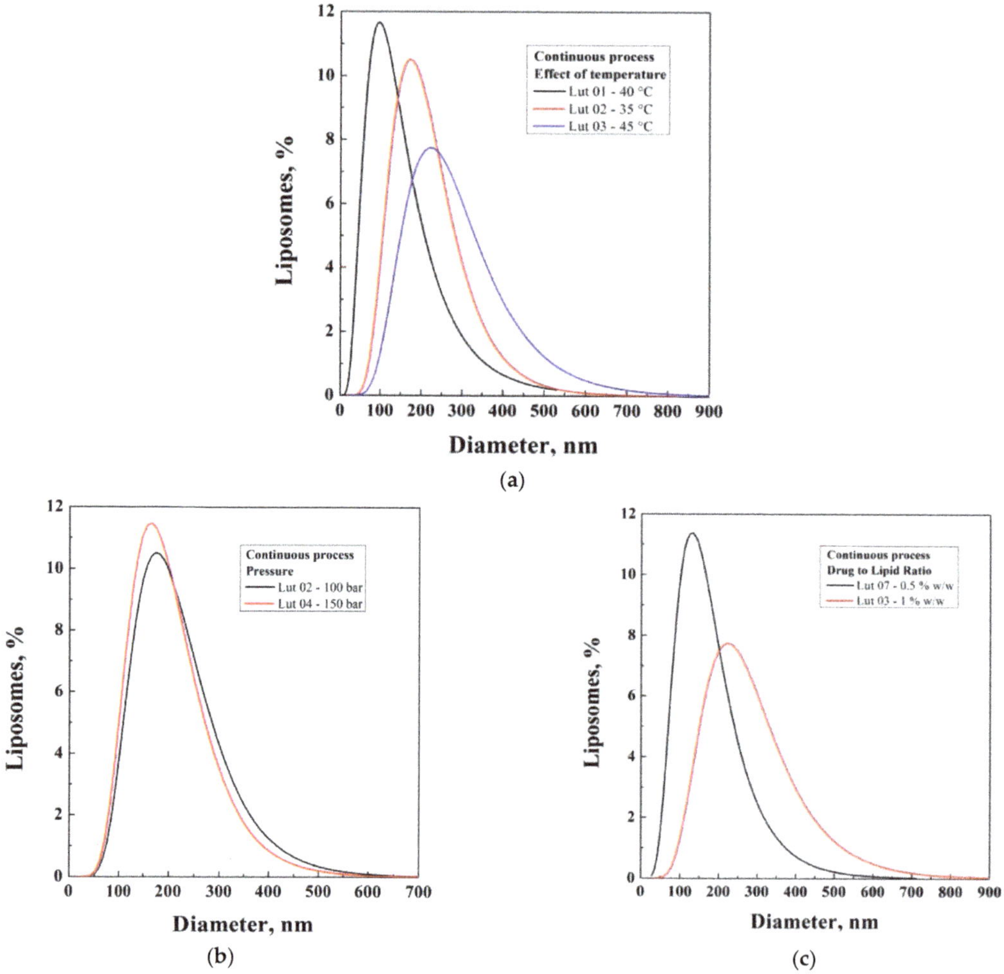

Figure 3. Influence of temperature on liposome mean dimensions for a pressure of 100 bar and a drug to lipid ratio of 1% mass (**a**); influence of operating pressure on liposome size distribution for a drug to lipid ratio of 1% w/w and a temperature of 35 °C (**b**); influence of lutein to lipid ratio on liposome size distribution for a pressure of 100 bar and a temperature of 45 °C (**c**).

3.3. Scanning Electron Microscope Observations

Liposomes produced using SuperLip were observed using a Scanning Electron Microscope, following the protocol described in Methods section.

Liposomes shown in Figure 4 are representatives of samples produced at 100 and 150 bar, among 35 °C and 45 °C, and DLR among 0.5% and 1%, thus resulting in the production of vesicles whose mean diameter is included among 200 and 300 nm, as confirmed by DLS analysis reported in Table 3. In all cases, liposomes have a spherical shape and smooth surface. As it is possible to see in all these SEM figures, the aggregation of lipids is due to the drying that was performed during liposomes preparation for SEM characterization, as reported in the preparation protocol of the Methods section.

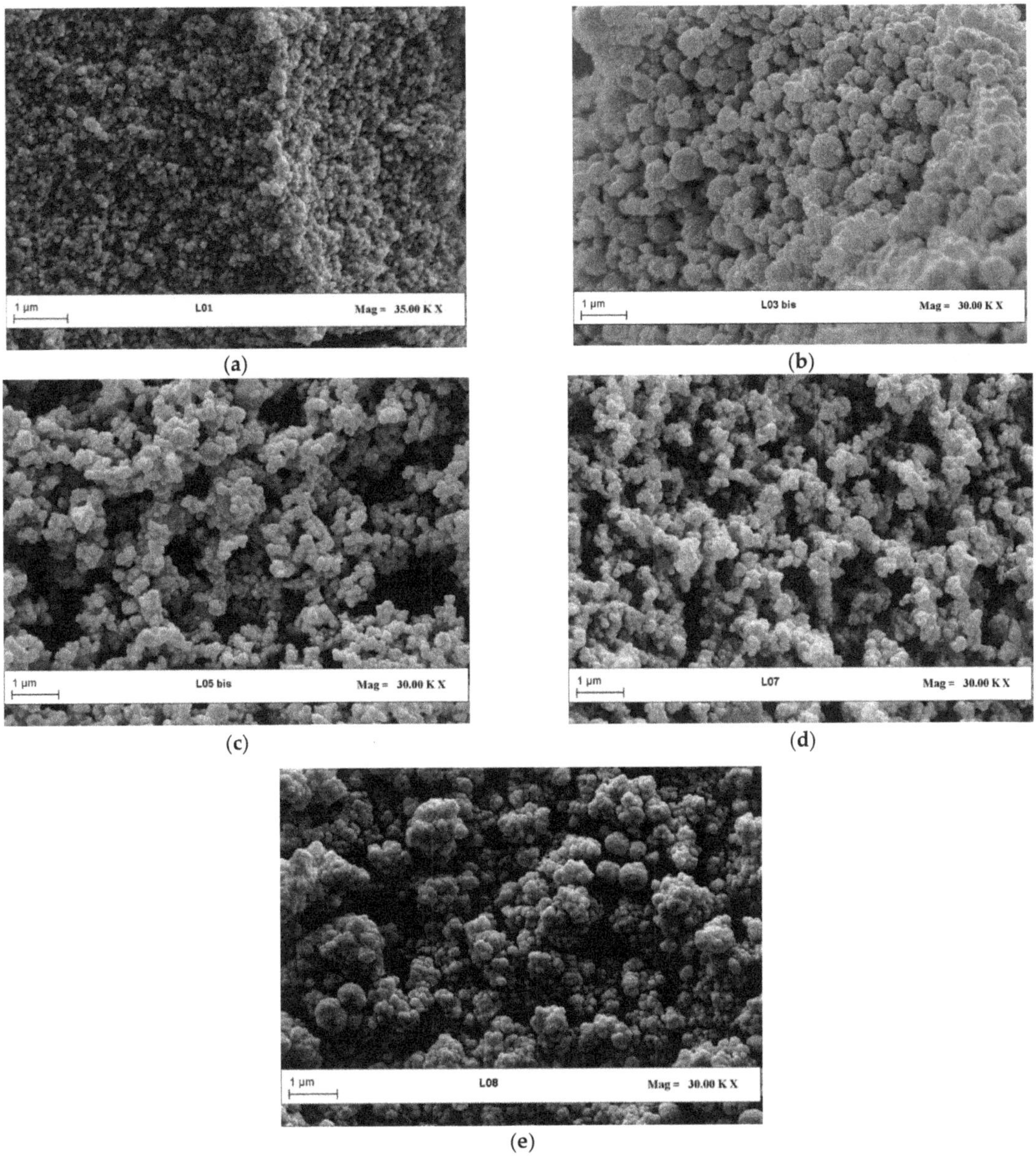

Figure 4. Scanning Electron Microscope observations of liposomes produced using SuperLip. (**a**) Lut1 (100 bar, 35 °C, and 1% DLR), (**b**) Lut3 (100 bar, 45 °C, and 1% DLR), (**c**) Lut 5 (150 bar, 45 °C, and 1% DLR), (**d**) Lut7 (100 bar, 45 °C, 0.5%), (**e**) Lut8 (150 bar, 35 °C, 0.5% DLR).

3.4. Drug Release Tests and Modelling

At the end of this study, Drug Release (DR) tests were performed from liposomes produced during experiments Lut 1, Lut 2, and Lut 3. These samples are referred to vesicles produced with SuperLip at 100 bar, 1% *w*/*w* DLR, and a working temperature of 35 °C, 40 °C, and 45 °C, respectively. The choice is due to the fact that in this set, the average EE was the highest (86%, 97%, and 96%). Even if drug release tests were performed at a constant temperature of 37 °C, the effect of the temperature used for producing samples was studied during drug release tests and shown in Figure 5.

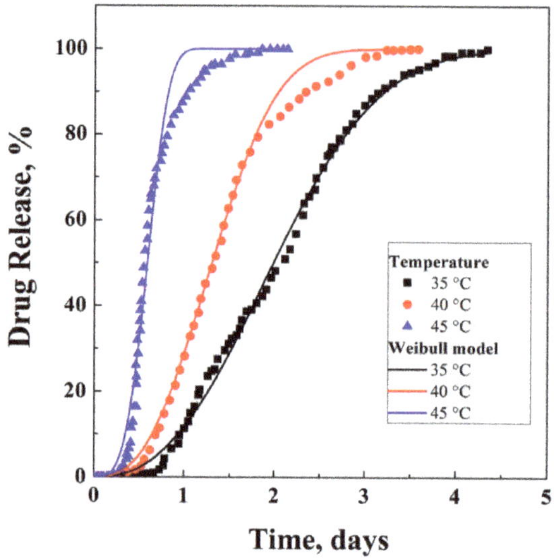

Figure 5. Drug release test comparison from liposomes, as an effect of liposomes produced at different temperatures.

The drug release of the experiment produced at 35 °C ended in about 4.5 days, with a significant delay time of 0.8 days before liposomes started to release lutein. A similar behavior was registered for drug release from liposomes produced at 40 °C, where the initial delay was about 0.6 days and it released all the drug content in 3 days. The third test, related to vesicles produced at 45 °C, was faster than previously analyzed ones, ending in less than 2 days.

Therefore, an evident and expected effect was detected: by increasing the production temperature, drug release was enhanced. A possible explanation of this trend is that the increase in the temperature from 35 °C to 45 °C reduced the strength of the lipidic barrier of liposomes, favoring a faster drug release. This indication is also confirmed by the fact that liposomes can release their content after production at an increased temperature. Moreover, the higher production temperature resulted in the production of liposomes with a less compact structure; this fact favored drug release performed in the receiving liquid medium.

As indicated above, drug release raw data were modelled according to Weibull model [50–52], to which literature refers as being generally used for drug release. In Figure 5, raw data obtained from drug release of liposomes prepared at 35, 40, and 45 °C, respectively, were fitted. The main parameters and the R^2 obtained for the model are reported in Table 4.

Table 4. Weibull model parameters at each process temperature (as reported in Equation (2)).

Temperature	A	d	K	R^2	Temperature
35 °C	100	2.504	0.435	0.997	35 °C
40 °C	100	2.693	0.670	0.995	40 °C
45 °C	100	3.596	1.549	0.975	45 °C

As it is possible to see from Figure 5 and Table 4, raw data of drug release from liposomes prepared at 35 °C showed R^2 value of 99.7%. The only set parameter was the amplitude, which was blocked at 100%, indicating the asymptotic total drug loss from liposomes. Drug release performed from liposomes prepared at 40 °C did not result in a R^2 value of 99.7%, even if the fitting parameters were significantly different. In the case of drug release from liposomes produced at 45 °C, the R^2 value was reduced to 97.5%. In this last case, the main controlling phenomenon was the drug diffusion from the inner core to the external bulk, but the less compact structure of liposomes contributed to allow a faster release of lutein. For this reason, the shape of the curve changed at 45 °C, and in this case, the model did not fit the raw data as well as in the case of lower temperatures.

4. Discussion

Drug carriers have been developed to overcome several problems of drug degradation and boor bioavailability that occurred frequently during administration to human tissues. The idea of protecting these drugs and active principles was successful, since an external shell or double layer (made of phospholipids or polymers) was able to avoid destructive interferences by external stimuli such as light or heat. This was particularly effective for antioxidant compounds, which are particularly sensitive to the conditions of human blood (temperature, pH, immune response). In particular, drugs were often recognized by the immune systems as intrusive and toxic, and was systematically attacked and eliminated by phagocytes and macrophages. Moreover, the simple drug delivery through the gastrointestinal tract was particularly ineffective, due to the high acidic conditions (pH < 2). All these drawbacks were solved by the use of drug carriers; in particular, liposomes were the most effective carriers, since they were characterized and constituted by the same backbone of human cells.

From one side, an intelligent solution was achieved; however, another problem became particularly critical for liposomes, and it was related to the method of production. Liposomes are spherical vesicles, but the mean dimensions, the stability over time, and the encapsulation efficiencies were critical parameters to control and monitor during and after the production. As was highlighted in the literature, the conventional methods for the production of liposomes suffered of low reproducibility and low encapsulation efficiencies of molecules. Several papers reported in the literature were finalized at the creation of liposomes in a batch mode, but they often needed post-processing steps of refining, such as filtration, sonication, or extrusion, in order to obtain nanometric mean dimensions. This created mechanical stress to the production of liposomes, often resulting in a loss of a significant fraction of the drug entrapped in the inner core or in the double lipidic layer. Moreover, some kinds of phospholipids are well-known temperature-sensitive materials, especially in relation to their phase transition behavior from gel phase to crystalline phase. The temperature of transition also depends on the type of phospholipid employed; this could impact to drug release behavior [53,54].

The specific problem of production was resolved by the idea to work at high pressures, using carbon dioxide in supercritical conditions. CO_2 is a cheap, non-toxic solvent, that has the ability to work in the place of the traditional solvent extraction. CO_2, working in a mixture with an organic solvent, can create an expanded liquid; this mixture has the ability to extract a specific solute, or induce an assisted atomization or even precipitation. Exploited by different high-pressure technologies, these phenomena have the ability to create drug carriers, with the great advantage to eliminate the organic solvent from the final product. The organic solvent is eliminated by carbon dioxide, while high-pressure conditions induce a fast and one-shot production of particles at nanometric dimensions. The process proposed in this paper and named SuperLip, after being employed several times in the literature, has also been capable of significantly reducing the ethanol residue in the final liposomes formulations. Moreover, liposomes produced using SuperLip were of nanometric dimensions, with high encapsulation efficiencies of hydrophilic and lipophilic compounds; that is why it was employed in this work for the secure and successful entrapment of lutein.

However, as it is possible to see from the results, the idea of this work was not only the encapsulation of lutein into liposomes, but also to try to tune drug release of lutein by varying the production operating parameters [55,56]. Among all the parameters studied in this paper, it was particularly interesting the effect of operating temperature on the successive drug release performed after the production. There is a substantial difference among the production of temperature-sensitive liposomes and the liposomes whose drug release is tuned by operating temperature. In the first case, liposomes are made of specific and expensive phospholipids, and are prepared in such a manner that they can open their structure and release their content only when a sudden external increase of temperature is applied. Instead, in this work, a different phenomenon was studied: the possibility to create a double lipidic layer with different resistance to mass transfer of lutein among the internal regions and the external bulk. In this case, the drug release was faster or slower, depending on the temperature of the process during production, and without the addition of external stimuli during the subsequent drug release.

The effect of the producing temperature on the external barrier structure of liposomes was studied in this work. The possibility to regulate the properties of liposomes' surfaces and the double lipidic layer was a particularly ambitious goal, which was achieved by creating temperature-responsive surfaces of liposomes. This fact was demonstrated by the effect on drug release profiles obtained, as it is possible to see in the Results section. In detail, the release was faster by increasing the production temperature, which resulted into a reduced robustness of the external double lipidic layer.

5. Conclusions

This work was aimed at producing liposomes whose drug release could be tunable, by varying the temperature of production. The innovation proposed was not focused on the supercritical technique for the production of liposomes, which has been demonstrated many times under different pharmaceutical applications, but it aimed at producing liposomes whose drug release could be regulated, in order to find specific therapeutic necessities of the human body.

Liposomes obtained in this work were produced one-shot at nanometric level, with mean diameters between 137 and 267 nm, and lutein encapsulation efficiency up to 98%, using a protocol validated by the literature and based on the elimination of the supernatant. The effect of drug to lipid ratio, pressure, and temperature was studied on mean size and EE, and mean dimensions were not significantly affected by the variation of these parameters.

The use of Weibull equation for the modelling of drug release raw data was investigated for the first time for liposomes produced using SuperLip. Liposomes produced with a higher temperature can be used in cases in which the drug release needs to be addressed in less time. This concept was also confirmed by the modelling part of this work, which completely fitted the raw data of release profiles at 35 °C and 40 °C, while losing fitting precision on raw data obtained at 45 °C.

Therefore, for vesicles produced at higher temperatures, a further comment could be that a model should also contain a dependency on temperature parameter. The study and production of these kinds of temperature-responsive liposomes using high-pressure continuous methods, could open new roots to the definitions of temperature-responsive vesicles; this could be a preliminary work that had the ambition to optimize and regulate processing parameters, but it will require further in vivo studies performed in sterile conditions by researchers from departments of medicine.

Author Contributions: Conceptualization, P.T. and M.M.; methodology, P.T. and M.M.; software, P.T.; investigation, P.T.; data curation, P.T.; writing—original draft preparation, P.T. and E.R.; writing—review and editing, P.T. and E.R.; supervision, E.R. All authors have read and agreed to the published version of the manuscript.

Funding: This research received no external funding.

Institutional Review Board Statement: Not applicable.

Informed Consent Statement: Not applicable.

Data Availability Statement: Not applicable.

Conflicts of Interest: The authors declare no conflict of interest.

Abbreviations

DDS	Drug Delivery Systems
EE	Encapsulation Efficiency
PSD	Particle Size Distribution
PDI	Polydispersity Index
SD	Standard Deviation
MD	Mean Diameter
DLS	Dynamic Light Scattering
SEM	Scanning Electron Microscope
DR	Drug Release
HPV	High-Pressure Vessel
S	Separator
PC	L-α-phosphatidylcholine

References

1. Allen, T.M. Drug Delivery Systems: Entering the Mainstream. *Science* **2004**, *303*, 1818–1822. [CrossRef] [PubMed]
2. Lavik, E.B.; Kuehn, M.H.; Kwon, Y. Novel drug delivery systems for glaucoma. *Eye* **2011**, *25*, 578–586. [CrossRef] [PubMed]
3. Crommelin, D.J.A.; Storm, G. Liposomes: From the Bench to the Bed. *J. Liposome Res.* **2003**, *13*, 33–36. [CrossRef] [PubMed]
4. Tsai, W.-C.; Rizvi, S. Simultaneous microencapsulation of hydrophilic and lipophilic bioactives in liposomes produced by an ecofriendly supercritical fluid process. *Food Res. Int.* **2017**, *99*, 256–262. [CrossRef]
5. Fatouros, D.G.; Antimisiaris, S.G. Effect of Amphiphilic Drugs on the Stability and Zeta-Potential of Their Liposome Formulations: A Study with Prednisolone, Diazepam, and Griseofulvin. *J. Colloid Interface Sci.* **2002**, *251*, 271–277. [CrossRef]
6. Mallick, S.; Choi, J.S. Liposomes: Versatile and Biocompatible Nanovesicles for Efficient Biomolecules Delivery. *J. Nanosci. Nanotechnol.* **2014**, *14*, 755–765. [CrossRef]
7. Alavi, M.; Asare-Addo, K.; Nokhodchi, A. Lectin Protein as a Promising Component to Functionalize Micelles, Liposomes and Lipid NPs against Coronavirus. *Biomedicine* **2020**, *8*, 580. [CrossRef]
8. Mozafari, M.R. Liposomes: An overview of manufacturing techniques. *Cell. Mol. Biol. Lett.* **2005**, *10*, 711–719. Available online: https://europepmc.org/article/med/16341279 (accessed on 5 June 2021).
9. Meure, L.A.; Foster, N.R.; Dehghani, F. Conventional and Dense Gas Techniques for the Production of Liposomes: A Review. *AAPS PharmSciTech* **2008**, *9*, 798–809. [CrossRef]
10. William, B.; Noémie, P.; Brigitte, E.; Géraldine, P. Supercritical fluid methods: An alternative to conventional methods to prepare liposomes. *Chem. Eng. J.* **2020**, *383*, 123106. [CrossRef]
11. Trucillo, P.; Campardelli, R.; Reverchon, E. Liposomes: From Bangham to Supercritical Fluids. *Processes* **2020**, *8*, 1022. [CrossRef]
12. Patil, Y.P.; Jadhav, S. Novel methods for liposome preparation. *Chem. Phys. Lipids* **2014**, *177*, 8–18. [CrossRef] [PubMed]
13. Deamer, D.W. From "Banghasomes" to liposomes: A memoir of Alec Bangham, 1921–2010. *FASEB J.* **2010**, *24*, 1308–1310. [CrossRef]
14. Maherani, B.; Arab-Tehrany, E.; Mozafari, M.R.; Gaiani, C.; Linder, M. Liposomes: A Review of Manufacturing Techniques and Targeting Strategies. *Curr. Nanosci.* **2011**, *7*, 436–452. [CrossRef]
15. Sultana, Y.; Aqil, M. Liposomal Drug Delivery Systems: An Update Review. *Curr. Drug Deliv.* **2007**, *4*, 297–305. [CrossRef]
16. Liu, G.; Hou, S.; Tong, P.; Li, J. Liposomes: Preparation, Characteristics, and Application Strategies in Analytical Chemistry. *Crit. Rev. Anal. Chem.* **2020**, 1–21. [CrossRef]
17. Has, C.; Sunthar, P. A comprehensive review on recent preparation techniques of liposomes. *J. Liposome Res.* **2020**, *30*, 336–365. [CrossRef]
18. Trucillo, P.; Reverchon, E. Production of PEG-coated liposomes using a continuous supercritical assisted process. *J. Supercrit. Fluids* **2021**, *167*, 105048. [CrossRef]
19. Trucillo, P.; Ferrari, P.; Campardelli, R.; Reverchon, E.; Perego, P. A supercritical assisted process for the production of amoxicillin-loaded liposomes for antimicrobial applications. *J. Supercrit. Fluids* **2020**, *163*, 104842. [CrossRef]
20. Trucillo, P.; Campardelli, R.; Reverchon, E. A versatile supercritical assisted process for the one-shot production of liposomes. *J. Supercrit. Fluids* **2019**, *146*, 136–143. [CrossRef]
21. Campardelli, R.; Trucillo, P.; Reverchon, E. A Supercritical Fluid-Based Process for the Production of Fluorescein-Loaded Liposomes. *Ind. Eng. Chem. Res.* **2016**, *55*, 5359–5365. [CrossRef]
22. Ciaglia, E.; Montella, F.; Trucillo, P.; Ciardulli, M.; Di Pietro, P.; Amodio, G.; Remondelli, P.; Vecchione, C.; Reverchon, E.; Maffulli, N.; et al. A bioavailability study on microbeads and nanoliposomes fabricated by dense carbon dioxide technologies using human-primary monocytes and flow cytometry assay. *Int. J. Pharm.* **2019**, *570*, 118686. [CrossRef]

23. Steiner, B.M.; McClements, D.J.; Davidov-Pardo, G. Encapsulation systems for lutein: A review. *Trends Food Sci. Technol.* **2018**, *82*, 71–81. [CrossRef]
24. Manayi, A.; Abdollahi, M.; Raman, T.; Nabavi, S.M.; Habtemariam, S.; Daglia, M. Lutein and cataract: From bench to bedside. *Crit. Rev. Biotechnol.* **2015**, *36*, 829–839. [CrossRef]
25. Chung, R.W.; Leanderson, P.; Lundberg, A.K.; Jonasson, L. Lutein exerts anti-inflammatory effects in patients with coronary artery disease. *Atherosclerosis* **2017**, *262*, 87–93. [CrossRef]
26. Maia, M.; Furlani, B.A.; Souza-Lima, A.A.; Martins, D.S.; Navarro, R.M.; Belfort, R. LUTEIN. *Retina* **2014**, *34*, 262–272. [CrossRef]
27. Tan, C.; Xia, S.; Xue, J.; Xie, J.; Feng, B.; Zhang, X. Liposomes as Vehicles for Lutein: Preparation, Stability, Liposomal Membrane Dynamics, and Structure. *J. Agric. Food Chem.* **2013**, *61*, 8175–8184. [CrossRef]
28. Xia, S.; Tan, C.; Zhang, Y.; Abbas, S.; Feng, B.; Zhang, X.; Qin, F. Modulating effect of lipid bilayer–carotenoid interactions on the property of liposome encapsulation. *Colloids Surf. B Biointerfaces* **2015**, *128*, 172–180. [CrossRef]
29. Xia, F.; Hu, D.; Jin, H.; Zhao, Y.; Liang, J. Preparation of lutein proliposomes by supercritical anti-solvent technique. *Food Hydrocoll.* **2012**, *26*, 456–463. [CrossRef]
30. Agarwal, S.; Rao, A.V. Carotenoids and Chronic Diseases. *Drug Metab. Drug Interact.* **2000**, *17*, 189–210. [CrossRef] [PubMed]
31. Dos Santos, P.P.; Andrade, L.D.A.; Flores, S.; Rios, A.D.O. Nanoencapsulation of carotenoids: A focus on different delivery systems and evaluation parameters. *J. Food Sci. Technol.* **2018**, *55*, 3851–3860. [CrossRef] [PubMed]
32. Nagao, A. Absorption and Function of Dietary Carotenoids. *Nutr. Cardiovasc. Dis.* **2009**, *61*, 55–63.
33. Tan, C.; Zhang, Y.; Abbas, S.; Feng, B.; Zhang, X.; Xia, S. Modulation of the carotenoid bioaccessibility through liposomal encapsulation. *Colloids Surf. B Biointerfaces* **2014**, *123*, 692–700. [CrossRef]
34. Zhao, L.; Temelli, F.; Curtis, J.M.; Chen, L. Encapsulation of lutein in liposomes using supercritical carbon dioxide. *Food Res. Int.* **2017**, *100*, 168–179. [CrossRef] [PubMed]
35. Campardelli, R.; Trucillo, P.; Reverchon, E. Supercritical assisted process for the efficient production of liposomes containing antibiotics for ocular delivery. *J. CO_2 Util.* **2018**, *25*, 235–241. [CrossRef]
36. Singh, P.; Bodycomb, J.; Travers, B.; Tatarkiewicz, K.; Travers, S.; Matyas, G.R.; Beck, Z. Particle size analyses of polydisperse liposome formulations with a novel multispectral advanced nanoparticle tracking technology. *Int. J. Pharm.* **2019**, *566*, 680–686. [CrossRef] [PubMed]
37. Junghans, A.; Sies, H.; Stahl, W. Macular Pigments Lutein and Zeaxanthin as Blue Light Filters Studied in Liposomes. *Arch. Biochem. Biophys.* **2001**, *391*, 160–164. [CrossRef]
38. Chakravarty, K.; Dalal, D. Mathematical modelling of liposomal drug release to tumour. *Math. Biosci.* **2018**, *306*, 82–96. [CrossRef]
39. Patel, G.P.; Crank, C.W.; Leikin, J.B. An Evaluation of Hepatotoxicity and Nephrotoxicity of Liposomal Amphotericin B (L-AMB). *J. Med. Toxicol.* **2010**, *7*, 12–15. [CrossRef]
40. Jain, A.; Jain, S.K. In vitro release kinetics model fitting of liposomes: An insight. *Chem. Phys. Lipids* **2016**, *201*, 28–40. [CrossRef] [PubMed]
41. Bozzuto, G.; Molinari, A. Liposomes as nanomedical devices. *Int. J. Nanomed.* **2015**, *10*, 975–999. [CrossRef] [PubMed]
42. Rodrigues, A.R.O.; Gomes, I.; Almeida, B.G.; de Araújo, J.P.E.; Castanheira, E.M.S.; Coutinho, P.J.G. Magnetic liposomes based on nickel ferrite nanoparticles for biomedical applications. *Phys. Chem. Chem. Phys.* **2015**, *17*, 18011–18021. [CrossRef] [PubMed]
43. Qv, X.-Y.; Zeng, Z.-P.; Jiang, J.-G. Preparation of lutein microencapsulation by complex coacervation method and its physicochemical properties and stability. *Food Hydrocoll.* **2011**, *25*, 1596–1603. [CrossRef]
44. Abdel-Aal, E.-S.M.; Rabalski, I. Composition of Lutein Ester Regioisomers in Marigold Flower, Dietary Supplement, and Herbal Tea. *J. Agric. Food Chem.* **2015**, *63*, 9740–9746. [CrossRef]
45. Becerra, M.O.; Contreras, L.M.; Lo, M.H.; Díaz, J.M.; Herrera, G.C. Lutein as a functional food ingredient: Stability and bioavailability. *J. Funct. Foods* **2020**, *66*, 103771. [CrossRef]
46. Sawant, R.R.; Torchilin, V.P. Liposomes as 'smart' pharmaceutical nanocarriers. *Soft Matter* **2010**, *6*, 4026–4044. [CrossRef]
47. Kono, K.; Hayashi, H.; Takagishi, T. Temperature-sensitive liposomes: Liposomes bearing poly (*N*-isopropylacrylamide). *J. Control. Release* **1994**, *30*, 69–75. [CrossRef]
48. Yuba, E. Development of functional liposomes by modification of stimuli-responsive materials and their biomedical applications. *J. Mater. Chem. B* **2020**, *8*, 1093–1107. [CrossRef] [PubMed]
49. Huang, Z.; Li, X.; Zhang, T.; Song, Y.; She, Z.; Li, J.; Deng, Y. Progress involving new techniques for liposome preparation. *Asian J. Pharm. Sci.* **2014**, *9*, 176–182. [CrossRef]
50. Lu, T.; Hagen, T.L.T. A novel kinetic model to describe the ultra-fast triggered release of thermosensitive liposomal drug delivery systems. *J. Control. Release* **2020**, *324*, 669–678. [CrossRef]
51. Papadopoulou, V.; Kosmidis, K.; Vlachou, M.; Macheras, P. On the use of the Weibull function for the discernment of drug release mechanisms. *Int. J. Pharm.* **2006**, *309*, 44–50. [CrossRef] [PubMed]
52. Kosmidis, K.; Argyrakis, P.; Macheras, P. A Reappraisal of Drug Release Laws Using Monte Carlo Simulations: The Prevalence of the Weibull Function. *Pharm. Res.* **2003**, *20*, 988–995. [CrossRef] [PubMed]
53. Liu, J.; Conboy, J.C. Direct Measurement of the Transbilayer Movement of Phospholipids by Sum-Frequency Vibrational Spectroscopy. *J. Am. Chem. Soc.* **2004**, *126*, 8376–8377. [CrossRef] [PubMed]
54. Koynova, R.; Caffrey, M. Phases and phase transitions of the phosphatidylcholines. *Biochim. Biophys. Acta* **1998**, *1376*, 91–145. [CrossRef]

55. Wang, J.; Ayano, E.; Maitani, Y.; Kanazawa, H. Tunable Surface Properties of Temperature-Responsive Polymer-Modified Liposomes Induce Faster Cellular Uptake. *ACS Omega* **2017**, *2*, 316–325. [CrossRef]
56. Zhan, W.; Gedroyc, W.; Xu, X.Y. Towards a multiphysics modelling framework for thermosensitive liposomal drug delivery to solid tumour combined with focused ultrasound hyperthermia. *Biophys. Rep.* **2019**, *5*, 43–59. [CrossRef]

processes

MDPI

Article

Supercritical Fluid Extraction of Compounds of Pharmaceutical Interest from *Wendita calysina* (Burrito)

Lucia Baldino * and Ernesto Reverchon

Department of Industrial Engineering, University of Salerno, Via Giovanni Paolo II, 132, 84084 Fisciano (SA), Italy

* Correspondence: lbaldino@unisa.it; Tel.: +39-089964232

Received: 25 June 2020; Accepted: 18 August 2020; Published: 21 August 2020

Abstract: *Wendita calysina* (commonly known as burrito) is an indigenous Paraguayan medical plat, whose essential oil (EO) is characterized by some pharmaceutical properties. In particular, the main component is D-carvone, with anticancer action and antimicrobial properties against various microorganisms. In this work, supercritical CO_2 (SC-CO_2) was used for the extraction of volatile compounds from this plant, selecting different operative conditions to optimize the extract yield and purity. Pressure was varied from 80 to 250 bar, and two CO_2 flow rates (0.8 kg/h and 1.2 kg/h) were tested. The highest EO percentage in the extract was obtained operating at 100 bar and 40 °C, using ground *Wendita calysina* leaves of 250 μm. CO_2 flow rate did not influence the extraction yield, indicating that an internal mass transfer resistance governs this process. The largely prevailing compound identified in the extract was D-carvone, with a mean percentage up to 90% *w*/*w*.

Keywords: *Wendita calysina*; burrito; supercritical fluid extraction; fractional separation; D-carvone; anticancer; antimicrobial

1. Introduction

The adoption of a correct diet together with regular physical exercise are recognized as useful for the prevention of some human diseases [1–3]. For instance, the daily consumption of foods rich in antioxidants can reduce free radical degradation of cells and tissues in an organism [2,4]. Herbal infusions are a possible sources of these bioactive molecules [5,6]; green tea is one of the major examples, widely used particularly in oriental regions due to its medical effects in the prevention and treatment of various diseases, such as cancer, cardiovascular disease, diabetes and obesity [7].

Infusions of *Wendita calysina* (Geraniaceae) leaves, a Paraguayan herbaceous plant, generally known as burrito, are also recognized as healthy for humans, thanks to the well-known antimicrobial, anti-inflammatory, antioxidant and anti-rheumatic properties [8,9]. Marzocco et al. [9] demonstrated that *Wendita calysina* water-soluble extract can be used in the therapy of acute inflammation. Mesa et al. [10] reported an ultrasonic extraction procedure from *Wendita calysina* leaves using a water/ethanol mixture (40/60% *v*/*v*), followed by spray-drying. These authors determined the antiplatelet action of this extract to be used as food supplement and phytomedicine. *Wendita calysina* is rich in D-carvone, a monoterpenic ketone with antimicrobial, anticonvulsant, and antioxidant properties [11]. In particular, it shows antimicrobial properties against various microorganisms, like *Listeria monocytogenes* [12], *Escherichia coli*, *Salmonella typhimurium*, *Photobacterium leiognathi* [13], *Fusarium subglutinans*, *Fusarium cerealis*, *Fusarium verticillioides*, *Fusarium proliferatum*, *Fusarium oxysporum*, *Fusarium sporotrichioides*, *Aspergillus tubingensis*, *Aspergillus carbonarius*, *Alternaria alternata* and *Penicillium sp.* [14]. Aydin et al. [15] demonstrated that carvone has also anticancer potential in vitro, which improves the general healthy *Wendita calysina* plant effects [8–10].

Maceration is the most frequently used traditional technique to produce extracts from a vegetable matter [16]; an organic solvent and, in some cases, water, affine with the compounds of interest, is used at a specific ratio with the starting vegetable matter. This procedure can last from some hours to some days, and the final extract is recovered after solvent evaporation. However, maceration is not a selective extraction technique and the obtained extracts have to be generally post-processed to remove the organic solvent residues and/or impurities [16].

Supercritical fluid extraction (SFE) is a green alternative technique to extract high-value compounds from vegetable matter, avoiding the use of toxic organic solvents that pollute the final product and the environment [17–19]. Supercritical fluids and, in particular, supercritical CO_2 (SC-CO_2), are characterized by favorable physico-chemical properties (i.e., gas-like diffusivity, liquid-like density, and near to zero surface tension) and reduced cost [20]. Numerous vegetal materials were processed by SFE [21,22]. Parisotto et al. [23] found that the antitumor activity of *Cordia verbenacea* extracts obtained by SFE was superior in the reduction of tumor cells viability and proliferation with respect to the extract obtained by classical organic solvent extraction with ethanol. Radojković et al. [24] performed SFE process on *Morus* leaves working at 300 bar and 40 °C. They obtained a linoleic and linolenic acid-rich extract to be used as dietary supplement and food product. Souza et al. [25] determined the effects of temperature and pressure on the extraction yield and chemical composition of extracts from *Aloysia gratissima* leaves obtained by SFE. Operative temperature was varied from 40 to 60 °C; whereas pressure was from 150 to 200 bar. The chemical analysis showed the presence of guaiol, pinocamphone, caryophyllene oxide and spathulenol as the major chemical components of this plant extract that contributed to a significant reduction of carrageenan-induced paw edema and myeloperoxidase activity.

SFE technology can also allow a fractional separation of the extracts, to obtain high purity compounds [19]. This result is possible when process conditions are opportunely selected. Operative pressure and temperature influence SC-CO_2 solvent power through CO_2 density; therefore, depending on the solubility and on the position inside the vegetable matter of the compounds of interest, these parameters should be selected, together with CO_2 flow rate and vegetable matter particle size, with the aim of minimizing the co-extraction of undesired compounds [26]. Another possibility is to perform the fractional extraction, changing the SC-CO_2 solvent power during processing. In particular, depending on the molecular weight (and on the solubility) of the various components, a reduced SC-CO_2 solvent power can be selected in a first step to extract essential oil compounds (low molecular weight compounds); then, in a second step, operative pressure is increased, to extract the other higher molecular weight compounds [26].

Taking into account these considerations, to our knowledge, this work performed SFE of *Wendita calysina* leaves for the first time, to obtain a high purity essential oil with enhanced healthy properties for pharmaceutical applications. Systematic gas chromatography–mass spectrometry (GC-MS) analyses were carried out on the extracts, to optimize the parameters related to SFE processing and on the starting vegetable matter, to produce a high added-value, solvent free product.

2. Materials and Methods

2.1. Plant Material

Dried *Wendita calysina* Martius leaves (moisture content lower than 10% *w/w*), produced by Coprosa Ltd.a, Paraguay (Cooperativa multiactiva de Produccìon Servicios publicos, Consumo, Ahorro y Credito, San Andres), were supplied by Commercio Alternativo s.r.l. (Ferrara, Italy). Before processing, *Wendita calysina* leaves were ground using a cooled milling apparatus (Waring Commercial Blender); mean particle size was determined using mechanical sieving.

2.2. Supercritical Fluid Extraction

Supercritical CO_2 extraction was performed in a lab-scale apparatus equipped with a 400 cm^3 internal volume extractor. CO_2 (purity 99%) was purchased from Morlando Group s.r.l. (Sant'Antimo, NA, Italy). An amount of 100 g of the vegetable matter (mean particle size of 250 μm) was mixed with

85 g of 3-mm glass beads to avoid caking and channeling phenomena in the fixed bed of particles inside the extractor. Extract fractions were recovered using two separation vessels in series, with an internal volume of 200 cm^3. Cooling of the first separator was achieved using a thermostated bath (Julabo, mod. F38-EH, Seelbach, Germany). The second separator allowed the continuous discharge of the product using a valve located at the bottom of the vessel. A membrane high-pressure pump (Lewa, mod. LDB1 M210S, Mazzo di Rho, Italy) pumped liquid CO_2 at the desired flow rate. CO_2 was heated up to the extraction temperature in a thermostated oven, and its flow rate was monitored by a calibrated rotameter (ASA, mod. N.5-2500, Serval 115022, Sesto San Giovanni, Italy). Temperatures along the extraction apparatus were measured by thermocouples and regulated using PID controllers (mod. 305, Watlow, Corsico (MI), Italy); whereas pressure was measured by test gauges (mod. MP1, OMET, Lecco, Italy). A schematic representation of an SFE plant can be found in [27]. Each experiment was performed in duplicate.

2.3. GC–MS Analyses of Compounds from Wendita Calysina Extracts

Wendita calysina extracts obtained by SFE were characterized by gas chromatography–mass spectrometry (GC-MS). The apparatus was a Varian (mod. Saturn 2100T, San Fernando, CA, USA); separation was achieved using a fused-silica capillary column (mod. DB-5, J&W, Folsom, CA, USA) with a 30-m length, 0.25-mm internal diameter and 0.25-μm film thickness. GC conditions were oven temperature at 40 °C for 5 min; then, programmed heating from 40 to 250 °C at 2 °C/min. After 60 min at 250 °C, temperature was increased to 270 °C (speed 2 °C/min) and, finally, from 270 to 290 °C, at 1 °C/min. Helium was used as the carrier gas (1 mL/min). Samples were run in dichloromethane using a dilution factor of 0.0008% *w/w*. All the components were identified according to the retention time and molecular weight. The same procedure was adopted for the identification of waxes.

3. Results

The aim of this work was the extraction of a high purity volatile oil from *Wendita calysina* leaves by SFE. Since this was the first time this technology was applied to *Wendita calysina*, a preliminary series of experiments was performed using different CO_2 flow rates, to determine the kind of mass transfer resistance governing this process. CO_2 flow rate was varied from 0.8 to 1.2 kg/h; whereas the plant mean particle size was fixed at 250 μm, obtained using the procedure described in Materials and Methods. SFE was performed at 250 bar and 40 °C (ρ_{CO2} = 0.880 g/cm^3); the first separator was set at −14 °C and the second separator was operated at 25 °C and 15 bar. The selection of these values of temperature in the two separators was based on previous works [17,19,28]; the aim was to induce waxes (i.e., paraffins) precipitation in the first separator due to the solubility reduction of these components in SC-CO_2 at low temperature, and the extract recovery in the second separator at conditions near to the ambient ones to avoid thermal stress on the liquid extract.

The extraction yield, measured as the ratio between the weight of the extract collected at regular time intervals and the starting vegetable matter loaded in the extractor, was similar using the two CO_2 flow rates. This information indicates that an internal mass transfer resistance governs this process [20]: i.e., the compounds of interest are located inside the vegetable matter, therefore, the effect of CO_2 flow rate is negligible. Once the kind of extraction mechanism was ascertained, all the following experiments were performed using a CO_2 flow rate of 0.8 kg/h and *Wendita calysina* leaves particle size of 250 μm.

The second part of the work was focused on the obtainment of a higher essential oil (EO) purity. Operative pressure was varied from 80 to 250 bar and temperature was fixed at 40 °C, to avoid thermal stresses on the vegetable material, and with the aim of determining the effect of SC-CO_2 solvent power. In all experiments, separators' conditions were the same as the preliminary experiments.

SFE was carried out in two steps (fractional extraction): the first one was performed at 80 bar and 40 °C (ρ_{CO2} = 0.281 g/cm^3), the second one at 150 bar and 40 °C (ρ_{CO2} = 0.780 g/cm^3). Extraction was stopped when a negligible increment of the extract was obtained. A final total yield of 0.763%

w/w and an EO yield of 0.717% *w/w* were measured. However, working at 80 bar and 40 °C, due to the reduced SC-CO_2 density, a too low EO extraction yield was obtained (≈0.03% *w/w*); therefore, these operative conditions were discarded. Moreover, observing the last 100 min of processing, the distance between the total extract yield curve and the EO yield curve tended to enlarge due to the extraction of non-terpenic compounds: i.e., higher molecular weight compounds were co-extracted that do not belong to *Wendita calysina* essential oil. Waxes extraction yield was 1.91% *w/w*, measured in the first separator at the end of the supercritical processing.

A further SFE experiment was performed at 250 bar e 40 °C, to determine the effect of a larger SC-CO_2 density on the total and EO yield. However, as expected, increasing SC-CO_2 density and, thus, its solvent power, the final total yield (0.794% *w/w*) was larger than the one measured working at 150 bar and 40 °C (0.763% *w/w*); but, also the difference between the final total yield and the EO yield (0.636% *w/w*) was still larger, due to the co-extraction of other undesired families of compounds (see Figure 1). Therefore, a lower EO purity characterized this extract. In Figure 1, this trend is clearly visible, even if the extraction kinetics were faster (420 min). Waxes collected in the first separator had a yield of 2.02% *w/w*.

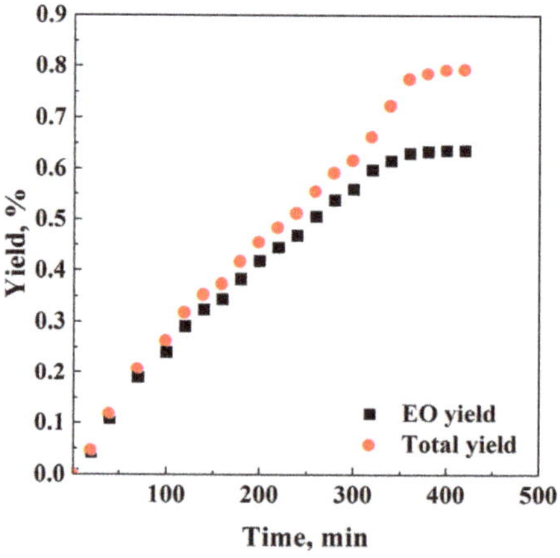

Figure 1. Representation of supercritical fluid extraction (SFE) total yield and essential oil (EO) yield versus time: 250 bar and 40 °C.

Taking into account these results, another series of experiments was performed at intermediate SC-CO_2 solvent power: i.e., 100 bar and 40 °C (ρ_{CO2} = 0.623 g/cm^3), to balance a fast extraction and a high EO purity in the extract. Figure 2 reports that a very similar final total yield (0.697% *w/w*) and EO yield (0.693% *w/w*) were obtained: i.e., the two curves practically overlap. This result confirms that adopting the opportune SFE operative conditions, it is possible to selectively extract the compounds of interest. Waxes yield was 1.56% *w/w* in this case.

Summarizing these results, SFE confirmed its flexibility in extracting the compounds of interest changing SC-CO_2 solvent power. This peculiarity, coupled with fractional separation, favored the obtainment of a high purity EO in the extract, without the use of toxic organic solvents.

Extracts collected in the first and in the second separator were systematically characterized by GC-MS, as described in Materials and Methods. The extracts composition was in all cases similar and mainly formed by monoterpenes and sesquiterpenes (at the second separator): identification of compounds related to SFE performed at 100 bar and 40 °C is reported in Table 1. In particular, D-carvone, a monoterpene responsible for the characteristic odor of caraway, was the major compound forming the *Wendita calysina* essential oil, as demonstrated by the GC-MS trace in Figure 3. Its mean amount in the extract represented up to 90% of the EO composition by supercritical processing.

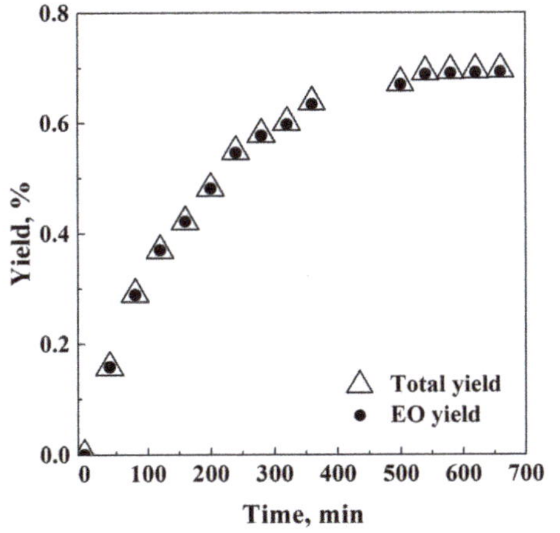

Figure 2. Representation of SFE total yield and EO yield versus time: 100 bar and 40 °C.

Table 1. Gas chromatography–mass spectrometry (GC-MS) analysis on SFE extracts obtained at 100 bar and 40 °C.

Collected Fraction	Compounds	% *w/w*
Collection #1–3 Time 80′	D-carvone	≈100
Collection #4–6 Time 200′	D-carvone	≈100
Collection #7–9 Time 320′	Linalool	1.18
	Para-cymene	0.72
	Camphene	2.14
	Para-cymenene	1.73
	D-carvone	87.87
	Caryophyllene	2.17
	Cuparene	1.67
	N.I. compounds	2.53

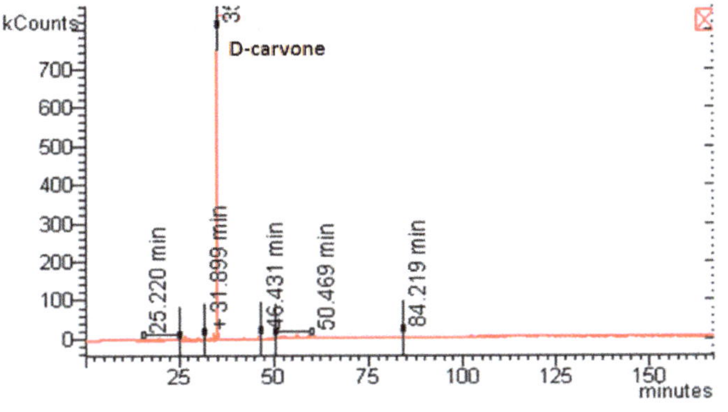

Figure 3. Example of a GC-MS trace related to the extract collected in the II separator.

Waxes collected in the first separator were white, odorless and mainly formed by Nonacosane (C_{29}) 1.227%, Triacontane (C_{30}) 12.616%, traces of Hentriacontane (C_{31}) and Dotriacontane (C_{32}) 86.150%, as shown in Figure 4. These results demonstrate the efficiency of SFE fractional separation since, in this separator, only waxes (paraffinic compounds) were precipitated; other peaks, related to other compounds were not detected. Moreover, waxes extraction yield increased from 1.56% *w*/*w* (@100 bar) to 2.02% *w*/*w* (@250 bar) when a larger SC-CO_2 solvent power was tested.

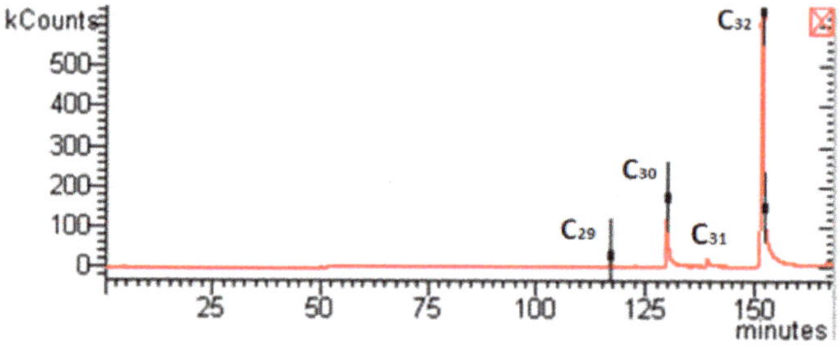

Figure 4. Example of a GC-MS trace related to waxes collected in the I separator at −14 °C and 100 bar.

4. Conclusions

Extracts produced by SFE can have a larger added value, since the co-extraction of undesired families of compounds can be minimized. A selective extraction is possible by obtaining a high purity, solvent-free extract. Operating at 100 bar and 40 °C, SFE applied to *Wendita calysina* leaves was successful in obtaining an extract with a content of D-carvone up to 90% *w*/*w*. This result was possible thanks to the addition of a fractional separation scheme that allowed waxes precipitation in the first separator, and pure extract collection in the second one. This product can be used in the pharmaceutical field thanks to its antimicrobial, anticancer, anti-inflammatory, antioxidant and anti-rheumatic properties.

A subsequent study will be focused on the testing of pharmaceutical properties of *Wendita calysina* leaves extract, in order to determine its efficacy and the doses required.

Author Contributions: Conceptualization, L.B. and E.R.; methodology, E.R.; formal analysis, L.B.; investigation, L.B.; resources, E.R.; writing—original draft preparation, L.B.; writing—review and editing, L.B. and E.R.; supervision, E.R. All authors have read and agreed to the published version of the manuscript.

Funding: This research received no external funding.

Conflicts of Interest: The authors declare no conflict of interest.

References

1. Sur, P.; Chandhuri, T.; Vedasiromoni, J.R.; Gomes, A.; Ganguly, D.K. Antiinflammatory and antioxidant property of saponins of tea [*Camellia sinensis* (L) O. Kuntze] root extract. *Phytother. Res.* **2001**, *15*, 174–176. [CrossRef] [PubMed]
2. Wongkham, S.; Laupattarakasaem, P.; Pienthaweechai, K.; Areejitranusorn, P.; Wongkham, C.; Techanitiswad, T. Antimicrobial activity of *Streblus asper* leaf extract. *Phytother. Res.* **2001**, *15*, 119–121. [CrossRef] [PubMed]
3. Yao, S.Z.; Tan, H.W.; Zhang, H.; Su, X.L.; Wei, W.Z. Bulk acoustic wave bacterial growth sensor applied to analysis of antimicrobial properties of tea. *Biotechnol. Prog.* **1998**, *14*, 639–644. [CrossRef] [PubMed]
4. Jin, D.; Hakamata, H.; Takahashi, K.; Kotani, A.; Kusu, F. Determination of quercetin in human plasma after ingestion of commercial canned green tea by semi-micro HPLC with electrochemical detection. *Biomed. Chromatogr.* **2004**, *18*, 662–666. [CrossRef]
5. Marongiu, B.; Porcedda, S.; Piras, A.; Rosa, A.; Deiana, M.; Dessi, M.A. Antioxidant activity of supercritical extract of *Melissa officinalis* subsp *officinalis* and *Melissa officinalis* subsp *inodora*. *Phytother. Res.* **2004**, *18*, 789–792. [CrossRef]
6. Wu, S.J.; Ng, L.T.; Lin, C.C. Antioxidant activities of some common ingredients of traditional Chinese medicine, *Angelica sinensis*, *Lycium barbarum* and *Poria cocos*. *Phytother. Res.* **2004**, *18*, 1008–1012. [CrossRef]
7. Arab, H.; Maroofian, A.; Golestani, S.; Shafaee, H.; Sohrabi, K.; Forouzanfar, A. Review of the therapeutic effects of *Camellia sinensis* (green tea) on oral and periodontal health. *J. Med. Plants Res.* **2011**, *5*, 5465–5469.

8. Piccinelli, A.L.; De Simone, F.; Passi, S.; Rastrelli, L. Phenolic Constituents and antioxidant activity of *Wendita calysina* leaves (Burrito), a folk paraguayan tea. *J. Agric. Food Chem.* **2004**, *52*, 5863–5868. [CrossRef]
9. Marzocco, S.; Piccinelli, A.L.; Rastrelli, L.; Mazzon, E.; Cuzzocrea, S.; Autore, G. Inhibition of inducible nitric oxide synthase in vitro and in vivo by a water-soluble extract of *Wendita calysina* leaves. *Naunyn-Schmiedeberg's Arch. Pharmacol.* **2007**, *375*, 349–358. [CrossRef]
10. Mesa, M.G.; Piccinelli, A.L.; Antonia, M.; Valiente, A.; Pinto, A.; Fazio, A.; Rastrelli, L. Inhibition of human platelet aggregation in vitro by standardized extract of *Wendtia calycina*. *Rev. Bras. Farmacogn. Braz. J. Pharmacogn.* **2011**, *21*, 884–888. [CrossRef]
11. Bicas, J.L.; Neri-Numa, I.A.; Ruiz, A.L.; De Carvalho, J.E.; Pastore, G.M. Evaluation of the antioxidant and antiproliferative potential of bioflavors. *Food Chem. Toxicol.* **2011**, *49*, 1610–1615. [CrossRef] [PubMed]
12. Friedman, M.; Henika, P.R.; Mandrell, R.E. Bactericidal activities of plant essential oils and some of their isolated constituents against *Campylobacter jejuni, Escherichia coli, Listeria monocytogenes,* and *Salmonella enterica*. *J. Food Prot.* **2002**, *65*, 1545–1560. [CrossRef]
13. Helander, I.M.; Alakomi, H.L.; Latva-Kala, K.; Mattila-Sandholm, T.; Pol, I.; Smid, E.J.; Gorris, L.G.M.; von Wright, A. Characterization of the action of selected essential oil components on gram-negative bacteria. *J. Agric. Food Chem.* **1998**, *46*, 3590–3595. [CrossRef]
14. Morcia, C.; Malnati, M.; Terzi, V. In vitro antifungal activity of terpinen-4-ol, eugenol, carvone, 1,8-cineole (eucalyptol) and thymol against mycotoxigenic plant pathogens. *Food Addit. Contam. Part A* **2012**, *29*, 415–422.
15. Aydın, E.; Türkez, H.; Keleş, M.S. Potential anticancer activity of carvone in N2a neuroblastoma cell line. *Toxicol. Ind. Health* **2015**, *31*, 764–772. [CrossRef] [PubMed]
16. Alizadehdakhel, A.; Golzary, A. An environmental friendly process for extraction of active constituents from herbal plants. *Environ. Energy Econ. Res.* **2020**, *4*, 69–81.
17. Baldino, L.; Reverchon, E.; Della Porta, G. An optimized process for SC-CO_2 extraction of antimalarial compounds from *Artemisia annua* L. *J. Supercrit. Fluids* **2017**, *128*, 89–93. [CrossRef]
18. Baldino, L.; Della Porta, G.; Sesti Osseo, L.; Reverchon, E.; Adami, R. Concentrated oleuropein powder from olive leaves using alcoholic extraction and supercritical CO_2 assisted extraction. *J. Supercrit. Fluids* **2018**, *133*, 65–69. [CrossRef]
19. Baldino, L.; Adami, R.; Reverchon, E. Concentration of *Ruta graveolens* active compounds using SC-CO_2 extraction coupled with fractional separation. *J. Supercrit. Fluids* **2018**, *131*, 82–86. [CrossRef]
20. Baldino, L.; Reverchon, E. Challenges in the production of pharmaceutical and food related compounds by SC-CO_2 processing of vegetable matter. *J. Supercrit. Fluids* **2018**, *134*, 269–273. [CrossRef]
21. Sharif, K.M.; Rahman, M.M.; Azmir, J.; Mohamed, A.; Jahurul, M.H.A.; Sahena, F.; Zaidul, I.S.M. Experimental design of supercritical fluid extraction—A review. *J. Food Eng.* **2014**, *124*, 105–116. [CrossRef]
22. Gallego, R.; Bueno, M.; Herrero, M. Sub- and supercritical fluid extraction of bioactive compounds from plants, food-by-products, seaweeds and microalgae—An update. *TrAC Trends Anal. Chem.* **2019**, *116*, 198–213. [CrossRef]
23. Parisotto, E.B.; Michielin, E.M.Z.; Biscaro, F.; Ferreira, S.R.S.; Wilhelm Filho, D.; Pedrosa, R.C. The antitumor activity of extracts from *Cordia verbenacea* D.C. obtained by supercritical fluid extraction. *J. Supercrit. Fluids* **2012**, *61*, 101–107. [CrossRef]
24. Radojković, M.; Zeković, Z.; Mašković, P.; Vidović, S.; Mandić, A.; Mišanb, A.; Đurović, S. Biological activities and chemical composition of *Morus* leaves extracts obtained by maceration and supercritical fluid extraction. *J. Supercrit. Fluids* **2016**, *117*, 50–58. [CrossRef]
25. Souza, M.A.; Guzatti, J.G.G.; Martello, R.H.; Schindler, M.S.Z.; Calisto, J.F.F.; Morgan, L.V.; Aguiar, G.P.S.; Locateli, G.; Scapinello, J.; Müller, L.G.; et al. Supercritical CO_2 extraction of *Aloysia gratissima* leaves and evaluation of anti-inflammatory activity. *J. Supercrit. Fluids* **2020**, *159*, 104753. [CrossRef]
26. Baldino, L.; Scognamiglio, M.; Reverchon, E. Extraction of rotenoids from *Derris elliptica* using supercritical CO_2. *J. Chem. Technol. Biotechnol.* **2018**, *93*, 3656–3660. [CrossRef]
27. Baldino, L.; Reverchon, E. *Artemisia annua* organic solvent extract, processed by supercritical CO_2. *J. Chem. Technol. Biotechnol.* **2018**, *93*, 3171–3175. [CrossRef]
28. Baldino, L.; Scognamiglio, M.; Reverchon, E. Supercritical fluid technologies applied to the extraction of compounds of industrial interest from *Cannabis sativa* L. and to their pharmaceutical formulations: A review. *J. Supercrit. Fluids* **2020**, *165*, 104960. [CrossRef]

processes

MDPI

Article

Compound Identification and In Vitro Cytotoxicity of the Supercritical Carbon Dioxide Extract of Papaya Freeze-Dried Leaf Juice

Kooi-Yeong Khaw [1,2,*], Paul Nicholas Shaw [1,*], Marie-Odile Parat [1], Saurabh Pandey [1] and James Robert Falconer [1,*]

1 School of Pharmacy, The University of Queensland, Brisbane, QLD 4072, Australia; m.parat@pharmacy.uq.edu.au (M.-O.P.); s.pandey2@uq.edu.au (S.P.)
2 Biofunctional Molecule Exploratory Research Group (BMEX) School of Pharmacy, Monash University Malaysia, Jalan Lagoon Selatan, Bandar Sunway 47500, Malaysia
* Correspondence: khaw.kooiyeong@monash.edu (K.-Y.K.); n.shaw@uq.edu.au (P.N.S.); j.falconer@uq.edu.au (J.R.F.); Tel.: +60-355146135 (K.-Y.K.); +61-733464636 (P.N.S.); +61-733461852 (J.R.F.)

Received: 10 May 2020; Accepted: 19 May 2020; Published: 20 May 2020

Abstract: *Carica papaya* leaves are used as a remedy for the management of cancer. Freeze-dried *C. papaya* leaf juice was extracted using a supercritical fluid extraction system. Compound identification was carried out using analytical techniques including liquid chromatography coupled to high-resolution quadrupole time-of-flight mass spectrometry (LC–QToF-MS) and gas chromatography–mass spectrometry (GC–MS). The cytotoxic activities of the $scCO_2$ extract and its chemical constituents were determined using the 3-(4,5-dimethylthiazol-2-yl)-2,5-diphenyltetrazolium bromide (MTT) assay on squamous cell carcinoma (SCC25) and human keratinocyte (HaCaT) cell lines. The chemical constituents were quantified by QToF-MS. The supercritical carbon dioxide ($scCO_2$) extract of papaya freeze-dried leaf juice showed cytotoxic activity against SCC25. Three phytosterols, namely, β-sitosterol, campesterol, and stigmasterol, together with α-tocopherol, were confirmed to be present in the $scCO_2$ extract. Quantitative analysis showed that β-sitosterol was the major phytosterol present followed by α-tocopherol, campesterol, and stigmasterol. β-Sitosterol and campesterol were active against SCC25 (half maximal inhibitory concentration (IC_{50}) ≈ 1 μM), while stigmasterol was less active (~33 μM) but was biologically more selective against SCC25. Interestingly, an equimolar mixture of phytosterols was not more effective (no synergistic effect was observed) but was more selective than the individual compounds. The compounds identified are likely accountable for at least part of the cytotoxicity and selectivity effects of *C. papaya*.

Keywords: supercritical fluids; mass spectroscopy; ultra-high-pressure liquid chromatography; cytotoxic compound; phytosterol; stigmasterol

1. Introduction

Carica papaya is a herbaceous plant indigenous to tropical Mexico, Central America, and northern South America [1]. It is widely cultivated in the tropical and subtropical countries for its nutritional edible fruit [2]. Apart from the fruit of *C. papaya*, the leaf is used as a food and medicine. Anecdotal evidence indicates that the leaf is used in Australia in the form of a decoction to treat cancer [3]. In other parts of the world, the decoction of the leaf is used as a tea to treat high blood pressure, diabetes, digestion disorder, and jaundice, as well as dengue fever, rheumatic complaints, and elephantoid growths [4–7].

Several chemical constituents from *C. papaya* were identified [7]. The leaf of papaya is reported to contain alkaloids, tocopherol, flavonoids, tannins, phytosterols, saponin, phenolic compounds, and chlorogenic acid [8,9]. Epidemiology studies showed that phytochemicals from plants are beneficial in reducing the risk of dementia, stroke, diabetes, cardiovascular disease, and cancers [10]. Some studies demonstrated that leaf extracts from *C. papaya* were selectively cytotoxic to skin cancer in vitro [11,12]. Nguyen and colleagues reported that phenoside A from papaya leaf juice was potently cytotoxic to the cancerous SCC25 cell line. However, it was also cytotoxic against the non-cancerous HaCaT cell line [12]. Active and selective anti-cancer compounds from leaf juice remain to be fully elucidated.

This research gap prompted our interest to discover the bioactive chemical constituents of *C. papaya* leaves with selective activity against skin cancer. Supercritical fluid extraction (SFE) offers an alternative method, whereby different chemical constituents are selectively extracted from those components yielded by conventional methods. This approach may, therefore, reduce the analytical workload in identifying chemical constituents of interest. Several analytical methods were used to identify and quantify chemical constituents from papaya leaves including LC–MS, GC–MS, and nuclear magnetic resonance (NMR) spectroscopy [11,13,14]. For example, linoleic and linolenic acids were identified from the ethyl acetate fraction of *C. papaya* leaves by GC–MS [15].

Over the course of our continuing efforts to characterize anti-cancer compounds from *C. papaya*, we aimed to identify bioactive chemical constituents from the $scCO_2$ extract of *C. papaya*. The tentatively identified compounds were obtained commercially and compared with those present in the $scCO_2$ extract (for their liquid chromatography retention times, accurate mass, and MS/MS fragmentations). The identified compounds were analyzed and quantified. Furthermore, the cytotoxicity of the identified active compounds alone and in combination were evaluated.

2. Materials and Methods

2.1. Materials

DL-α-Tocopherol (purity >96%), stigmasterol (purity >95%), and 3-(4,5-dimethylthiazol-2-yl)-2,5-diphenyltetrazolium bromide were purchased from Sigma-Aldrich (Castle Hill, NSW, Australia). β-Sitosterol (purity >95%) was from Chromadex (USA) and campesterol (purity >95%) was from Novachem (Heidelberg West, VIC, Australia). Ethanol, HPLC-grade methanol, and HPLC-grade acetone were purchased from Merck (Darmstadt, Germany). Dulbecco's modified Eagle's medium (DMEM), DMEM/F12, trypsin, penicillin/streptomycin, and fetal bovine serum (FBS) were obtained from Invitrogen (Life Technologies, Mulgrave, VIC, Australia). Freeze-dried leaf juice was prepared from Australian *C. papaya* leaves (grown organically) gifted by Tropical Fruit World Pty Ltd., (Duranbah, New South Wales, Australia) according to the protocol described previously [16].

2.2. Supercritical Fluid Extraction (SFE)

SFE of *C. papaya* leaf juice was performed using a laboratory-scale extraction system. Earlier experiments determined that freeze-dried leaf juice was a better starting material in comparison with fresh leaves or freeze-dried leaves when considering both extraction yield and selective toxicity to cancer cells. The SFE system comprised a liquid carbon dioxide (CO_2) reservoir, high-pressure syringe pump (Teledyne Isco 260D), a vertical 60-mL stainless-steel (SS-316) extraction vessel (University of Nottingham), backpressure regulator, heating jacket (WatLow, USA), overhead stirrer (200 rpm) and fixed straight blade paddle, and SS precipitation chamber; details are described in previous work [16]. Glass wool (Merck, Darmstadt, Germany) was placed inside the extraction vessel and covered by a stainless-steel mesh to prevent any entrainment of the sample. We previously optimized the conditions of SFE that provided the extract exhibiting most cytotoxicity toward cancer cells [16]. The extraction was operated with the following conditions: pressure 250 bar; temperature 35 °C; freeze-dried leaf juice 5 g; extraction time 3 h. Following completion of the extraction, the supercritical fluid ($scCO_2$) extract was separated to the precipitation chamber via a capillary nozzle (0.0625 inch; 1.5875 mm) and the CO_2 gas was discharged to the atmosphere. The extract was weighed and stored at −20 °C for further analysis. The experiments were repeated three times.

2.3. Gas Chromatography–Mass Spectrometry

Analysis of the $scCO_2$ extract was carried out using a Shimadzu GCMS-TQ8040. Separation was obtained on an Rtx-5ms column (30 m × 0.25 mm, 0.25 μm film thickness; Restek, USA) with helium as the carrier gas at a constant linear velocity of 46.6 cm/s. The injection volume was 1 μL with a split ratio of 10. The initial column temperature was held at 160 °C for 1 min and then increased to 300 °C at a rate of 10 °C/min. The final column temperature was maintained at 300 °C for another 10 min. The temperatures of the injector and the detector were 240 °C and 200 °C, respectively. The interface temperature was set to 300 °C. Mass acquisition was performed in the range of 42–500 *m/z* using electron impact ionization at 70 eV. The components detected in the sample were identified by performing spectral database matching against the National Institute of Standards and Technology (NIST) library (v14).

2.4. Sample Preparation for UHPLC–QToF-MS Analysis

The standard stock solutions of DL-α-tocopherol, stigmasterol, β-sitosterol, and campesterol were prepared by dissolving accurately weighed standards in HPLC-grade methanol to give a concentration of 1 mg/mL. Each stock solution was filtered through a 0.22-μm polyvinylidene fluoride (PVDF) sterile filter (Merck Millipore, Germany) and stored in a freezer at −20 °C prior to analysis.

2.5. UHPLC–QToF-MS Analysis

The chromatography analysis of DL-α-tocopherol and phytosterols was performed on an Agilent 1290 UHPLC system (Agilent Technologies, Santa Clara, CA, USA) equipped with an Agilent 6520 high-resolution accurate mass quadruple time of flight (QToF) mass spectrometer. Chromatographic separation was conducted by a 2.0 × 150 mm, 100 Å, 2.6 μm C18 analytical column (Phenomenex, USA). Ultra-purified MilliQ water was mobile phase A, while HPLC-grade methanol was mobile phase B. The gradient elution conditions were as follows: 50% B for the first 5 min; 50% B increasing to 90% B from 5–40 min; 90% B increasing to 100% B from 40–60 min; 100% B decreasing to 50% B from 60–75 min. The sample injection volume was 5 μL with the flow rate of 0.2 mL/min. Mass spectral acquisition was monitored by MassHunter software (version B.02.01 SP3 –Agilent). The operating conditions for the mass spectrometer included a scan rate of 0.8 cycles/per second with the following conditions: nebuliser pressure 30 psi, *m/z* scan 100–1700, drying gas flow 5.0 L/min, gas temperature 300 °C, fragmenting voltage 175 V, and skimmer voltage 65.0 V.

2.5.1. Calibration Curves, Linearity Ranges, Limits of Detection (LOD), and Limits of Quantification (LOQ)

In this study, four biomarker analytes were evaluated at concentrations ranging from 1.56 to 50 μg/mL for the determination of the linear dynamic ranges. Each individual analyte at a fixed concentration was injected three times and the resultant peak heights obtained. Calibration curves were constructed by plotting peak heights against the analyte concentrations prepared, and the linearity of response to the four compounds was evaluated by linear regression analysis.

2.5.2. Quantification of DL-α-Tocopherol and Phytosterols by LC–QToF-MS

A 5-μL volume of the sample was injected into the LC–MS system. The concentration and profile of DL-α-tocopherol and phytosterols in the extracts were obtained using optimized LC–QToF-MS parameters. Peak identifications were performed by matching the retention times and accurate masses with the standard analytes. The sample was quantified using the external standard method.

2.6. Sample Preparation for the Cell Viability Assay

DL-α-Tocopherol, campesterol, stigmasterol, and β-sitosterol were dissolved in ethanol at concentrations of 10 mg/mL or 25 mM for the combined phytosterol assay, while the $scCO_2$ extract was solubilized in ethanol at a concentration of 50 mg/mL. All samples were sterile-filtered by a 0.22-μm polyvinylidene fluoride (PVDF) filter (Merck Millipore, Germany), resulting in stock solutions that were diluted with serum-free medium to the indicated final concentrations prior to performing the experiments.

2.7. Cell Culture Conditions

Human oral squamous cell carcinoma (SCC25) cells were obtained from ATCC® CRL-1628™, Manassas, VA, USA. The cells were maintained in DMEM/F12 medium added with 10% *v/v* heat-inactivated FBS, penicillin (100 units/mL), streptomycin (100 μg/mL), and hydrocortisone (0.4 μg/mL). Non-cancerous human keratinocyte (HaCaT) cells were a generous gift from Professor Fusenig. The cells were grown in DMEM containing 10% FBS, penicillin (100 units/mL), and streptomycin (100 μg/mL). All cell lines were cultured in a humidified atmosphere with 5% CO_2 at 37 °C. The cultures were passaged every third day, at which point they were approximately 70%–90% confluent.

Cell Viability Assay

The 3-(4,5-dimethylthiazol-2-yl)-2,5-diphenyltetrazolium bromide (MTT) assay was performed according to the method previously optimized for anticancer bioactive discovery [11,12]. In brief, 6000 cells (SCC25) per well or 3000 cells (HaCaT) were seeded in each well of a 96-well microtiter™ microplate. The cells were allowed to attach for 24 h prior to addition of samples. The culture medium was replaced with samples to be tested diluted in serum-free medium. The samples were incubated for 48 h at 37 °C, and the serum-free medium was replaced with a 0.5 mg/mL MTT solution. After a 2-h incubation, the medium was replaced with 100 μL of dimethyl sulfoxide (DMSO) on an orbital shaker for 20 min. The absorbance values were measured at 595 nm using an Lmark plate reader (BioRad, Hercules, California, USA). Wells containing no cells was being used as blanks whose absorbance was subtracted. Results are expressed as the percentage of viable cells with control, with untreated cells taken as 100%. The half maximal inhibitory concentration (IC_{50}) values of the compounds were estimated using non-linear regression analysis implemented in Prism 7 (GraphPad software Inc., San Diego, CA, USA).

2.8. Statistical Analysis

All statistical analysis was performed using Prism 7 (GraphPad software Inc., San Diego, CA, USA). All data are presented as means ± standard error of mean (SEM). Two-way ANOVA with a Sidak post hoc test was employed to compare the differences between the two cell lines.

3. Results

3.1. Supercritical Carbon Dioxide Extraction

Freeze-dried leaf juice of *C. papaya* was extracted using a $scCO_2$ extraction system. The extraction yield of $scCO_2$ freeze-dried leaf juice was 3.2% ($n = 3$).

3.2. Chemical Analysis of $scCO_2$ Extract with GC–MS

GC–MS results showed that at least 11 volatile compounds were present in the $scCO_2$ extract of freeze-dried leaf juice of papaya (Figure A1).

The mass spectral database of these compounds was matched (>95%) with the NIST library. Lipophilic compounds including fatty acids, vitamin E, and the phytosterols were tentatively assigned, and they are listed in Table 1. The results showed that DL-α-tocopherol (25.15%) and β-sitosterol (40.96%) were the major compounds detected, followed by campesterol (9.87%) and stigmasterol (8.86%).

Table 1. Volatile compounds identified by GC–MS.

Peak	Retention Time (Min)	Peak Area	[a] Peak Area (%)	Tentative Compound Identification
1	8.673	863,427	1.71	Linolenic acid
2	13.820	542,784	1.07	Squalene
3	15.417	2,380,812	4.71	γ-Tocopherol
4	15.884	624,002	1.23	β-Sitosterol acetate
5	16.034	12,690,750	25.15	DL-α-Tocopherol
6	17.000	4,980,249	9.87	Campesterol
7	17.293	4,474,003	8.86	Stigmasterol
8	17.89	20,663,841	40.96	β-Sitosterol
9	18.08	1,819,771	3.60	α-Amyrin
10	18.74	1,403,996	2.78	Brassicasterin
11	19.51	1,289,165	2.53	3-Oxocholest-4-en-27-yl acetate

[a] Percentage of peak area relative to the total peak area.

3.3. *Compound Identification by Comparison with Authentic Standards*

Based on the results from the database searching and the tentative identification of compounds from GC–MS analysis, direct comparison of the four compounds with four commercially available standards (DL-α-tocopherol, β-sitosterol, stigmasterol, and campesterol) with the scCO_2 extract of freeze-dried leaf juice was performed using LC–MS.

Table 2 shows the comparison of the masses and retention times of the respective standards and features derived from the scCO_2 extract of freeze-dried leaf juice detected in positive ion mode. All protonated phytosterols likely lost a molecule of water and, hence, the ions detected by the mass spectrometer are of the form $[M + H - H_2O]^+$. All compounds were matched to the retention times, accurate masses, and MS/MS fragmentations of the reference compounds (Figure 1a–d).

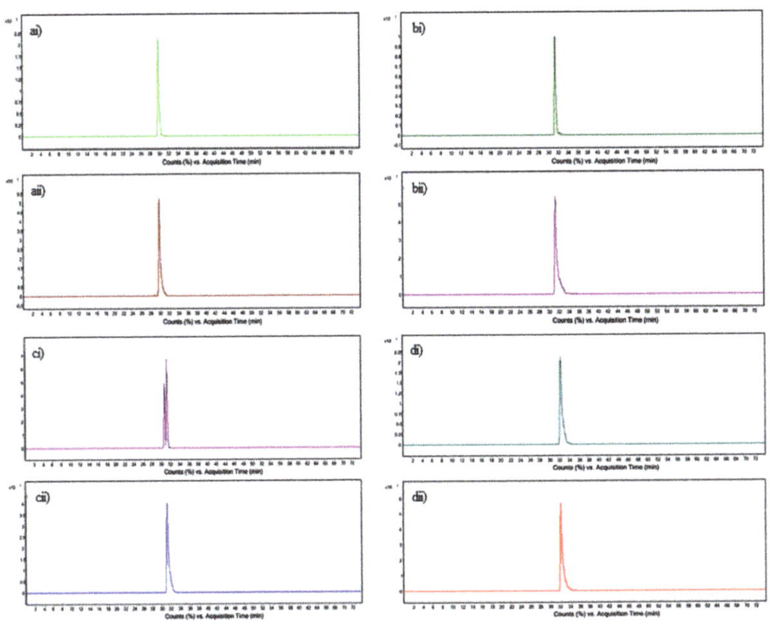

Figure 1. **(ai)** Extracted ion chromatogram (EIC) of *m/z* 431.3884 for the scCO_2 extract. **(aii)** EIC of *m/z* 431.3823 for the DL-α-tocopherol standard; **(bi)** EIC of *m/z* 383.3662 for the scCO_2 extract; **(bii)** EIC of *m/z* 383.3683 for the campesterol standard; **(ci)** EIC of *m/z* 395.3668 for the scCO_2 extract; **(cii)** EIC of *m/z* 395.3669 for the stigmasterol standard; **(di)** EIC of *m/z* 397.1361for the scCO_2 extract; **(dii)** EIC of *m/z* 397.1347 for the β-sitosterol standard.

Table 2. Phytochemicals determined by LC–electrospray ionization (ESI)-ToF MS in the $scCO_2$ extract of freeze-dried leaf juice (positive mode).

Standard	Mass of Standard $[M+H-H_2O]^+$ m/z	Retention Time of Standard (min)	Product ion Standard m/z	Mass of Feature $[M+H-H_2O]^+$ m/z	Retention Time of Feature (min)	Product ion of Feature m/z	Molecular Formula
DL-α-Tocopherol	431.3823	30.049	165.120	431.3884	30.043	165.139	$C_{29}H_{50}O_2$
Campesterol	383.3683	31.429	161.180	383.3662	31.464	161.187	$C_{28}H_{48}O$
Stigmasterol	395.3669	31.457	255.187	395.3668	31.497	255.220	$C_{29}H_{48}O$
β-Sitosterol	397.1347	32.407	161.185	397.1361	32.409	161.188	$C_{29}H_{50}O$

3.4. Quantification of Tocopherol and Phytosterols by UPLC–QToF-MS

The responses of all compounds possessed good linearity where DL-α-tocopherol, campesterol, and β-sitosterol were linear over the concentration range of 1.56–50 μg/mL (r^2 = 0.9992, 0.9987, and 0.9991, respectively), and stigmasterol was linear over the range 3.12–50 μg/mL. The LODs were calculated using a signal-to-noise ratio of 3. The LOD was 0.39 μg/mL for stigmasterol campesterol, and β-sitosterol, and 0.19 μg/mL for DL-α-tocopherol. The LOQs were calculated at a signal-to-noise ratio of 10. The LOQ was 1.56 μg/mL for stigmasterol, campesterol, and β-sitosterol, and 0.78 μg/mL for DL-α-tocopherol (Table 3).

Table 3. Linearity, limit of detection (LOD), and limit of quantification (LOQ) of tocopherol and phytosterols.

Compound	Calibration Curve	R^2	Linear Range (μg/mL)	LODs (μg/mL)	LOQs (μg/mL)
DL-α-Tocopherol	$y = 2668.2x + 4301.4$	0.9992	50–1.56	0.19	0.78
Stigmasterol	$y = 956.46x + 2623.6$	0.9979	50–3.12	0.39	1.56
Campesterol	$y = 1008.1x + 1213.6$	0.9987	50–1.56	0.39	1.56
β-Sitosterol	$y = 1718.1x + 1316.6$	0.9991	50–1.56	0.39	1.56

The concentrations of tocopherol and phytosterols from *C. papaya* were evaluated. The content of each compound was calculated as μg/100 mg of original leaf. The concentrations of tocopherol and phytosterols were in the range of 0.33 to 1.91 μg/100 mg of original leaf material. β-Sitosterol was present at the highest concentration, followed by DL-α-tocopherol, campesterol, and stigmasterol (Table 4).

Table 4. Concentrations of tocopherol and phytosterols in *Carica papaya* leaves.

Sample	Content (μg/100 mg ± SEM of Original Leaf) n = 3 Independent Extractions			
	DL-α-Tocopherol	Stigmasterol	Campesterol	β-Sitosterol
$scCO_2$ extract	1.07 ± 0.36	0.33 ± 0.10	0.55 ± 0.08	1.91 ± 0.45

3.5. Cytotoxicity of $scCO_2$ Extract and Identified Compounds

The cytotoxic effects of $scCO_2$ extract, DL-α-tocopherol, stigmasterol, campesterol, and β-sitosterol were evaluated against cancerous SCC25 and non-cancerous HaCaT cell lines using the MTT assay. The concentrations used ranged from 1 to 100 μg/mL for the individual compounds and 30 to 500 μg/mL for $scCO_2$ extract. Figures 2 and 3 show SCC25 HaCaT cell viabilities upon exposure to extracts or pure compounds. The $scCO_2$ extract and phytosterols, but not DL-α-tocopherol, showed statistically significant cytotoxic effects against SCC25 and HaCaT cell lines in a dose-dependent manner. Approximately 60% of SCC25 cells survived when exposed to a concentration of 1 μg/mL campesterol or β-sitosterol, while 1 μg/mL stigmasterol only reduced cell viability to 77.1%. Approximately 60% of SCC25 cells survived after treatment with 250 μg/mL of the $scCO_2$ leaf juice extract. Interestingly, when comparing the selectivity of the phytosterols at 1 μg/mL, stigmasterol resulted in the most significant difference in viabilities between the two cell lines (13.4% difference), whereas campesterol and β-sitosterol demonstrated statistically significant selectivity, but with only 3.4% and 5.6% differences, respectively, at the same concentrations. On the other hand, the $scCO_2$ extract showed significant selectivity between the two cell lines ($p < 0.001$, two-way ANOVA) with the post hoc test comparison showing the significance at each concentration tested.

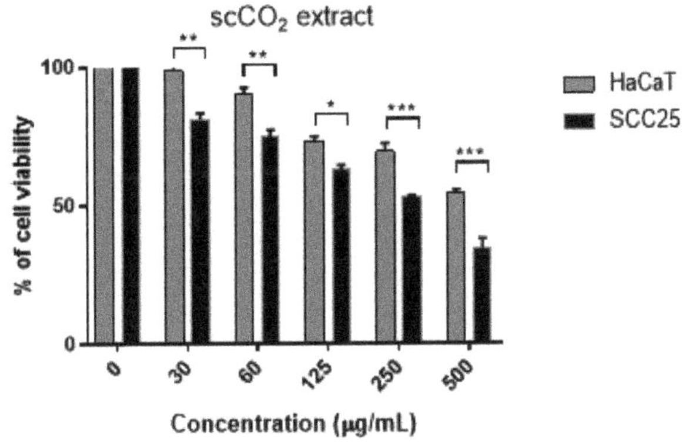

Figure 2. Effect of $scCO_2$ extract on the viability of SCC25 and HaCaT cell lines. Results are expressed as means ± standard error of the mean (SEM) ($n = 3$ independent experiments). Statistical significance was determined by two-way ANOVA with the Sidak post hoc test, comparing the survival of HaCaT vs, scc25 (* $p < 0.05$, ** $p < 0.01$, *** $p < 0.001$).

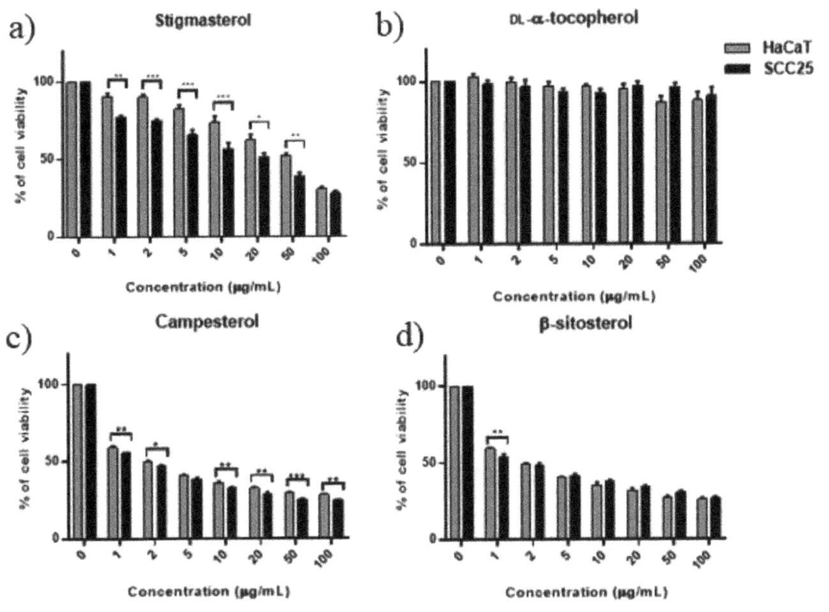

Figure 3. Effect of stigmasterol (**a**), DL-α-tocopherol (**b**), campesterol (**c**), and β-sitosterol (**d**) on the cell viabilities of SCC25 and HaCaT cell lines. Results are expressed as means ± SEM ($n = 3$ independent experiments). Statistical significance was determined by two-way ANOVA with the Sidak post hoc test, comparing the survival of HaCaT vs. SCC25 (* $p < 0.05$, ** $p < 0.01$, *** $p < 0.001$).

Table 5 summarizes the IC_{50} values of the compounds and $scCO_2$ extract toward each cell line. Campesterol was very active but not selective (had the same IC_{50} against SCC25 and HaCaT cells).

Stigmasterol and β-sitosterol exhibited some selectivity toward SCC25, and stigmasterol was less toxic than the other two sterols tested. The $scCO_2$ extract was cytotoxic against SCC25 and HaCaT cell lines at concentrations of 88.07 to 120.60 μg/mL, respectively, confirming its selectivity.

Table 5. Half maximal inhibitory (IC_{50}) values of tested compounds and extract for SCC25 and HaCaT cells.

Cell Lines	IC_{50} (95% Confidence Interval)								
	DL-α-Tocopherol	Stigmasterol		Campesterol		β-Sitosterol		Equimolar Sterol * Mixture	$scCO_2$ Extract
		μg/mL	μM	μg/mL	μM	μg/mL	μM	μM	
SCC25	ND	14.35 (11.8–18.2)	34.8	0.63 (0.51–0.74)	1.57	0.41 (0.2–0.56)	0.98	5.98 (3.9–8.1)	88.07 (67.9–111)
HaCaT	ND	23.41 (19.1–28.7)	57.7	0.63 (0.53–0.73)	1.57	0.78 (0.65–0.91)	1.88	15.60 (11.2–21.6)	120.60 (95.7–152.3)

* Equal concentrations of stigmasterol, campesterol, and β-sitosterol. ND means not detected. Parentheses show the 95% confidence interval for each compound.

3.6. *Effect of Combined Phytosterols against SCC25 and HaCaT Cell Lines*

To determine if there was a synergistic interaction between the compounds of interest, the cytotoxic effects of an equimolar mixture of stigmasterol, campesterol, and β-sitosterol on SCC25 and HaCaT cell lines were evaluated at a range of concentrations from 2.5 to 250 μM (equivalent to 1–100 μg/mL). In this experiment, 1 μM of the mixture represents a 0.333 μM concentration of each individual sterol. Figure 4 reveals the effects of the mixture against SCC25 and HaCaT cell lines. The mixture of phytosterols showed cytotoxicity with an IC_{50} of 5.98 and 15.60 μM toward SCC25 and HaCaT cells, respectively. The selectivity of the mixture was the highest at a concentration of 5 μM with 89% viable HaCaT and 68% viable SCC25 cells (** $p < 0.01$, two-way ANOVA with the Sidak post hoc test). Two-way ANOVA analysis indicated that, overall, the treatment effect was different between the two cell lines ($p < 0.0001$).

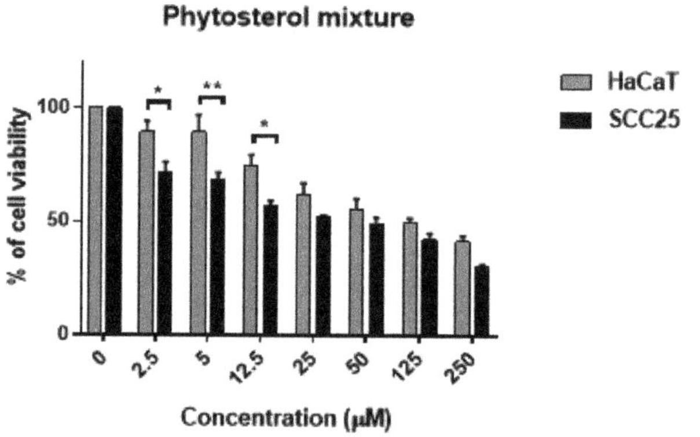

Figure 4. Effect of an equimolar mix of the three phytosterols on SCC25 and HaCaT cell lines. Statistical significance was determined by two-way ANOVA with the Sidak post hoc test, comparing the survival of HaCaT vs. SCC25 (* $p < 0.05$, ** $p < 0.01$).

The IC_{50} of the mixture of phytosterols indicates that the mixture of compounds is not more potent than campesterol or β-sitosterol on either cell line (Table 5), thereby excluding significant synergistic effects. It is interesting to note, however, that, with an IC_{50} ~3 times lower on cancer cells than non-cancer cells, the mixture showed better selectivity than any of the individual sterols or the extract.

4. Discussion

Several groups reported the potential anticancer properties of *C. papaya* leaves prepared by different extraction methods [12,17,18]. However, the phytochemicals responsible for cytotoxicity and selectivity are yet to be identified. An orthogonally different extraction method such as supercritical CO_2 extraction method is necessary to ease the identification of compounds of interest from extracts of simplified composition [19]. There are no scientific studies in the literature which characterize the $scCO_2$ extracts from freeze-dried leaf juice of *C. papaya* and, therefore, the present study provides the first insights into the chemotherapeutic potential of identified compounds from $scCO_2$ extract from freeze-dried leaf juice of *C. papaya*.

Among the phytosterols confirmed to be present in the $scCO_2$ extract in our study (stigmasterol, campesterol, and β-sitosterol), β-sitosterol was found to be the most abundant compound, followed by campesterol and stigmasterol. This quantitative result is consistent with the phytosterol biosynthesis pathway, which suggests that the most abundant end-product of plant sterol synthesis is β-sitosterol, followed by campesterol and stigmasterol [20].

The search for potential chemotherapeutic drugs involves screening for compounds selectively cytotoxic toward cancerous cells, sparing non-cancerous cells. In the present study, although campesterol and β-sitosterol were potently cytotoxic to SCC25 (with ~1 μM IC_{50}), only stigmasterol demonstrated statistically and biologically significant selectivity over a range of concentrations. For campesterol, the selective cytotoxicity is statistically significant but of an amplitude unlikely to be of biological relevance. By comparing the structural features of all three phytosterols, a double bond in the side chain (C-22) of stigmasterol might be hypothesized to selectively affect cancer cells; however, more data are needed to confirm this claim. Another study showed that stigmasterol demonstrates chemo-preventive properties against dimethylbenz[a]anthracene-induced carcinoma in a mouse model at concentrations of 200 mg/kg and 400 mg/kg body weight [21]. This supports the further evaluation of stigmasterol as a therapeutic agent, as well as the possibility that the traditional use of papaya leaves may be derived from such activity.

While a previous study showed that phenoside A was more cytotoxic toward non-cancerous HaCaT cells, this study for the first time proves that the cytotoxicity and selectivity of the *C. papaya* leaf extract was due to phytosterol derivatives. However, more studies are needed to discover potential compounds that possess better selectivity toward SCC25. Some strategies including dereplication and single-compound purification from the crude extract may lead to fruitful outcomes. The scaffold of stigmasterol is for medicinal chemists to investigate its structural activity relationship and selectivity. Some of the successful cases for semi-synthetic compounds including dihydroartemisinin (arteminisin analogue) are used together with holotranferrin to treat breast cancer [22]. Interestingly, the combination of sterols demonstrated greater selectivity against SCC25 than individual sterols. This study paves the way for further in vivo study for future chemotherapy purposes and important clinical implications.

A synergistic effect is the result of the interaction of multiple agents exerting a greater effect than the total of their individual effects [23]. For example, a mixture of berberine and evodiamine demonstrated greater inhibition of the human hepatocellular carcinoma cells SMMC-7221 than single treatment by berberine or evodiamine [24]. We evaluated the possibility that the three most abundant sterols could act synergistically in selectively impairing cancer cell viability. The results suggest that no benefit results from combining equimolar concentrations of β-sitosterol, campesterol, and stigmasterol in terms of half maximal effective concentration (EC_{50}). However, it is interesting to note that the mixture showed better selectivity than each compound applied individually. A study by Csupor-Löffler and co-workers reported that the mixture of β-sitosterol plus stigmasterol (at an unknown ratio) from the roots of *Conyza canadensis* [22] was five times more cytotoxic against skin carcinoma (A431) cells than non-cancerous human fetal fibroblasts (MRC-5). The IC_{50} values of each individual compound against A431 and MRC-5 were not determined.

5. Conclusions

In this study, four compounds from $scCO_2$ freeze-dried leaf juice extract were identified by UPLC–QToF-MS-based chemometric and GC–MS analysis. Phytosterols were reported for the first time from papaya freeze-dried leaf juice for their cytotoxic activities against skin cancer. Stigmasterol was selective toward cancerous SCC25 cells in comparison to non-cancerous HaCaT cells. β-Sitosterol was found to be the most abundant, followed by campesterol and stigmasterol. The present study provides evidence for further studies on the mechanism of action and in vivo efficacy of papaya leaf extracts in skin cancer.

Author Contributions: J.R.F., M.-O.P., P.N.S., and K.-Y.K. conceptualized the study; K.-Y.K. performed the experiments, analyzed the data, and wrote the manuscript; J.R.F., M.-O.P., S.P., and P.N.S. critically reviewed the manuscript. All authors have read and agreed to the published version of the manuscript.

Funding: This research was not funded.

Acknowledgments: We wish to thank Tropical Fruit World Pty Ltd., 29 Duranbah Road, Duranbah, NSW 2487 for providing fresh papaya leaves (website: tropicalfruitworld.com.au). We would like to thank Elvis Chua for his research contribution toward the GC–MS analysis.

Conflicts of Interest: The authors declare no conflicts of interest.

Appendix A

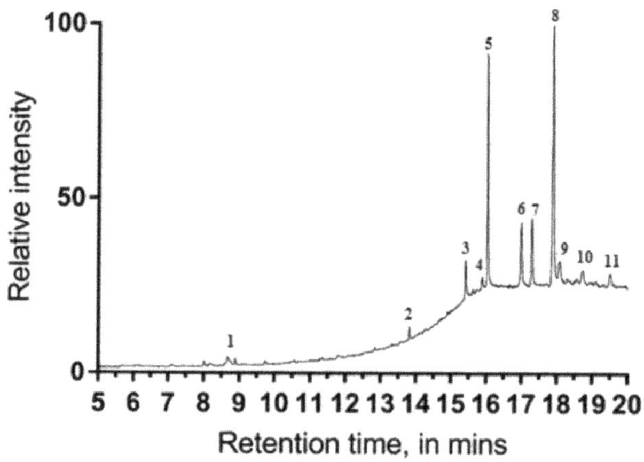

Figure A1. GC–MS analysis of $scCO_2$ extract.

References

1. Antunes Carvalho, F.; Renner, S.S. A dated phylogeny of the papaya family (Caricaceae) reveals the crop's closest relatives and the family's biogeographic history. *Mol. Phylogenet. Evol.* **2012**, *65*, 46–53. [CrossRef] [PubMed]
2. O'Hare, T.J.; Williams, D.J. Papaya as a Medicinal Plant. In *Genetics and Genomics of Papaya*; Ming, R., Moore, P.H., Eds.; Springer: New York, NY, USA, 2014; pp. 391–407. [CrossRef]
3. Nguyen, T.T.T.; Shaw, P.N.; Parat, M.-O.; Hewavitharana, A.K. Anticancer activity of *Carica papaya*: A review. *Mol. Nutr. Food Res.* **2013**, *57*, 153–164. [CrossRef] [PubMed]
4. Ahmad, N.; Fazal, H.; Ayaz, M.; Abbasi, B.H.; Mohammad, I.; Fazal, L. Dengue fever treatment with *Carica papaya* leaves extracts. *Asian Pac. J. Trop. Biomed.* **2011**, *1*, 330–333. [CrossRef]
5. Gunjan, M.; Karna, L.; Dayalan, K.; Sasigaran, P. A review and search of phytomedicine used by traditional people in Malaysia (Ipoh, Perak). *Int. J. Phytoterapy Res.* **2012**, *2*, 26–41.
6. Kapoor, S.; Saraf, S. Topical herbal therapies an alternative and complementary choice to combat acne. *Res. J. Med. Plant* **2011**, *5*, 650–669. [CrossRef]

7. Lim, T.K. *Carica papaya, Edible Medicinal and Non-Medicinal Plants: Volume 1, Fruits*; Springer: Dordrecht, The Netherlands, 2012; pp. 693–717. [CrossRef]
8. Pandey, S.; Cabot, P.J.; Shaw, P.N.; Hewavitharana, A.K. Anti-inflammatory and immunomodulatory properties of *Carica papaya*. *J. Immunotoxicol.* **2016**, *13*, 590–602. [CrossRef]
9. Vij, T.; Prashar, Y. A review on medicinal properties of *Carica papaya* Linn. *Asian Pac. J. Trop. Dis.* **2015**, *5*, 1–6. [CrossRef]
10. Boeing, H.; Bechthold, A.; Bub, A.; Ellinger, S.; Haller, D.; Kroke, A.; Leschik-Bonnet, E.; Müller, M.J.; Oberritter, H.; Schulze, M.; et al. Critical review: Vegetables and fruit in the prevention of chronic diseases. *Eur. J. Nutr.* **2012**, *51*, 637–663. [CrossRef]
11. Nguyen, T.; Parat, M.-O.; Hodson, M.; Pan, J.; Shaw, P.; Hewavitharana, A. Chemical Characterization and in vitro Cytotoxicity on Squamous Cell Carcinoma Cells of *Carica Papaya* Leaf Extracts. *Toxins* **2016**, *8*, 7. [CrossRef]
12. Nguyen, T.T.; Parat, M.-O.; Shaw, P.N.; Hewavitharana, A.K.; Hodson, M.P. Traditional Aboriginal Preparation Alters the Chemical Profile of *Carica papaya* Leaves and Impacts on Cytotoxicity towards Human Squamous Cell Carcinoma. *PLoS ONE* **2016**, *11*, e0147956. [CrossRef]
13. Gogna, N.; Hamid, N.; Dorai, K. Metabolomic profiling of the phytomedicinal constituents of *Carica papaya* L. leaves and seeds by ^{1}H NMR spectroscopy and multivariate statistical analysis. *J. Pharm. Biomed. Anal.* **2015**, *115*, 74–85. [CrossRef] [PubMed]
14. Lieb, V.M.; Esquivel, P.; Cubero Castillo, E.; Carle, R.; Steingass, C.B. GC–MS profiling, descriptive sensory analysis, and consumer acceptance of Costa Rican papaya (*Carica papaya* L.) fruit purees. *Food Chem.* **2018**, *248*, 238–246. [CrossRef] [PubMed]
15. Melariri, P.; Campbell, W.; Etusim, P.; Smith, P. Antiplasmodial Properties and Bioassay-Guided Fractionation of Ethyl Acetate Extracts from *Carica papaya* Leaves. *J. Parasitol. Res.* **2011**, *2011*, 104954. [CrossRef] [PubMed]
16. Khaw, K.-Y.; Parat, M.-O.; Shaw, P.N.; Nguyen, T.T.T.; Pandey, S.; Thurecht, K.J.; Falconer, J.R. Factorial design-assisted supercritical carbon-dioxide extraction of cytotoxic active principles from *Carica papaya* leaf juice. *Sci. Rep.* **2019**, *9*, 1716. [CrossRef] [PubMed]
17. Pandey, S.; Walpole, C.; Cabot, P.J.; Shaw, P.N.; Batra, J.; Hewavitharana, A.K. Selective anti-proliferative activities of *Carica papaya* leaf juice extracts against prostate cancer. *Biomed. Pharmacother.* **2017**, *89*, 515–523. [CrossRef]
18. Otsuki, N.; Dang, N.H.; Kumagai, E.; Kondo, A.; Iwata, S.; Morimoto, C. Aqueous extract of *Carica papaya* leaves exhibits anti-tumor activity and immunomodulatory effects. *J. Ethnopharmacol.* **2010**, *127*, 760–767. [CrossRef]
19. Khaw, K.-Y.; Parat, M.-O.; Shaw, P.N.; Falconer, J.R. Solvent Supercritical Fluid Technologies to Extract Bioactive Compounds from Natural Sources: A Review. *Molecules* **2017**, *22*, 1186. [CrossRef]
20. Hartmann, M.-A. 5 Sterol metabolism and functions in higher plants. In *Lipid Metabolism and Membrane Biogenesis*; Daum, G., Ed.; Springer: Berlin/Heidelberg, Germany, 2004; pp. 183–211. [CrossRef]
21. Ali, H.; Dixit, S.; Ali, D.; Alqahtani, S.M.; Alkahtani, S.; Alarifi, S. Isolation and evaluation of anticancer efficacy of stigmasterol in a mouse model of DMBA-induced skin carcinoma. *Drug Des. Dev. Ther.* **2015**, *9*, 2793–2800. [CrossRef]
22. Csupor-Löffler, B.; Hajdú, Z.; Zupkó, I.; Molnár, J.; Forgo, P.; Vasas, A.; Kele, Z.; Hohmann, J. Antiproliferative Constituents of the Roots of Conyza canadensis. *Planta Med.* **2011**, *77*, 1183–1188. [CrossRef]
23. van Vuuren, S.; Viljoen, A. Plant-Based Antimicrobial Studies—Methods and Approaches to Study the Interaction between Natural Products. *Planta Med.* **2011**, *77*, 1168–1182. [CrossRef]
24. Wang, X.N.; Han, X.; Xu, L.N.; Yin, L.H.; Xu, Y.W.; Qi, Y.; Peng, J.Y. Enhancement of apoptosis of human hepatocellular carcinoma SMMC-7721 cells through synergy of berberine and evodiamine. *Phytomedicine* **2008**, *15*, 1062–1068. [CrossRef] [PubMed]

processes

MDPI

Review

Mechanism of Particle Agglomeration for Single and Multi-Nozzle Atomization in Spray Drying: A Review

Ireneusz Zbicinski *, Krzysztof Ciesielski and Bangguo Ge

Department of Environmental Engineering, Lodz University of Technology, 213 Wolczanska Str., 93-005 Lodz, Poland; krzysztof.ciesielski@p.lodz.pl (K.C.); bangguo.ge@dokt.p.lodz.pl (B.G.)
* Correspondence: ireneusz.zbicinski@p.lodz.pl

Abstract: This paper reviews experimental works on the effects of single nozzle location and multi-nozzle atomization on the mechanism of particle agglomeration in spray drying. In addition to the naturally occurring primary agglomeration, forced and secondary agglomeration is observed as an effect of different nozzle positions or multiple-nozzle atomization in spray drying. Particle size diameters in the spray drying process for atomization from a single nozzle located at the top of the tower are larger than at the bottom of the tower because of the lower ambient air temperatures and longer residence time in the agglomeration zone. The trend of reduction in particle size is observed in all analyzed works when the nozzle is moved down towards the air inlet, due to droplets' exposure to higher air temperatures and shorter residence time in the drying chamber. Conditions of droplet–droplet, dry–dry or sticky–dry collisions leading to the development of coalescence, agglomeration and rebound zones for multiple-nozzle atomization are described and discussed. Typically, log normal PSD was found for single-nozzle spraying whereas for multi nozzle arrangement, bi-modal particle size distribution was found both for drying in lab and industrial scale.

Keywords: spray drying; agglomeration; multiple nozzles; single nozzle location

Citation: Zbicinski, I.; Ciesielski, K.; Ge, B. Mechanism of Particle Agglomeration for Single and Multi-Nozzle Atomization in Spray Drying: A Review. *Processes* **2022**, *10*, 727. https://doi.org/10.3390/pr10040727

Academic Editors: Roberta Campardelli and Paolo Trucillo

Received: 28 March 2022
Accepted: 7 April 2022
Published: 9 April 2022

1. Introduction

One of the common methods for producing dry loose material without prior grinding is spray drying. Spray drying involves atomization of a solution, slurry, or paste to form the droplets and contact with hot drying medium, usually air. In spray drying chambers, the feed is converted into micro-sized droplets via atomization, which facilitates a robust heat and mass transfer. Spray drying is a fast (10 sec-1min) convective drying process, in which hot air provides energy for evaporation of solvent in the form of liquid drops [1]. Over the years, spray drying has evolved as an industry-friendly drying technology with a wide range of applications in many sectors, including the food, pharmaceutical, and chemical industries [2].

Producing a powder with controlled particle size distribution (PSD) as well as particle shape and structure is still of paramount interest in the dairy, instant food, pharmaceutical, and detergent industries, and even in the production of medical and biological products [3,4].

The most common configuration of spray drying is co-current and counter-current process. In the co-current spray towers, both phases flow through the chamber in the same direction (downward), which is suitable for temperature-sensitive products. In the counter-current towers, the spray and the air are introduced at opposite ends of the dryer, with the atomizer positioned at the top and the air entering at the bottom [5]. Counter-current spray dryers offer high throughput, control of product bulk density, and an option for application of multilevel atomization [6].

Particles produced by spray drying can be smaller than 50 μm in diameter, leading to poor flowability and slower reconstitution or lump formation during rehydration, affecting the quality of products. To overcome these problems, the agglomeration process can be applied to powders produced by the spray drying method. The required particle size can be achieved by particle agglomeration and control of particle size and properties, which is commonly implemented in the industry. There are various methods for particle agglomeration: direct agglomeration during drying [7–9], fluid-bed agglomeration [10], steam-jet agglomeration, and different pressure agglomeration processes [3].

Many theoretical attempts have been made to determine mechanism of the agglomeration process, such as by direct modelling [11–13], stochastic models [14–16], and particle populations balance models [17,18]; however, only a few agglomeration models have been validated due to difficulties in the industrial scale measurements and a lack of data on material properties (glass transition or stickiness), tower geometry, and flow hydrodynamics.

This paper reviews experimental works and experimentally validated theoretical models on the effect of nozzle system configuration on the mechanism of particle agglomeration in spray drying.

2. Agglomeration in Spray Drying

2.1. Study on Particle Agglomeration Mechanism in Spray Drying

Particle agglomeration is a complex and rapid process in spray drying affected by many factors, such as drying temperature, hydrodynamics of phases, material characteristics, and the configuration of the nozzles system [19].

Agglomeration can be described as the sticking of particulate solids, which is caused by short-range physical or chemical forces among the particles because of physical or chemical modifications of the surface of the solid [20]. Agglomeration can be defined as a process during which primary particles stick together so that bigger porous secondary particles (conglomerates) are formed, in which the original particles are still identifiable, Figure 1. In this way, the characteristics of a single particle are maintained, whereas the bulk powder properties are improved by the development of the larger agglomerates [19].

Agglomeration in the spray dryer chamber results from collisions between two or more particles [21]. With drying, stable solid bridges between particles are developed, leading to formation of a new bigger structure. Not every particle contact results in agglomeration; at least one of the colliding particles should have a sticky surface for adhesion [22]. To some extent, both primary and secondary agglomeration occur naturally in the spray chamber through random collisions of moving particles in the chamber [23].

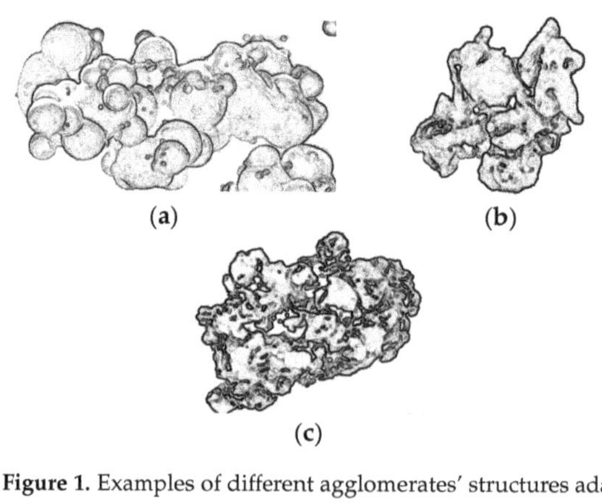

Figure 1. Examples of different agglomerates' structures adapted from Refs. [19,24,25].

When the surface reaches the sticky state, collisions with other particles (sticky or dry) could lead to agglomeration, depending on velocity, particle momentum, collision angle, and time of contact between particles [26,27].

Extensive analysis of the agglomeration mechanism in industrial spray drying towers was elaborated by Huntington [28]. The author distinguished the following mechanisms of agglomeration: droplet/droplet agglomeration within a spray cone, droplet/semi-dried particle, and particle/particle, taking place during particle flow as well as on the dryer wall. Huntington [28] claims that in counter-current industrial swirl dryers, agglomeration increases with increasing tower load, where the mass median diameter (MMD) grows by a factor of about three from droplet to the powder, whereas for nil-swirl, lightly loaded towers, MMD may double. Huntington, Ref. [28] concludes that to prevent production of excessive coarse product agglomeration process in spray drying, towers must be controlled.

Recently, Fröhlich [27], analyzed the effect of solid content in the feed solution on agglomerate formation. Drying of maltodextrin in pilot plant spray dryer showed that an increase in total solid content from 30 to 50% (w/w) resulted in an increased particle size from 44.7 to 64.6 μm. The authors attribute this effect to a changed particle viscosity, as increased solid content resulted in less coalescence of the agglomerates, concluding that the agglomeration in the atomization zone is controlled air hydrodynamics in the chamber.

Control of particle size and the agglomeration process can be achieved by proper configuration of the atomizing nozzles producing droplets [29]. In addition to the naturally occurring primary agglomeration, forced, secondary agglomeration is observed in many industrial applications as an effect of using different nozzle positions or multiple-nozzle atomization [9].

2.2. *Particle Agglomeration and Product Quality*

In the process of spray drying, particle agglomeration changes the structure and size of the particles (spherical or semi-spherical), thus affecting some properties of the final product, such as bulk density, porosity, solubility, and fluidity. Refs. [8,24], investigated agglomeration in spray drying towers for different numbers and locations of the nozzles to study and understand the mechanism of the particle agglomeration process. The agglomerates themselves evolve from grapelike structures at the finest fractions to complex shapes for large sizes (Figure 1). Through the agglomeration process, the primary particles become granule-shaped with improved dissolution characteristics, which quickly re-hydrate because of irregular structure and high porosity [3,22]. After agglomeration, the morphology of the particles is substantially changed, which improves the properties of the powder's functionalities (flow properties, dust generation, explosion risks, storage, mixing capacity, dispersion, solubility, and controlled release).

3. Effect of Nozzle System Configuration on Particle Agglomeration in Spray Drying

The rate of agglomeration and the size of the resulting agglomerates, in addition to the physicochemical properties of the atomized solution, are also influenced by the operating process parameters, the geometry of the tower, the method of phase contact (co-current, counter-current, or mixed) [30], and the configuration of the nozzle system. In this review, we focus on the analysis of the effect of nozzle arrangement for particle agglomeration in spray drying.

To determine of the influence of the nozzle system configuration on particle agglomeration more comprehensively, two aspects should be considered: one is the effect of a single nozzle at different locations in the drying chamber on particle agglomeration (nozzle height is generally relative to the hot air inlet), and the other is the effect of applying a multi-nozzle system (two or more nozzles atomizing simultaneously) on particle agglomeration.

3.1. *The Effect of a Single Nozzle Position on Particle Agglomeration in Spray Drying*

The position of the nozzle in a spray tower affects heat transfer and the velocity of the particles, which can influence particle agglomeration.

The first extensive experiments on agglomeration in spray drying to include in situ PDA laser measurements of particle size distribution in a tower were carried out by

Zbicinski and Piatkowski [31], and Piatkowski [32]. The agglomeration process was studied in the counter-current 8 m high and 0.5 m diameter spray drying tower (Figure 2).

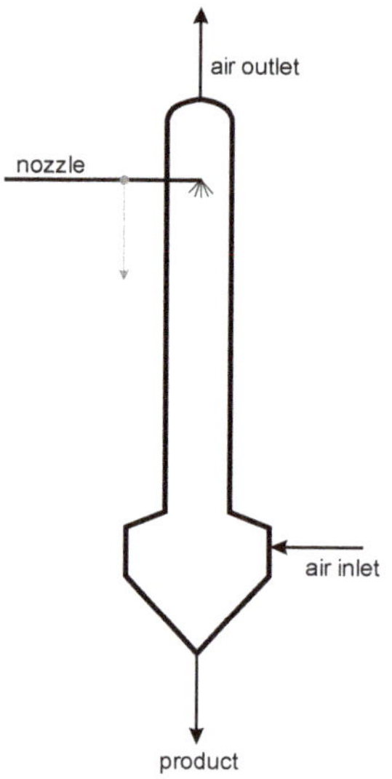

Figure 2. Counter-current spray drying tower.

The position of the nozzle in the tower was changed against the distance of the air inlet, which allowed analyses of the effect of nozzle location on drying and agglomeration. The nozzle was installed in three different positions from the air inlet: 2.4 m, 4.7 m, and 6.7 m.

Figure 3 shows a comparison of average particle diameters along the drying tower for three different nozzle locations and different process parameters; the final average diameter of agglomerates changed from 150 to 450 μm. The most intensive agglomeration and the highest average particle growth was observed when the nozzle was installed at 4.7 m and 6.7 m from the air inlet.

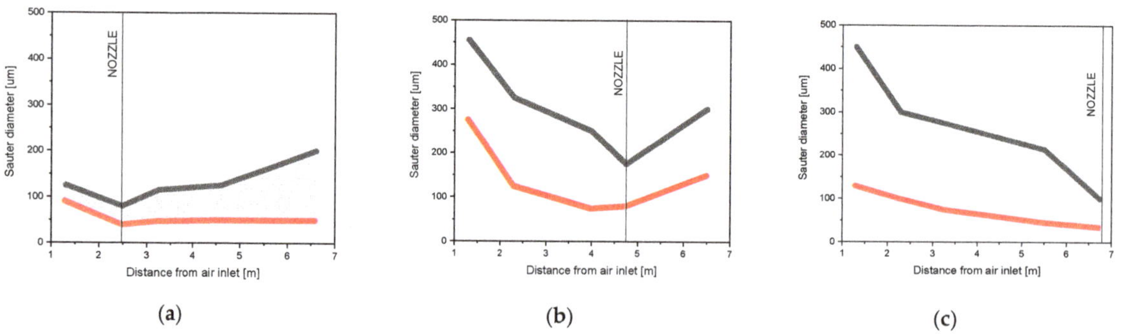

Figure 3. Area of changes of an average particle diameter along the drying tower for different nozzle positions (**a–c**), slurry and airflow rate, swirl flow, atomization ratio, and air temperatures.

Results of the experiments proved that the position of the nozzle in relation to the hot air inlet had a significant influence on the agglomeration process. Phenomena decisive for the course of agglomeration take place in the atomization zone, i.e., the area from the nozzle to about 1.5 m below the nozzle. By changing the location of the nozzle, the temperature and air velocity profiles—which are in direct contact with the sprayed particles—are changed.

For the nozzle installed at 2.4 m from the air inlet, the final particle diameters were relatively small because the high air temperature in the atomization zone hindered agglomeration, and the final particle diameters for these conditions in most cases did not exceed 150 µm. The poor agglomeration of the particles was the result of the rapid evaporation of moisture and the establishment of a low adhesive particle structure, so that contact between the particles did not lead to the formation of the agglomerates.

The displacement of the nozzle from 2.4 m to 4.7 m from the air inlet caused the material to be sprayed into the air with different velocity profiles and, most significant of all, to lower air temperature by about 20 °C. Under these conditions, the agglomeration process was more effective, and the maximum particle diameters for several drying conditions reached approx. 450 µm and were even 10 times larger than the average initial droplet diameters released from the atomizer.

Another change of the nozzle position (to the level of 6.7 m from the air inlet) did not significantly change the size of agglomerates; an average diameter size obtained under these conditions was up to 400 µm.

PDA analysis showed negative value of the particle velocity in the drying chamber, which confirmed the existence of particle agglomeration recycling zones in counter-current spray drying [32].

Similar research was carried out by Francia [8,24], who investigated particle growth in a pilot plant counter-current swirl dryer for detergents, operating with one nozzle at three locations (8.2D, 5.9D, 3.5D). Figure 4 shows a comparison of the product mass-based particle size distribution for different nozzle locations. In each case for single nozzles, mono-modal size distributions were obtained. When the spray nozzle was located at the top position (a distance of 8.2D from the air inlet), the size of the product was significantly higher (between 600 to 850 µm), which implied intensive growth into the large fractions. For the nozzle located at distance of 5.9D from the air inlet, the growth of the size of the product was reduced. The trend of reduction of the particle size continued when the nozzle was moved further down to the lowest distance from the air inlet (3.5D), because of droplets' exposure to higher air temperatures. In a high-temperature environment, the outer crust of droplets was produced earlier, the surface was less prone to stick and with lower residence time, and agglomeration was suppressed. When spraying from higher nozzles locations, due to longer residence time and lower ambient air temperatures, the particle aggregation increases.

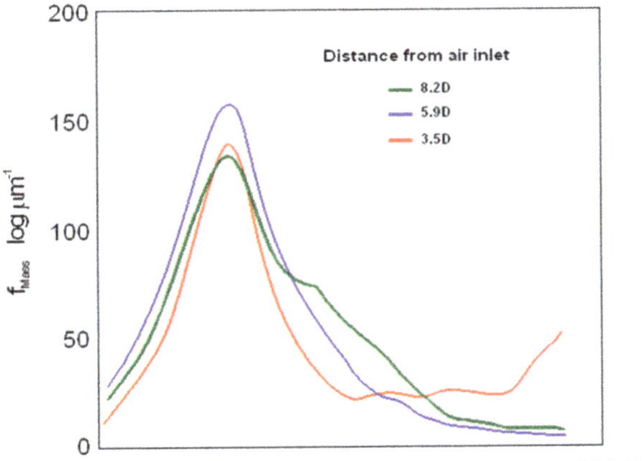

Figure 4. Comparison of the product mass based size distribution obtained under different nozzle locations Ref. [7].

However, when the nozzle was situated at the lowest distance from the air inlet (3.5D), the proportion of particles in the powder with size > 850 μm was larger than at higher distances from the air inlet (>35% of the production, versus 24% and 14%, respectively), which means that the mechanism of particle growth might be different to inter-particle contacts. The authors of [8] suggest that interaction of particles with the wall deposits may lower disruptive stresses of the agglomerates and promote re-entrainment of large granules to the chamber.

Recently, Piatkowski et al. [33] experimentally studied particle size distribution, particle velocities, and air flow pattern in a counter-current pilot plant drying tower (8 m in height and 0.5 m in diameter) above and below the nozzles located at the top and close to the air inlet. A schematic of the pilot plant tower and location of the nozzles and levels of PDA measurements are shown in Figure 5.

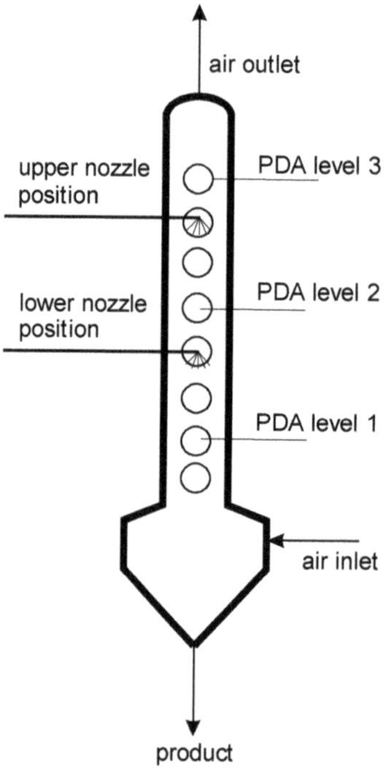

Figure 5. Schematic of pilot plant tower.

PDA measurements carried out below and above the nozzle trace changes in particle size distribution and particle velocities at the axis and near the wall (0.22 m from the axis). Downward air flow in the axis area and upward flow near the wall of the tower cause collisions and promote agglomeration of the particles. Figure 6a,b shows the particle size distribution of the product for the nozzles at the top and the bottom positions in the tower. Figure 6a shows that PSD above the nozzle is smaller, both at the axis (D32 = 25.4 μm) and at the wall (D32 = 78 μm), in comparison with particles measured below the nozzle (D32 = 69.5 μm) and at the wall (D32 = 96.6 μm), which is a result of intensive agglomeration in the recirculation zones below the nozzle, whereas smaller and lighter particles were entrained with upward airflow to the cyclones. The presence of a large number of smaller particles at the axis results from airflow hydrodynamics, as swirl flow in the chamber moves bigger particles to the wall of the tower.

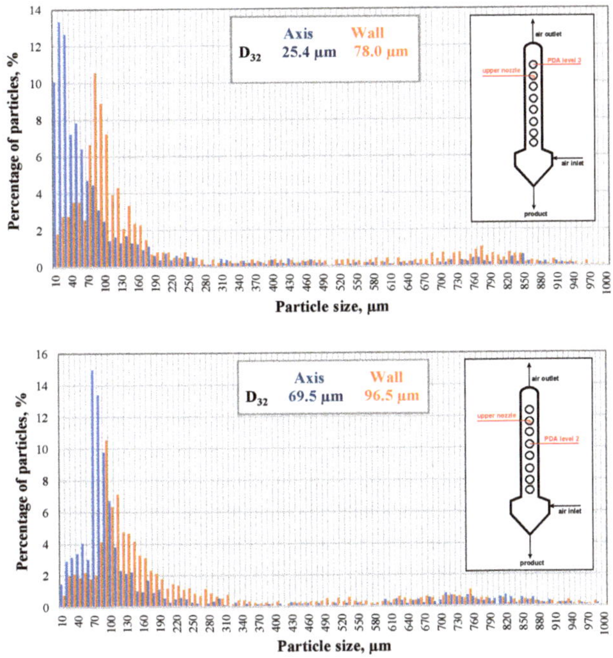

(**a**)

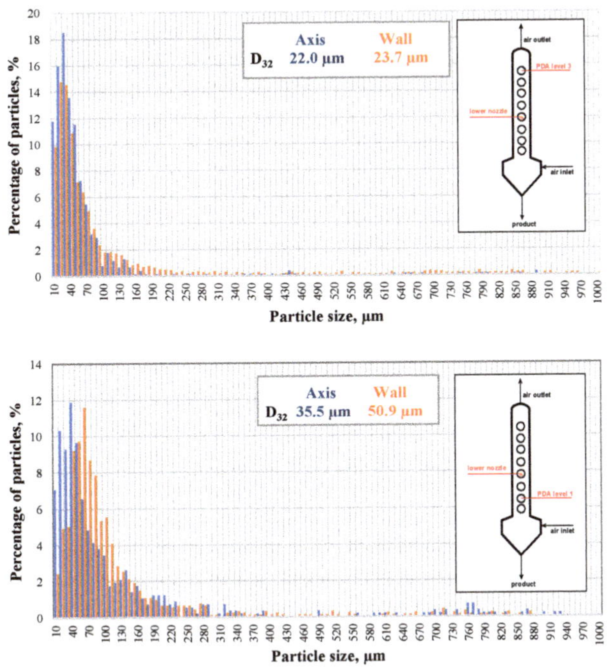

(**b**)

Figure 6. ((**a**) Nozzle at the top (**b**) Nozzle at the bottom) PSD at the axis and at the wall above and below the nozzle (adapted from Ref. [33]).

A similar PSD was observed when the nozzle was situated lower in the dryer, near the air inlet, Figure 6b. The particles above the nozzle are smaller both at the axis (D32 = 22.0 μm) and at the wall (D32 = 23.7 μm) in comparison with particles found below the nozzle (D32 = 35.5 μm) and at the wall (D32 = 50.9 μm).

Comparing PSD of the particles for the nozzle located at the top and at the bottom of the tower, we observe smaller particle diameters at the bottom and bigger for the top nozzle positions. Reduction in particle size when moving the nozzle down to the air inlet is a result of droplets' exposure to higher air temperatures, which hampers the agglomeration process [8,24].

The authors also showed that for a nozzle situated at the top of the dryer, where the air temperature in the atomization zone is lower, intensive air circulation promotes agglomeration of the particles in the form of large uniform structures (Figure 7a). For the lower position of the nozzle (closer to air inlet), the high air temperature in the atomization zone hinders agglomeration in the form of spheres and develops complex 3D structures consisting of many smaller particles (Figure 7b).

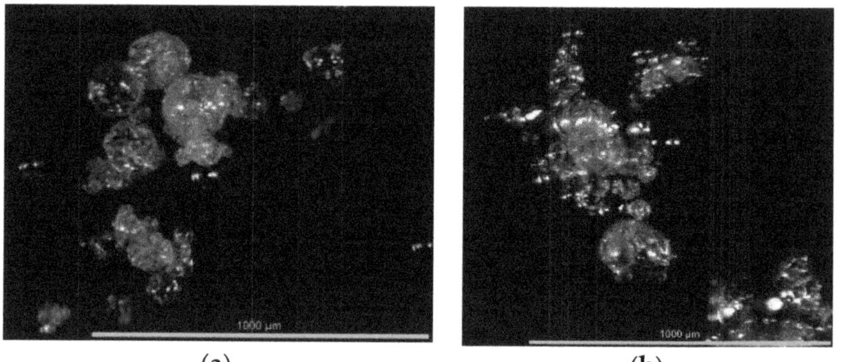

(a) (b)

Figure 7. Product photos from the stereoscopic microscope for the nozzle at the top position (**a**) and bottom position (**b**) (adapted from Ref. [33]).

Distance from the nozzle to the air inlet has a decisive effect on the agglomeration of particles in the counter-current system; nozzle location and appropriate selection of drying process parameters enable control of particle structure in the spray drying process.

3.2. Particle Agglomeration in Multiple-Nozzle System Atomization

The concept of multiple-nozzle spraying means using single nozzles on different levels or several nozzles on each level in the drying chamber [34]. Only a few researchers have carried out experiments on agglomeration for a multiple-nozzle system; mostly mathematical models are used to simulate particle agglomeration without experimental verification [17,35]

Francia et al. [25] studied agglomeration in a counter-current dryer between multiple spraying levels. In contrary to the application of single nozzles, a multi-nozzle arrangement generates bi-modal particle size distributions (PSD). Figure 8 presents product size distributions between 300–425 μm, with a typical log-normal shape for single-nozzle operation. The use of two sprays (3.5D and 8.2D from the air inlet) generates a bi-modal-shaped PSD, with a second mode between 850–1180 μm.

The authors estimated that inter-nozzle interactions account for the production of >6–11% of agglomerates > 600 μm.

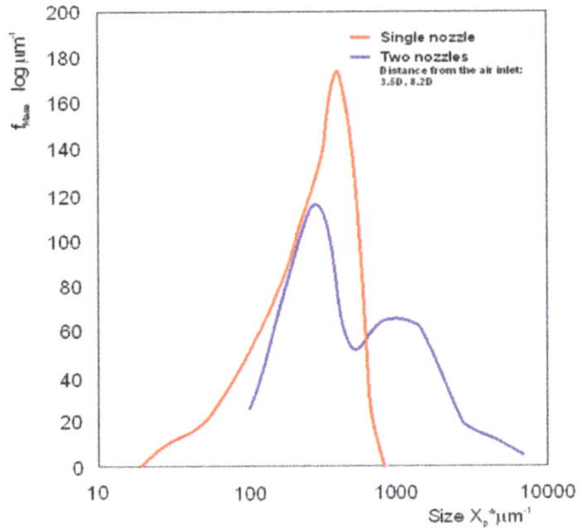

Figure 8. Product size distribution for single- and multi-nozzle operation (3.5D and 8.2D from air inlet).

Francia et al. [8] also examined the effect of air flow conditions (air temperature and velocity) on agglomeration in an industrial swirl spray tower for the drying of detergents, for a single- and double-nozzle arrangement. The authors confirmed that the particle size of powder responds to changes in the air flow in the drying chamber. High air velocity in the chamber makes droplets and particles swirl faster, which moves particles to the wall where the agglomerates are constantly forming and breaking. This mechanism occurs independently of dryer throughput and nozzle location, both for single and multiple nozzles, and inhibits the development of coarse aggregates > 850 µm.

Wawrzyniak et al. [36] applied PDA measurements to determine the effect of spray direction of two nozzles on the final properties of the products in a pilot plant drying tower (Figure 5). The upper nozzle was installed 5.05 m above the air inlet and sprayed downwards, whereas the lower nozzle at 2.8 m above the air inlet sprayed in two directions: downwards (Test A) and upwards (Test B) (Figure 9). The results confirmed bi-modal PSD for both configurations of the nozzles spraying in downward and upward direction (see Figure 8). The drying zone under the lower nozzle and agglomeration zone between the nozzles were detected when the two nozzles sprayed downwards (Test A). Below the lower nozzle, reduced agglomeration was determined, as the slurry was sprayed into the hot air, which made the particles less prone to stick, also due to short residence time. Only smaller and lighter particles were flowing upward to the agglomeration zone between the nozzles.

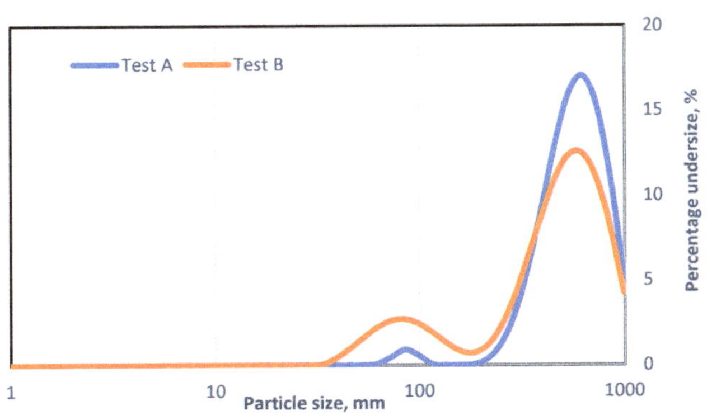

Figure 9. Product particle size distribution for two nozzles spraying downwards (Test A) and for lower nozzle spraying upwards (Test B) (adapted from Ref. [36]).

When the upper nozzle sprayed downward and lower nozzle sprayed upward (Test B), intensive agglomeration between the nozzles was observed. The particles sprayed by the lower nozzle quickly fall into a stream of moist particles from the upper nozzle.

Average powder PSD for Test A and B are similar, D50 = 542 μm and 460 μm, respectively; however, PSD for Test B has a more pronounced two-peak character (Figure 9).

Jubaer et al. [9] studied forced agglomeration to predict coalescence, agglomeration, and elastic rebound zones in a lab-scale counter-current spray dryer with two-nozzle atomization. Figure 10 shows schematic of a lab spray dryer with nozzle locations; the primary (main) nozzle is installed at the top and the secondary nozzles (positions 1 to 3) are inclined perpendicularly or at an angle of 45° to the dryer wall. The arrangements of the of the secondary nozzles simulates fine returns, which is a typical method in the industry for controlling and promoting the agglomeration process [23].

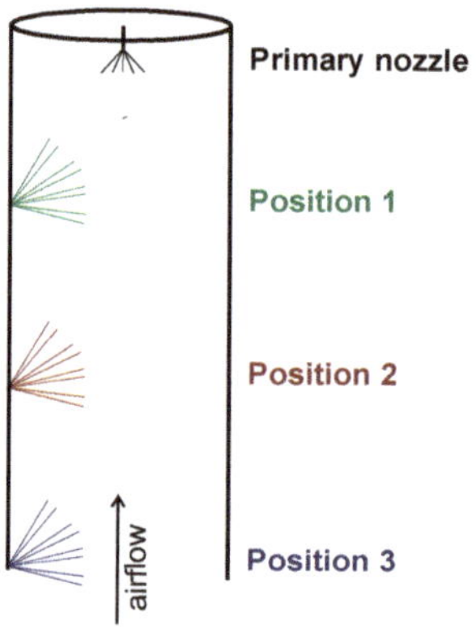

Figure 10. Schematic of lab spray dryer with nozzle locations; primary nozzle at the top and the secondary nozzles (positions 1 to 3, Ref. [9]).

During drying experiments, skim milk with 10% initial solid content was sprayed from primary (top nozzle) and one of the secondary nozzles. To obtain profound picture of agglomeration mechanism, CFD model was developed and validated on the basis of the experimental data of particle size distribution and SEM analyses of the powder morphology.

The trials showed significant differences in particle size and morphology for spray drying with different secondary nozzles locations.

Results of the work are summarized in Figure 11 which shows interaction of sprays developing droplet, sticky and dry zones in the drying chamber for different secondary nozzle configurations (Position 1, 2, 3, [9]). For the position 1 of secondary nozzle droplet-droplet collisions dominates due to short distance between secondary and primary nozzles. In the position 2 sticky particles from both sprays could collide forming sticky-sticky or sticky-dry collision zone (the most effective for agglomeration). For position 3 of the secondary nozzle, where most of the particles are dry; dry-dry or sticky-dry collisions zone occurs. Similar particles behavior was observed for the case when the secondary nozzles were positioned at an upward angle.

The Authors claim that the position 2 of the secondary nozzle located upward produce the most intensive forced agglomeration. Asymmetry introduced by the single secondary injection might affect air temperature and velocity in the chamber and increase the wall deposition.

Cumulative distribution of particle size for experiments with different nozzle configurations are shown in Figure 12. Analysis of Figure 12 confirms size enlargement of the particles for each trail with secondary nozzle atomization; however, the extent of the enlargement is different for different secondary nozzle locations. The most significant size enlargement was obtained when the secondary nozzle was located at Position 2 (both for perpendicular and upward spraying angle) due to the highest probability of sticky–sticky and sticky–dry particle contacts. The particle size distribution at Position 1 is slightly smaller than that of Position 2, as agglomeration there develops only by coalescence. The smallest diameter enlargement was observed for a secondary nozzle situated at Position 3 due to droplets' exposure to higher air temperatures, which is in line with conclusions made by Francia [24]. The most intensive agglomeration was found for nozzle spraying upward at Position 2.

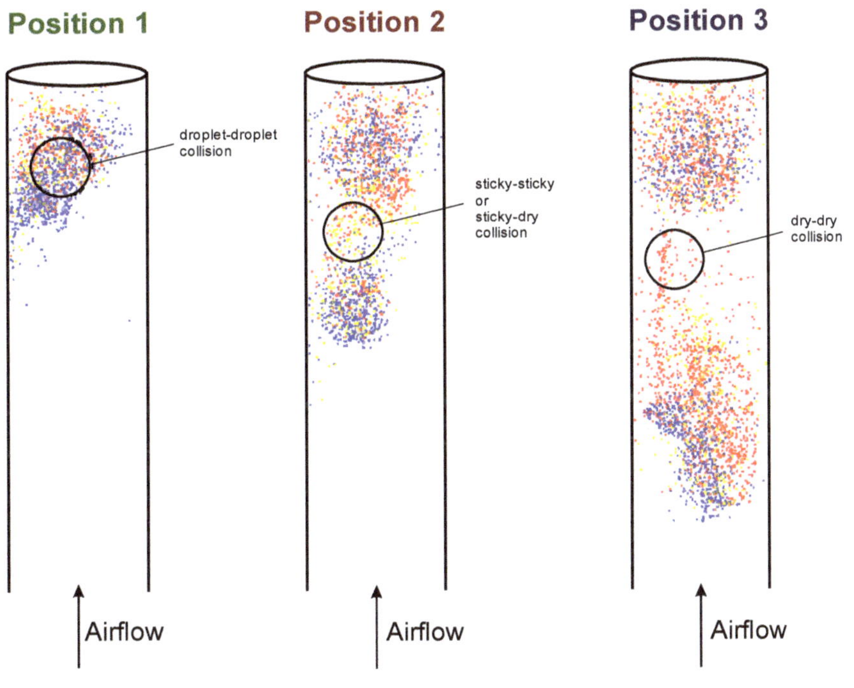

Figure 11. Location of droplet, sticky, and dry collision zones for different nozzle configurations (Position 1, 2, 3, Ref. [9]).

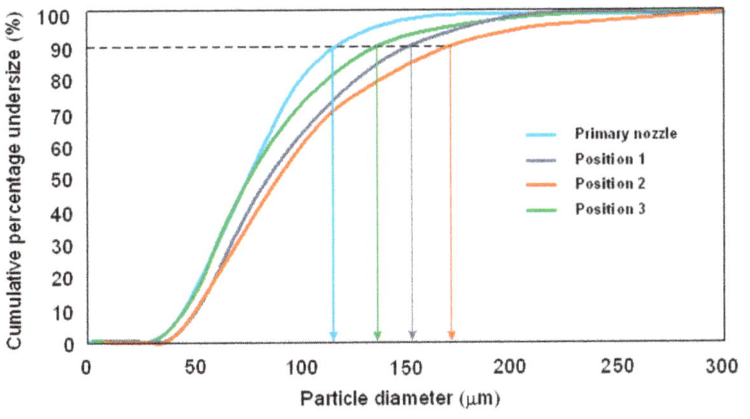

Figure 12. Comparison of particle size distributions of the powder samples collected from the experimental trials for all tested nozzle positions Ref. [9].

Wawrzyniak et al. [37,38] analyzed the agglomeration process in a 37 m high and 6 m diameter nil-swirl industrial spray drying tower for detergent production. The slurry was sprayed by 12 nozzles an angle of 65° located on two levels of 18 m and at 10 m from the hot air inlets (Figure 13).

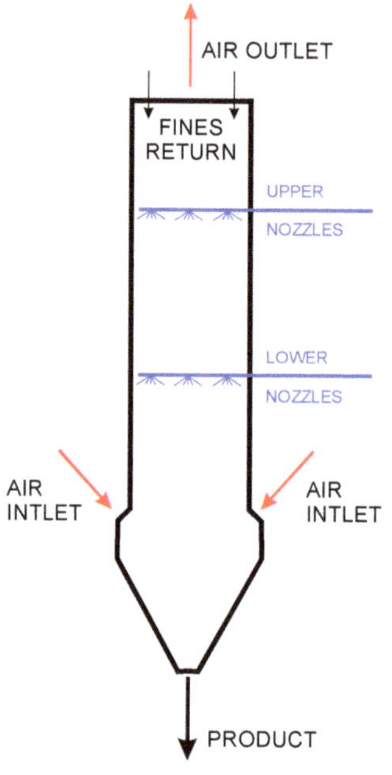

Figure 13. Schematic of industrial spray dryer for detergents with nozzle locations.

The lack of swirl due to the geometry of the tower and construction of the air inlet ring resulted in high instability of air flow in the tower. A robust agglomeration process was observed in the tower, as the initial average particle size increased from 285 µm to the final 770 µm in the product. The initial velocity of particles was 50 m/s. Cumulative initial and final and percentage undersizing PSD in the industrial tower are presented in Figure 14.

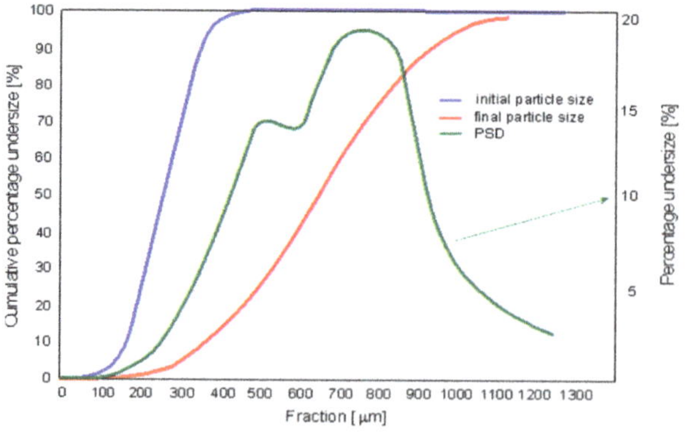

Figure 14. Cumulative initial and final and percentage undersized PSD in the industrial tower.

Experimentally determined percentage undersized PSD shows a bi-modal distribution as in [25,36].

To determine mechanism of agglomeration in the nil-swirl industrial dryer, CFD of particle flow was developed. The model was verified indirectly based on temperature and velocity measurements in the industrial tower using "negative heat source concept" described in [38]. By analyzing the residence time of particles in the drying tower and the diameter growth, the zones of particles agglomeration were determined. Residence time of particles in the dryer was found to be in range from 20 to 60 sec. Average agglomeration time was estimated as 20 sec. Two particle recirculation zones, above the upper and below the lower nozzles level in the dryer were detected (Figure 15). Small particles injected at the upper nozzles level followed air path lines, whereas bigger particles fell with the progress of the agglomeration and leave the dryer.

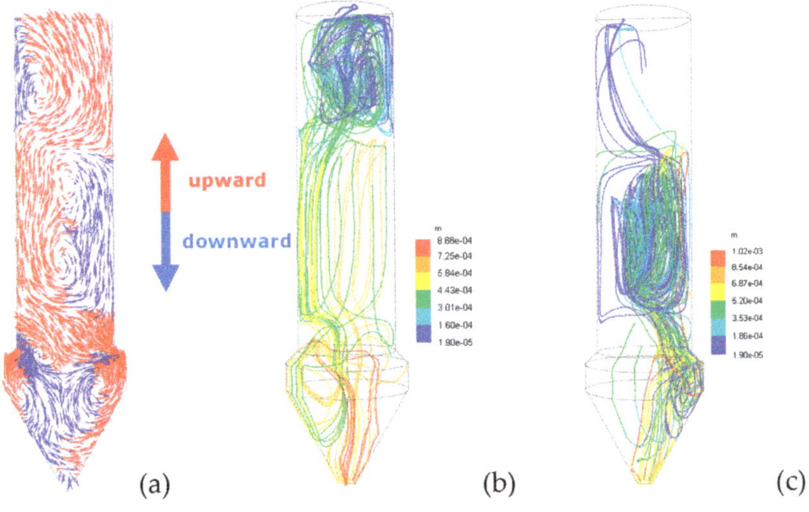

Figure 15. Comparison of drying air flow (**a**) and particle trajectories for upper (**b**) and lower (**c**) nozzles level, Ref. [38].

Diameters of larger fractions increase due to the collisions with wet particles produced by the nozzles from the upper level, but also due to the agglomeration with particles entrained with the air from the lower level. Finally, particles produced at the upper nozzle level are larger than the particles with similar initial diameters released from the lower level because of the longer residence time in the agglomeration zone.

Based on experiments in an industrial tower ([38]), Jaskulski et al. [39] confirmed the mechanism of agglomeration in the industrial tower (Figure 12) using a CFD agglomeration model accounting for collisions of droplets and particles and coupling of agglomeration with heat, mass, and momentum transfer for the industrial tower. Fine and dry fractions were entrained above the atomizing nozzle but did not agglomerate, as the lower moisture content hampered the formation of liquid bridges between the particles during collisions.

Similar conclusions regarding the mechanism of agglomeration were presented by Nijdam et al. [17], who researched the agglomeration process for one and two nozzles pointing toward each other and spraying water. The authors showed that the inertia of a droplet plays an important role in droplet movement in the spray drying chamber. Larger droplets tend to concentrate on the outer part of the spray, maintaining radial momentum further downstream from the nozzle, whereas smaller droplets follow the air flow more closely at the core of the spray. Both turbulent collisions and difference in droplet relative velocity can promote agglomeration. The authors confirmed that besides primary agglomeration, forced and secondary agglomeration occurs as an effect of different nozzle positions.

4. Conclusions

Intensive agglomeration is observed in industrial spray drying towers, which results in a significant increase in the particle diameter of the product. Powder containing small particles, commonly produced in spray drying, is characterized by poor flowability, slower reconstitution, or lump formation during rehydration. Agglomeration changes the morphology of the particles, improving the powder's functionalities such as particle shape, bulk density, porosity, solubility, and fluidity. Most works in the literature examine agglomeration in spray drying and the hydrodynamics of phase contact (co-current, counter-current, or mixed) and the effect of physicochemical properties of the atomized solution, the spray drying process parameters, and method of feed atomization in controlling the rate of agglomeration and size of the agglomerates. The literature review carried out in this paper reveals a strong effect of nozzle location and multi-nozzle atomization on the mechanism of particle agglomeration in spray drying.

The coarse agglomeration mode is mainly generated by spraying from the top of drying towers because of the longer residence time and lower ambient air temperature in the agglomeration zone. Agglomeration growths are inhibited when the nozzles are located near the bottom the tower due to contacting the high inlet air temperature and short residence time, which hampers the formation of liquid bridges between the particles.

For multiple-nozzle atomization, droplet–droplet, dry–dry or sticky–dry particle collisions form coalescence, agglomeration, and rebound zones. Log normal PSD, characteristic for single-nozzle atomization, turns into bi-modal for multi-nozzle arrangements.

Literature analysis proves that atomization from multiple nozzles promotes agglomeration by increasing collision rate, collision frequency and efficiency, increasing the probability of particle growth with the aid of air flow recirculation in the chamber.

In future work, the research of the effect of spray direction of the nozzles on the mechanism of particle agglomeration, final properties of the products and reduction of the wall deposition in co- and counter-current spray drying will be carried out.

Author Contributions: Conceptualization, methodology, resources, writing—original draft preparation, I.Z.; formal analysis, visualization, K.C.; resources, B.G. All authors have read and agreed to the published version of the manuscript.

Funding: This research received no external funding.

Institutional Review Board Statement: Not applicable.

Informed Consent Statement: Not applicable.

Data Availability Statement: Not applicable.

Conflicts of Interest: The authors declare no conflict of interest.

References

1. Gianfrancesco, A.; Turchiuli, C.; Dumoulin, E.; Palzer, S. Prediction of Powder Stickiness along Spray Drying Process in Relation to Agglomeration. *Part. Sci. Technol.* **2009**, *27*, 415–427. [CrossRef]
2. Selvamuthukumaran, M.; Tranchant, C.; Shi, J. Spraying DryingConcept, Application and Its Recent Advances in Food Processing. In *Handbook on Spray Drying Applications for Food Industries*; CRC Press: Boca Raton, FL, USA, 2019; ISBN 978-0-429-05513-3.
3. Palzer, S. Agglomeration of Pharmaceutical, Detergent, Chemical and Food Powders—Similarities and Differences of Materials and Processes. *Powder Technol.* **2011**, *206*, 2–17. [CrossRef]
4. Izonin, I.; Tkachenko, R.; Gregus, M.; Duriagina, Z.; Shakhovska, N. PNN-SVM Approach of Ti-Based Powder's Properties Evaluation for Biomedical Implants Production. *Comput. Mater. Contin.* **2022**, *71*, 5934–5947. [CrossRef]
5. Patel, R.P.; Patel, M.P.; Suthar, A.M. Spray Drying Technology: An Overview. *Indian J. Sci. Technol.* **2009**, *2*, 44–47. [CrossRef]
6. Rahse, W.; Dicoi, O. *Spray Drying in the Detergent Industry*; Conference Proceedings: Dotmund, Germany, 2001; pp. 83–87.
7. Francia, V.; Martín, L.; Bayly, A.E.; Simmons, M.J.H. Particle Aggregation in Large Counter-Current Spray Drying Towers: Nozzle Configuration, Vortex Momentum and Temperature. *Procedia Eng.* **2015**, *102*, 668–675. [CrossRef]
8. Francia, V.; Martín, L.; Bayly, A.E.; Simmons, M.J.H. Agglomeration during Spray Drying: Airborne Clusters or Breakage at the Walls? *Chem. Eng. Sci.* **2017**, *162*, 284–299. [CrossRef]
9. Jubaer, H.; Xiao, J.; Chen, X.D.; Selomulya, C.; Woo, M.W. Identification of Regions in a Spray Dryer Susceptible to Forced Agglomeration by CFD Simulations. *Powder Technol.* **2019**, *346*, 23–37. [CrossRef]

10. Yuksel, H.; Dirim, S.N. Application of the Agglomeration Process on Spinach Juice Powders Obtained Using Spray Drying Method. *Dry. Technol.* **2020**, *39*, 19–34. [CrossRef]
11. Hirt, C.W.; Nichols, B.D. Volume of Fluid (VOF) Method for the Dynamics of Free Boundaries. *J. Comput. Phys.* **1981**, *39*, 201–225. [CrossRef]
12. Mezhericher, M.; Levy, A.; Borde, I. Theoretical Drying Model of Single Droplets Containing Insoluble or Dissolved Solids. *Dry. Technol.* **2007**, *25*, 1025–1032. [CrossRef]
13. Li, M.X. *Analysis and Modeling of Droplet-Droplet Interaction and Particle-Droplet Interaction in Dispersions*; Particles and Process Engineering Department, Bremen University: Bremen, Germany, 2012.
14. Ho, C.; Sommerfeld, M. Modelling of Micro-Particle Agglomeration in Turbulent Flows. *Chem. Eng. Sci.* **2002**, *57*, 3073–3084. [CrossRef]
15. Kim, S.; Lee, D.J.; Lee, C.S. Modeling of Binary Droplet Collisions for Application to Inter-Impingement Sprays. *Int. J. Multiph. Flow* **2009**, *35*, 533–549. [CrossRef]
16. Taskiran, O.O.; Ergeneman, M. Trajectory Based Droplet Collision Model for Spray Modeling. *Fuel* **2014**, *115*, 896–900. [CrossRef]
17. Nijdam, J.J.; Guo, B.; Fletcher, D.F.; Langrish, T.A.G. Challenges of Simulating Droplet Coalescence within a Spray. *Dry. Technol.* **2004**, *22*, 1463–1488. [CrossRef]
18. Tsotsas, E.; Mujumdar, A.S. Modern Drying Technology Vol. 1 Computational Tools at Different Scales. *Dry. Technol.* **2008**, *26*, 812–814. [CrossRef]
19. Verdurmen, R.E.M.; Menn, P.; Ritzert, J.; Blei, S.; Nhumaio, G.C.S.; Sonne Sørensen, T.; Gunsing, M.; Straatsma, J.; Verschueren, M.; Sibeijn, M.; et al. Simulation of Agglomeration in Spray Drying Installations: The EDECAD Project. *Dry. Technol.* **2004**, *22*, 1403–1461. [CrossRef]
20. Dhanalakshmi, K.; Ghosal, S.; Bhattacharya, S. Agglomeration of Food Powder and Applications. *Crit. Rev. Food Sci. Nutr.* **2011**, *51*, 432–441. [CrossRef]
21. Turchiuli, C.; Gianfrancesco, A.; Palzer, S.; Dumoulin, E. Evolution of Particle Properties during Spray Drying in Relation with Stickiness and Agglomeration Control. *Powder Technol.* **2011**, *208*, 433–440. [CrossRef]
22. Palzer, S. Influence of Material Properties on the Agglomeration of Water-Soluble Amorphous Particles. *Powder Technol.* **2009**, *189*, 318–326. [CrossRef]
23. Hazlett, R.; Schmidmeier, C.; O'Mahony, J.A. Approaches for Improving the Flowability of High-Protein Dairy Powders Post Spray Drying—A Review. *Powder Technol.* **2021**, *388*, 26–40. [CrossRef]
24. Francia, V.; Martín, L.; Bayly, A.E.; Simmons, M.J.H. Agglomeration in Counter-Current Spray Drying Towers. Part A: Particle Growth and the Effect of Nozzle Height. *Powder Technol.* **2016**, *301*, 1330–1343. [CrossRef]
25. Francia, V.; Martín, L.; Bayly, A.E.; Simmons, M.J.H. Agglomeration in Counter-Current Spray Drying Towers. Part B: Interaction between Multiple Spraying Levels. *Powder Technol.* **2016**, *301*, 1344–1358. [CrossRef]
26. Gianfrancesco, A.; Turchiuli, C.; Dumoulin, E. Powder Agglomeration during the Spray-Drying Process: Measurements of Air Properties. *Dairy Sci. Technol.* **2008**, *88*, 53–64. [CrossRef]
27. Fröhlich, J.A.; Raiber, T.V.; Hinrichs, J.; Kohlus, R. Nozzle Zone Agglomeration in Spray Dryers: Influence of Total Solid Content on Agglomerate Properties. *Powder Technol.* **2021**, *390*, 292–302. [CrossRef]
28. Huntington, D.H. The Influence of the Spray Drying Process on Product Properties. *Dry. Technol.* **2004**, *22*, 1261–1287. [CrossRef]
29. Guo, B.; Fletcher, D.F.; Langrish, T.A.G. Simulation of the Agglomeration in a Spray Using Lagrangian Particle Tracking. *Appl. Math. Model.* **2004**, *28*, 273–290. [CrossRef]
30. Hussain, F.; Jaskulski, M.; Piatkowski, M.; Tsotsas, E. CFD Simulation of Agglomeration and Coalescence in Spray Dryer. *Chem. Eng. Sci.* **2022**, *247*, 117064. [CrossRef]
31. Zbicinski, I.; Piatkowski, M. Continuous and Discrete Phase Behavior in Countercurrent Spray Drying Process. *Dry. Technol.* **2009**, *27*, 1353–1362. [CrossRef]
32. Piatkowski, M. *Drying Kinetics of Counter-Current Spray Drying*; TUL: Lodz, Poland, 2011. (In Polish)
33. Piatkowski, M.; Wawrzyniak, P.; Jaskulski, M.; Zbicinski, I.; Zawadzki, D. *Experimental Analysis of the Effect of Nozzle Position on Product Properties in Counter-Current Pilot Plant Spray Drying Tower*; Conference Proceedings: Worcester, MA, USA, 2022.
34. Davis, R.P.; Haines, M.S.; Sagel, J.A. Multilevel Spray-Drying Method. U.S. Patent 3,629,951, 28 December 1971.
35. Jaskulski, M.; Wawrzyniak, P.; Zbicinski, I. *CFD Prediction of Powder Particle Size Distribution in the Industrial Scale Spray Drying Process*; Conference Proceedings: Gifu, Japan, 2016.
36. Wawrzyniak, P.; Piatkowski, M.; Jaskulski, M.; Zbicinski, I.; Jubaer, H. *Analysis of Product Properties in Counter-Current Spray Drying with Two Nozzles System*; Conference Proceedings: Worcester, MA, USA, 2022.
37. Wawrzyniak, P.; Podyma, M.; Zbicinski, I.; Bartczak, Z.; Rabaeva, J. Modeling of Air Flow in an Industrial Countercurrent Spray-Drying Tower. *Dry. Technol.* **2012**, *30*, 217–224. [CrossRef]
38. Wawrzyniak, P.; Podyma, M.; Zbicinski, I.; Bartczak, Z.; Polanczyk, A.; Rabaeva, J. Model of Heat and Mass Transfer in an Industrial Counter-Current Spray-Drying Tower. *Dry. Technol.* **2012**, *30*, 1274–1282. [CrossRef]
39. Jaskulski, M.; Wawrzyniak, P.; Zbicinski, I. CFD Simulations of Droplet and Particle Agglomeration in an Industrial Counter-Current Spray Dryer. *Adv. Powder Technol.* **2018**, *29*, 1724–1733. [CrossRef]

processes

MDPI

Review

Supercritical Antisolvent Process for Pharmaceutical Applications: A Review

Paola Franco and Iolanda De Marco *

Department of Industrial Engineering, University of Salerno, Via Giovanni Paolo II, 132, 84084 Fisciano (SA), Italy; pfranco@unisa.it

* Correspondence: idemarco@unisa.it

Received: 8 July 2020; Accepted: 1 August 2020; Published: 4 August 2020

Abstract: The supercritical antisolvent (SAS) technique has been widely employed in the biomedical field, including drug delivery, to obtain drug particles or polymer-based systems of nanometric or micrometric size. The primary purpose of producing SAS particles is to improve the treatment of different pathologies and to better the patient's compliance. In this context, many active compounds have been micronized to enhance their dissolution rate and bioavailability. Aiming for more effective treatments with reduced side effects caused by drug overdose, the SAS polymer/active principle coprecipitation has mainly been proposed to offer an adequate drug release for specific therapy. The demand for new formulations with reduced side effects on the patient's health is still growing; in this context, the SAS technique is a promising tool to solve existing issues in the biomedical field. This updated review on the use of the SAS process for clinical applications provides useful information about the achievements, the most effective polymeric carriers, and parameters, as well as future perspectives.

Keywords: supercritical antisolvent; micronization; coprecipitation; biomedical field; drug delivery

1. Introduction

Nowadays, the focus on novel polymer/drug formulations is continuously growing, in order to improve the therapeutic efficacy of an active compound and to augment the patient's compliance. In this context, the micronization processes provide a robust solution to issues related to the biomedical field, including drug delivery. The use of polymeric carriers to produce composite particles has multiple functions, such as the drug protection against its oxidation and/or deactivation caused by external agents (light, oxygen, temperature, pH) and the masking of unpleasant taste and/or odor. However, the primary motivation is often the modification of the release kinetics of the active principle embedded in the polymeric particles. In the biomedical field, a rapid or a prolonged release of the active compound can be necessary depending on the pathology, which may require a fast or long-term response. Many prescribed active principles have a low solubility in water, resulting in low bioavailability. The particle size reduction, which is achieved by micronizing the active compound, allows improving the dissolution rate of poorly water-soluble drugs [1–7]. However, drug dissolution can be further enhanced by developing polymer-based formulations. This approach is also useful when the active compound cannot be micronized alone [8–13], or when a highly water-soluble drug has to be released slowly. For example, in the case of infections, long-term antibiotic therapies are often prescribed. Unfortunately, many antibiotics are very soluble in water, and are characterized by a very short half-life; in these cases, prolonged-release drug delivery systems are desirable to reduce the number of administration and the unwanted effects, including the antibiotic resistance [14–17]. Instead, given the existence of many inflammatory conditions, in the case of non-steroidal anti-inflammatory drugs (NSAIDs), both a fast and a prolonged release can be required for the treatment of minor inflammations (e.g., headaches, toothache) or chronic disease (e.g., rheumatoid arthritis, osteoarthritis), respectively [18–20]. However, when NSAIDs are frequently taken, serious side effects may occur, such as nausea, vomiting, gastrointestinal bleeding, and peptic ulcer. All these issues can be solved by controlling the drug release with suitable polymers, which may have mostly hydrophobic or hydrophilic behavior, so producing alternative administration systems.

The typical release kinetics of an active principle from conventional pharmaceutical forms is sketched in Figure 1a, showing how the drug concentration can go below or above the therapeutic range, in an uncontrolled manner. Thus, in this case, high and repeated drug dosages are necessary, leading to serious side effects on the patient's health. In Figure 1b, different release kinetics from alternative drug delivery systems are represented, such as immediate, prolonged, delayed, or pulsed drug releases. In this way, the therapy is targeted for the specific application, and the drug concentration in the plasma is ensured within the therapeutic region, i.e., above the minimum effective concentration (MEC) required to have a therapeutic effect and below the maximum safe concentration (MSC) to avoid toxicities.

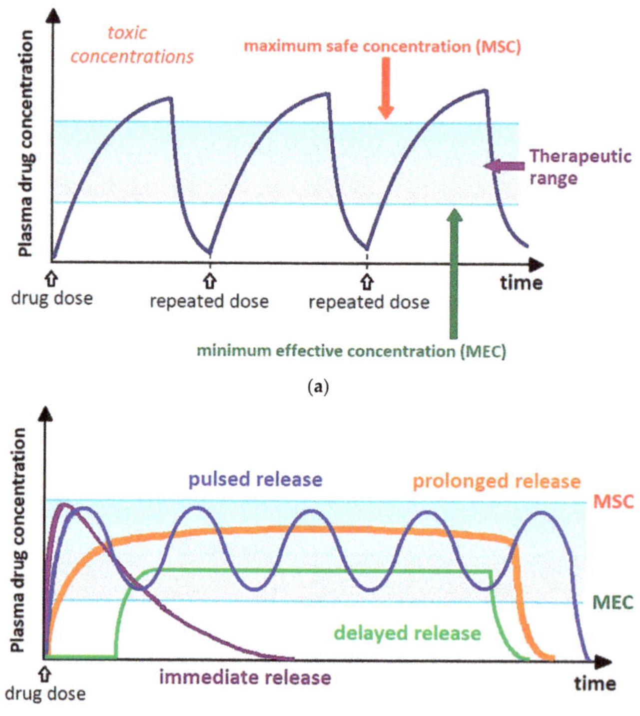

Figure 1. (**a**) Conventional pharmaceutical formulations; (**b**) modified-release drug delivery systems.

There are different technologies to micronize active compounds or to produce composite polymer/drug particles, such as spray drying, freeze-drying, jet-milling, solvent evaporation, or coacervation. Despite the advantages of using these conventional techniques, such as the possibility of scaling up the process without particular complications, in the case of the widely employed spray drying, these methods also present significant limitations [7,12,21–25]. Indeed, multistage processing is often involved to remove the toxic organic solvents used, whose residues are often not negligible in the final product. Moreover, the employed high process temperatures can cause the thermal degradation of the thermosensitive active principles. In addition, using conventional micronization techniques, it is difficult to control the morphology and the dimensions of the produced particles.

On the contrary, these drawbacks can be overcome using processes based on the use of supercritical fluids [7,12,21–25], among which the supercritical carbon dioxide ($scCO_2$) is the most commonly employed in the industrial field. Considerably reducing the use of organic solvents, the $scCO_2$-based micronization techniques permit to produce regular particles of minimal dimensions, which are difficult to obtain with conventional processes.

This goal is possible because, in supercritical conditions, fluids have some gas-like properties (e.g., high diffusivity, low viscosity) and other liquid-like properties (e.g., high density, high solvent power). In particular, the supercritical antisolvent (SAS) technique stands out for the production of nanoparticles and microparticles of one or more compounds in a controlled manner. The first papers on the SAS process date back to the early 90s. For example, Yeo et al. processed para-linked aromatic polyamides obtaining polycrystalline spherulites and microfibrils [26,27], Bertucco et al. obtained hyaluronic acid benzylic ester microspheres that can be used for controlled drug delivery [28], Reverchon et al. obtained nanoparticles and sub-microparticles of different superconductor precursors [29,30].

Regarding the pharmaceutical field, using the SAS process, up to now, micronized drug particles and several polymer/active compound composites have been proposed for different biomedical applications [31–35]. Indeed, SAS precipitated particles can be used for drug delivery by oral, parenteral, transdermal, or topical routes [2,36–42]. The versatile SAS particles can be used in powder form for the oral administration of composites or the preparation of aerosol formulations, mainly for the treatment of respiratory diseases; they can also be used for the production of medicated patches or active topical gels, by incorporating nanoparticles/microparticles into various types of dressings, supports or pomades, respectively. The dissolution kinetics of many active principles, both of synthetic and natural origin, have been successfully modified by selecting carriers suitable for the required therapy. Different biopolymers, which are biocompatible [11,22,43,44] and sometimes even biodegradable [32,45–49], have been studied to tune the release of the active compounds. In literature, there are few reviews about the SAS technique [50–53]. An older overview [53] was focused on the precipitation of single compounds in the form of microparticles and nanoparticles, mainly introducing the fundamentals of the SAS technique. In this study, the micronization of different kinds of materials was considered, without choosing a specific field of application. Abuzar et al. [50] focused their attention to poorly water-soluble drugs, recommending the SAS process as an innovative tool to enhance the bioavailability of these active principles. In another review [51], the SAS process has been indicated as a promising drug encapsulation method, but a limited number of studies compared to the existing ones have been analyzed. Moreover, the crucial role of the polymeric carrier to produce novel formulations for specific biomedical applications and drug dissolution rate is not highlighted. In the paper of Gonzalez-Coloma et al. [52], the attention was instead given to the supercritical antisolvent fractionation (SAF) of natural products. The SAF process is conceptually similar to the SAS one since, in any case, the supercritical fluid acts as an antisolvent; however, the difference is that, in the former technology, a complex mixture consisting of various compounds in a solvent has to be selectively fractionated, precipitating a compound or a series of compounds of interest. It has to be considered that none of the aforementioned reviews report the interesting results obtained in recent years, such as the new carriers identified, whose use has been optimized lately.

Considering the advantages associated with the use of the SAS technique and the goals achieved so far, it is clear that this technology can still make a significant contribution to improving the treatment of several diseases. Therefore, this updated review is focused on the application of the SAS process in the medical/pharmaceutical field, both regarding the micronization of active compounds and their coprecipitation with polymeric carriers. The advantages and disadvantages associated with the use of the SAS technique, the most effective polymeric carriers, and operating conditions, as well as the processed active principles and the potential applications, are indicated. This review provides guidelines for future prospects.

2. SAS Process: Fundamentals and Test Procedure

The SAS process is based on some main prerequisites. In this technique, the $scCO_2$ has the role of the antisolvent; therefore, it has to be completely miscible with the selected liquid solvent. On the contrary, the solute/solutes to be processed has/have to be soluble in the liquid solvent but insoluble in the binary mixture formed in the precipitator consisting of the solvent and the antisolvent. Hence, the precipitation/coprecipitation of solute/solutes occurs because of the fast diffusion of $scCO_2$ into the liquid solvent and the consequent supersaturation of the solute/solutes.

A schematic representation of the SAS process is reported in Figure 2. Briefly, a typical SAS experiment begins by pumping the CO_2 to reach the desired pressure in the precipitator, which is heated up to the selected temperature. Once the operating conditions are stabilized, the pure solvent is delivered to the precipitator, passing through a nozzle. Then, the liquid solution, which consists of the solute/solutes dissolved in the selected solvent, is injected. Due to the supersaturation, the solute/solutes precipitate on a filter, whereas the mixture solvent/antisolvent is recovered and separated downstream the precipitator, where a vessel to collect the liquid solvent is located. Once the solution is injected, the $scCO_2$ continues to flow to eliminate the solvent residues. At the end of this washing step, the precipitator is depressurized down to atmospheric pressure, and the precipitated powder can be collected.

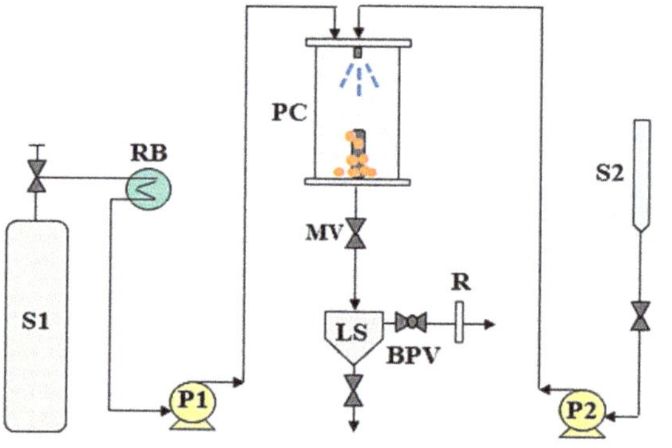

Figure 2. A scheme of the supercritical antisolvent (SAS) apparatus. P1, P2: Pumps; S1: CO_2 supply; S2: Liquid solution supply; RB: Refrigerating bath; PC: Precipitation chamber; LS: Liquid separator; MV: Micrometric valve; BPV: Back-pressure valve; R: Rotameter.

There are also some modified versions of the SAS technique, in which the $scCO_2$ acts equally as an antisolvent, such as the supercritical gas antisolvent (GAS) process and the solution enhanced dispersion by supercritical fluids (SEDS). In the SEDS process, the $scCO_2$ and the liquid solution are both injected through a coaxial nozzle. Instead, in the GAS process, the $scCO_2$ is introduced into the solution consisting of solute/solutes to be precipitated and the liquid solvent. The SAS process was also proposed in the literature with other different acronyms, such as the aerosol solvent extraction system (ASES).

3. SAS Micronization of Active Compounds

Different factors can influence the oral bioavailability of an active compound, including the first-pass metabolism, the drug permeability, and its solubility and dissolution rate in water [54]. In particular, poor solubility and slow dissolution rates are the leading causes of the low bioavailability for most of the active principles [55–57]. Consequently, a low drug amount, often at an insufficient concentration, is available at the specific site of action.

In the last decades, drug particles with nanometric or micrometric size have been applied to treat or prevent different pathologies or medical conditions. As previously mentioned, the micronization of an active compound revealed to be a possible approach to enhance its dissolution rate, since the particle size reduction leads to an increase of the specific surface area in contact with the aqueous solution.

Most of nonsteroidal anti-inflammatory drugs (NSAIDs) are a striking example of poorly-water soluble drugs. The micronization of active principles is also a valid solution to reach a target part of the body. Aerosol drug administration or anticancer therapies are striking applications. The drug micronization is also a valid solution when a specific range of particles' size is required for a certain disease; for example, in the case of pulmonary diseases, drug particles (e.g., antibiotics) with mean diameters lower than 1 μm and a controlled particle size distribution (PSD) are desirable.

For these purposes, the SAS technique is one of the most effective micronization techniques based on the use of supercritical CO_2. Up to now, several active compounds, such as antibiotics [25,58–64], NSAIDs [46,65], antioxidant compounds [1,6,7,38,66], anticancer drugs [3,4,67–69], and others [2,32,70–72], were micronized in the form of nanoparticles, sub-microparticles, and microparticles using the SAS process.

As in the case of the other supercritical fluids assisted technologies, SAS micronization allows overcoming the main drawbacks of conventional micronization techniques, such as spray drying, freeze-drying, jet-milling, solvent evaporation, and grinding [7,25]. For example, Rogers et al. [73] focused their review on a comparison between $scCO_2$-based processes, including SAS, and cryogenic spray-freezing technologies in the micronization of active principles. Both approaches were presented as potential solutions for the production of particles of active compounds that cannot be processed by conventional micronization techniques, such as spray drying and jet-milling. However, the authors highlighted the advantage of using an environmentally-friendly antisolvent as the $scCO_2$ (in the specific case of the SAS technique) compared to the organic and toxic antisolvents conventionally employed. Moreover, the gas-like and liquid-like properties of $scCO_2$, such as the high diffusivity and the high solvent power, allow obtaining dry particles without solvent residues in one step. In reverse, lyophilization or liquid antisolvent extraction are needed to dry powders prepared by the cryogenic micronization technologies. The study of Park et al. [25] demonstrated the superiority of the SAS process with respect to spray drying in the micronization of cefdinir (a model drug). Indeed, microparticles with a mean size equal to 2.32 ± 1.76 μm were obtained by spray drying, whereas nanoparticles with a narrow PSD (0.15 ± 0.07 μm) were produced by the SAS process. Moreover, the specific surface area of SAS-processed particles (55.79 ± 0.06 m^2/g) was higher than that of spray-dried particles (35.01 ± 0.63 m^2/g). The increase in the surface area led to a better contact with the water molecules, so the dissolution of the SAS-processed cefdinir was faster than the spray-dried drug particles. According to the results obtained by Ha et al. [7], SAS micronization also proved to be more effective than milling techniques. Indeed, SAS nanoparticles and sub-microparticles (size in the range: 0.15–0.50 μm) allowed to speed up the dissolution of resveratrol, with respect to crystals (18.75 μm) or crystals/irregular coalescent particles (1.94 μm) produced by Fritz milling and jet-milling, respectively. Therefore, it is ascertained that SAS micronization provides benefits in the final pharmaceutical product both in terms of morphology and dimensions of the particles that constitute it.

Similarly, in other works [1–7,38,66,72,74–77], the SAS micronization led to an enhancement of the drug dissolution rate; in some cases, a slight improvement was already observed even when smaller crystals and no regular particles were obtained by SAS with respect to the unprocessed drug. The SAS technique offers the possibility of obtaining both crystalline and amorphous materials by modulating the operating conditions of pressure and temperature. For example, at a fixed temperature, crystals mostly precipitate working at pressures below the mixture critical point (MCP) of the drug/solvent/$scCO_2$ system, i.e., in the biphasic region. The possibility of producing an amorphous material increases by increasing the pressure, up to a value much higher than the critical pressure, i.e., at fully-developed supercritical conditions. The attainment of amorphous powders involves numerous benefits, including the enhancement of the drug dissolution rate in an aqueous environment, due to a higher contact between the water molecules and the active principle with an increased specific surface area.

However, as emerged from several papers, the selection of the proper liquid solvent is crucial. Indeed, the success of SAS precipitation strongly depends on the affinity between the solvent and the supercritical antisolvent, i.e., the solubility of the liquid solvent in the supercritical CO_2 and the fast gas-like diffusion of the supercritical CO_2 in the solvent. These fundamental aspects assure the attainment of small particles. Many studies showed that dimethylsulfoxide (DMSO) is one of the best solvents for SAS micronization [1,3,6,17,46,58–62,65,67,71], as it often allows to produce particles with good morphology and small size, in addition to being the most used solvent. Respecting the basic prerequisites of SAS precipitation, up to now, DMSO allowed micronizing a wide variety of active principles in the form of nanometric and micrometric particles. Ethanol [5,7,63,64,69,78] and acetone [2,68] also emerged as good solvents for the SAS micronization of active compounds. In addition, some active principles were effectively micronized using methanol [7,25,72] and N-methyl-2-pyrrolidone (NMP) [38,58,60]. Moreover, acetone belongs to the category of solvents that have a narrow pressure range to switch from the two-phase to the single-phase mixing with $scCO_2$; using this kind of solvent, nanoparticles are generally obtained [79]. Instead, microparticles usually precipitate with DMSO, EtOH, and NMP, which are solvents that present a broad transition zone from the two-phase to the single-phase mixing with $scCO_2$. For the latter category of solvents, high operating pressures are necessary to produce nanoparticles. In some papers, SAS precipitation was also attempted by solubilizing drugs into solvent mixtures, which consisted of the solvents above, but satisfying results were reached only in few cases [1,7,13,59,66,80,81].

Another critical parameter for a successful micronization is the molar fraction of CO_2, which should allow the complete miscibility between the solvent and the antisolvent at the selected pressure and temperature. For this purpose, in most of the reported papers, CO_2 molar fractions higher than 0.95 were used. In addition, the operating pressure and the concentration of the active compound in the liquid solution also influence the morphology and the size of the precipitated powders. In particular, when spherical particles are produced, a decrease of the pressure or an increase of the drug concentration often results in an increase of particles' dimensions [5,38,58,61,62,69,72]. However, it is possible that large crystals significantly precipitate, increasing drug concentration. This result can be explained considering that the presence of the active compound can modify the high-pressure vapor-liquid equilibria (VLEs) of the solvent/antisolvent binary system.

Consequently, the MCP of the drug/solvent/$scCO_2$ ternary system shifts towards higher pressures with respect to the MCP of the solvent/$scCO_2$ binary system [82,83]. Therefore, the operating point could lie below the MCP, i.e., in the biphasic region. This effect is accentuated by increasing the drug concentration in the liquid solvent.

The morphology and the dimensions of SAS-processed particles obviously will influence the dissolution of the active compound. Indeed, as highlighted in various papers [7,38,66,72], a smaller particle size, which means an increased surface area, led to an increase in the drug dissolution rate.

SAS particles can be used for oral, parenteral, and topical drug administration; in the latter case, the micronized drug particles can be incorporated into various types of dressings or supports. Up to now, SAS-micronized drugs were proposed for different diseases or clinical conditions, including inflammations and pains [11,12,75,84], infections [2,4,25,58,59,61,63,64,74,77,85,86], diabetes [5,7,66], psychological disorders [87], asthma [32,71,88], tumors [1,3,4,31,38,67–69,78,89–92], cardiovascular pathologies [6,33,70,76], etc.

However, some active compounds are not processable by the SAS technique. Indeed, in many cases, SAS micronization substantially fails since crystals precipitate [11–13,74,87,90,93] or, even, the active compound can be extracted by the solvent/$scCO_2$ mixture [8,9] not respecting the fundamental prerequisites of the SAS process.

Nevertheless, different studies proved that it is possible to force the morphology of these compounds not suitable for SAS micronization through their coprecipitation with a proper selected polymeric carrier [5,8–13,44].

In Table 1, a summary of the active compounds processed by SAS micronization is reported. The active principle to be micronized, the morphology and size of precipitated powders, as well as the applications are indicated. The operating conditions that mainly influence the morphology of particles are also specified, i.e., the liquid solvent used, the pressure (P), the temperature (T), and the drug concentration in the injected solution (C).

Table 1. An overview of SAS micronization of active compounds. C: Crystals; NP: Nanoparticles; SMP: Sub-microparticles; MP: Microparticles; BL: Balloons; c: Coalescent; AGG: Aggregates; m.s.: Mean size; NSAID: Nonsteroidal anti-inflammatory drug; DLBS2347: Manilkara kauki L. Dubard's leaf extract; HCPT: 10-Hydroxycamptothecin; GBE: Ginkgo biloba extract; AC: Acetone; CHF: Chloroform; EtAc: Ethyl acetate; EtOH: Ethanol; DMSO: Dimethylsulfoxide; DCM: Dichloromethane; DMFA: N,N-dimethylformamide; MeOH: Methanol; NMP: N-methyl-2-pyrrolidone; iPrOH: Isopropanol; THF: Tetrahydrofuran.

Active Compound	Solvent	P [MPa]	T [°C]	C [mg/mL]	Morphology	Applications	Reference
Risperidone	CHF	5–20	40	25–100	C (size: 10–200 μm)	antipsychotic to treat bipolar and obsessive-compulsive disorders, schizophrenia	[87]
Sulfathiazole	AC, EtOH	10–14	35, 45, 55	3–13	C+MP (m.s.: 2.1–16.9 μm)	antimicrobial drug	[2]
Paracetamol	MeOH	10	40	Not reported	C (size: 4.2 μm)	analgesic and antipyretic to treat fever, headaches, and others	[75]
Cefonicid, Cefuroxime, Cefoperazone	DMSO	9–18	40–60	10–90	SMP, MP or BL of Cefonicid and Cefuroxime SMP or MP of Cefoperazone (size: 0.1–50 μm)	cephalosporins used before-surgery, to treat pneumonia, skin infections, urinary tract and post-operative infections	[61]
Ampicillin	NMP, EtOH, DMSO	8–15	40	20	SMP, MP, cSMP, irregular MP (m.s.: 0.26–1.2 μm)	antibiotic to treat respiratory, gastrointestinal and urinary tract infections	[58]
Amoxicillin	DMSO, DMSO/EtOH, DMSO/MeOH	10–25	40	0.005–0.02	SMPs, MPs (size: 0.2–1.6 μm)	antibiotic to treat infections of the skin, urinary and respiratory tracts	[59]
Amoxicillin	NMP, DMSO	15	40	20–100	SMP, MP (m.s.: 0.25–1.2 μm)		[60]
Ampicillin, Amoxicillin	DMSO	9	40	20	SMP (m.s.: 0.23 μm for Ampicillin, 0.26 μm for Amoxicillin)	antibiotics	[17]
Griseofulvin, Ampicillin, Amoxicillin, Tetracycline	NMP, DMSO, EtOH, DCM	10–18	40	20–120	C of Griseofulvin AGG of Ampicillin Film of Amoxicillin cSMP of Tetracycline (m.s.: 0.2–0.6 μm)	Antibiotics for various infections	[85]
Griseofulvin	THF, DCM	9.7	35	5	C, needles, irregular NP-SMP (m.s. 0.13–0.51μm)	Antibiotic and antifungal drug	[86]
Cefdinir	MeOH	12	45	20	NP (m.s.: 0.15 μm)	antibiotic to treat infections of skin, eyes and respiratory tract	[25]

Table 1. *Cont.*

Active Compound	Solvent	P [MPa]	T [°C]	C [mg/mL]	Morphology	Applications	Reference
Rifampicin	DMSO, NMP, MeOH, EtAc, DCM	9–14.5	40	5–70	C, SMP, MP (m.s.: 0.4–5.0 µm)	antibiotic to treat tuberculosis, meningitis, biliary tract infections	[62]
Minocycline hydrochloride	ethanol	7.5–13	35–50	1–20	cSMP/AGG (m.s.: 0.2–0.3 µm)	antibiotic to treat infections of skin, urinary and respiratory tracts	[63,64]
Moxifloxacin	DMSO, DMFA, MeOH, acetic acid	15	40	1–50	C (m.s.: 0.3–8.2 µm)	antibiotic to treat tuberculosis	[77]
Sulfamethoxazole	AC	10–12	35	88	C (m.s.: 42–5 µm)	antibiotic to treat urinary tract infections, otitis, shigellosis, interstitial pneumonia	[74]
Theophylline	DMSO	9	40	20	C	bronchodilator to treat asthma	[44]
Salbutamol	DMSO, MeOH, EtOH/H_2O	9.5–15	40	3–10	BL, cMP, rods (length: 1–3 µm; diameter: 0.2–0.4 µm)	drug to treat bronchial asthma	[71]
Fluticasone propionate	DCM	6.5–11	40–60	5–17	C (m.s.: 3.7–9.1 µm)	corticosteroid to prevent asthma symptoms	[88]
Dexamethasone, prednisolone, budesonide	EtOH	9–15	40	20	C	corticosteroids to treat ocular and pulmonary diseases, hepatitis, ulcers	[93]
Budesonide	DMC	7.9–13.9	35–45	0.002–0.01	AGG, MP (m.s. 1.4–2.0 µm)	corticosteroid to treat asthma, nasal polyps, bronchiectasis, pulmonary disease	[32]
Cilostazol	DCM, acetic acid	8–15	40–60	50–150	AGG (size: 1.0–4.5 µm)	vasodilator drug to treat vascular claudication	[76]
Telmisartan	EtOH/DCM	12	45	25	Irregular cSMP (m.s.: 740 µm)	drug for hypertension treatment	[70]
HCT	DMF, NMP, DMSO, THF, AC, 2-butanone	15	42	20	AGG, irregular particles	drug to treat hypertension	[33]
Atorvastatin	MeOH	10–18	40–60	25–150	NP, SMP (m.s.: 0.15–0.86 µm)	drug to treat hyper-cholesterolemia	[72]
Piroxicam	DCM	9.7	25	0.05	C	NSAID to treat arthritis, osteoarthritis, spondylitis	[12]
Diclofenac sodium	DMSO	9	40	20	NP (m.s.: 0.14 µm)	NSAID to treat arthritis, osteoarthritis, spondylitis	[46]
Diflunisal	AC, AC/DCM	14–15	35–40	0.02–0.04	C	NSAID for tuberculosis treatment	[11]

Table 1. *Cont.*

Active Compound	Solvent	P [MPa]	T [°C]	C [mg/mL]	Morphology	Applications	Reference
Ibuprofen sodium	EtOH	8–12	35–50	0.0002–0.0004	C	NSAID to treat fever, pains, various inflammations	[84]
Licorice	EtOH	15–20	40	10–14	AGG	thrombin inhibitor with antiulcer, antimicrobial, antidiabetic, hepatoprotective and anticancer activities	[89]
5-fluorouracil	DMSO, DMSO/DCM	15–18	40–50	0.1–0.2	AGG, cSMP (m.s.: 0.22–0.67 μm)	anticancer drug	[67]
DLBS2347	DMSO, EtOH, MeOH, EtAc, AC, DCM	8–20	40–60	Not reported	Film, C, NP-AGG	anticancer drug	[68]
Camptothecin	DMSO	10–25	35–68	1–5	cNP, SMP (m.s.: 0.4–0.9 μm)	anticancer drug	[4]
HCPT	DMSO	10–25	35–68	0.5–5	NP (m.s.: 0.18 μm)	anticancer drug	[3]
Taxol	EtOH	10–25	35–68	2.5–10	NP, MP (m.s.: 0.2–1.9 μm)	anticancer drug	[69]
GBE	EtOH	10–40	35–80	1–5	NP (m.s.: 0.1–0.2 μm)	antioxidant, antifungal and antitumor drug to treat diabetes cardiovascular diseases, cerebral insufficiency, dementia	[5]
Curcumin	EtOH, AC, AC/EtOH	9	40	2–10	AGG	polyphenol with antioxidant and anticancer properties	[80]
Mangiferin	DMSO, AC, NMP, DMSO/AC, DMSO/EtOH, NMP/AC, NMP/EtOH	8–15	40–50	8–14	NP, SMP, cSMP (size: 0.22–1.44 μm)	polyphenol with antioxidant, analgesic, anti-allergic, anticancer properties to treat diabetes, aging, periodontitis, neurodegenerative disease	[66]
Mangiferin	NMP	10–20	35–59	5–59	cSMP, SMP, MP (m.s. 0.56–1.04 μm)	antioxidant	[38]
Curcumin	AC/EtOH	9–12	40	20	Needles	polyphenol	[13]
Resveratrol	MeOH, EtOH, MeOH/DMC, EtOH/DMC	15	40	Not reported	NP, SMP (m.s.: 0.15–0.50 μm)	phenol to treat diabetes, cancer, cardiovascular and neurological disease	[7]
Folic acid	DMSO	15	40	20	AGG	vitamin B9 to prevent neural tube defects in infants, vascular diseases and megaloblastic anemia	[94]

Table 1. *Cont.*

Active Compound	Solvent	P [MPa]	T [°C]	C [mg/mL]	Morphology	Applications	Reference
Lycopene	DCM	7–15	35–45	0.13–0.5	C (m.s.: 10–80 μm)	carotenoid with antioxidant and anticancer properties	[90]
Lutein	EtAc	6.5–9	35–45	0.5–0.9	needles, cNP	carotenoid	[45]
β-carotene	AC/EtOH	8.5	40	4–8	C (m.s. 4–247 μm)	carotenoid with antioxidant and anticancer properties, also used to treat cardiovascular diseases and osteoporosis	[81]
β-carotene	DCM	8–12	40	32–61	C	carotenoid	[95]
β-carotene	DCM	8–20	40	4–8	C	carotenoid	[96]
Rosemary extracts	EtOH	8–12	25–50	Not reported	AGG, cSMP (size: 0.2–1 μm)	antioxidants with antimicrobial, anti-inflammatory and anticancer activities	[78]
Quercetin, Rutin	DMSO	9, 13	40	20	C	flavonoids with antioxidant and anticancer properties, also used to prevent cardiovascular disease	[97]
Quercetin	EtAc	10	35	1.4	needles (0.63 ± 0.06 μm)	flavonoid	[91]
Quercetin	EtOH	8–25	35–65	2–11	C, AGG, needles	flavonoid	[92]
Rutin	DMSO, DMSO/EtOH, DMSO/AC	8–20	40–60	20–85	cSMP, SMP, MP (m.s.: 0.3–1.9 μm)	flavonoid	[1]
Fisetin	EtOH/DCM	10	45	1	rods	flavonoid with antioxidant, neuroprotective, anticancer effects	[31]
Vitexin	DMSO	15–30	40–70	1–2.5	Irregular NP (m.s.: 0.13 μm)	flavonoid to prevent heart disease	[6]

4. SAS Coprecipitation of Active Compounds with Polymeric Carriers

Up to now, the SAS process has been widely exploited to produce composite polymer/active compound systems for various biomedical applications. In particular, composite particles were produced to treat inflammations [8,11,46,65,98,99], infections [17,100–106], asthma and allergies [22,32,44,93], diabetes [35,107,108], hypertension [34,70], and other diseases [39,94,109–113]. Different kinds of active principles, both with synthetic and natural origin, have been incorporated into polymeric particles. In particular, SAS particles loaded with natural active compounds have been often proposed as alternative therapies to conventional ones, e.g., for the prevention and treatment of tumors or cardiovascular diseases [31,49,91,97,114–116]. Carotenoids, phenols, and flavonoids belong to this category of compounds that offer numerous benefits for human health, given their antioxidant, anticancer, antimicrobial, and anti-inflammatory properties.

As occurs for the drug micronization, the coprecipitation of a polymer and an active principle via the SAS process offers many advantages if compared with conventional techniques [12,21–24]. The study of Wu et al. [12] proved the superiority of the SAS process with respect to spray drying in the attainment of polyvinylpyrrolidone (PVP)/piroxicam microparticles. Indeed, the PSD obtained with the SAS technique was narrower than that of the spray-dried particles. Moreover, the dissolution rate of piroxicam released from SAS particles is approximately twice as fast than that of the drug released from the spray-dried particles. In particular, about 2% of pure piroxicam dissolved in 5 min, whereas the percentage of piroxicam released from SAS-processed microparticles and the spray-dried microparticles was about 100% and 55%, respectively [12].

Similarly, Lee et al. [22] prepared inclusion complexes both by the freeze-drying method and the SAS technique. By coprecipitating cetirizine dihydrochloride (an antihistamine drug) with β-cyclodextrin (β-CD), large and irregular crystals were produced by the freeze-drying method, whereas spherical particles were obtained by the SAS technique. In the work of Won et al. [23], the SAS process also showed better performance with respect to the conventional solvent evaporation in the preparation of felodipine-loaded particles based on hydroxypropylmethyl cellulose (HPMC) and poloxamers. In particular, SAS particles proved to be more stable over time, in addition to exhibit a higher drug dissolution rate.

As previously mentioned, polymer/drug composites are produced for various purposes, but the main challenge is to suitably modify the dissolution kinetics of the active principles, aiming for excellent therapies. Depending on the medical application, different drug release kinetics can be required. The choice of the right polymeric carrier is strategic to release the drug at the desired time/speed and/or to a specific site of action. In this context, it was proven that the use of different carriers for SAS coprecipitation leads to different drug releases [46,65]. In particular, when a hydrophilic polymer is used as the carrier, the dissolution of the active principle contained in the SAS composite particles is enhanced. PVP belongs to this category of polymers; indeed, its use allowed to increase the dissolution rate of various poorly-water soluble active compounds [8,9,11,31,33,65,70,81,93,94,97,115–117]. In reverse, the drug release is prolonged using a polymer with a hydrophobic behavior, as occurs, for example, by selecting zein [17,46–49]. Similarly, a sustained or prolonged release of the active compound was observed from particles based on polylactic acid (PLA) and poly (L-lactic acid) (PLLA) [32,67,98,99,111,112,118–121], which, until now, has been mostly employed to deliver anticancer drugs.

In addition, it is also possible to promote a targeted drug release by producing SAS particles based on pH-sensible polymers, i.e., polymers that dissolve at specific pH values. The Eudragit polymers are a striking example [43,44,98,99,104,122].

Two significant examples of improvement of the drug release kinetics by SAS-produced particles are reported in Figure 3. In Figure 3a, the dissolution of a model NSAID (i.e., ketoprofen) was speeded up by selecting PVP as a hydrophilic carrier to ensure a rapid relief against inflammations, such as headache or toothache. In Figure 3b, the release of a model antibiotic (i.e., ampicillin) was prolonged by producing particles based on zein as a hydrophobic carrier, thus reducing the frequency of administration and the side effects due to antibiotics overuse.

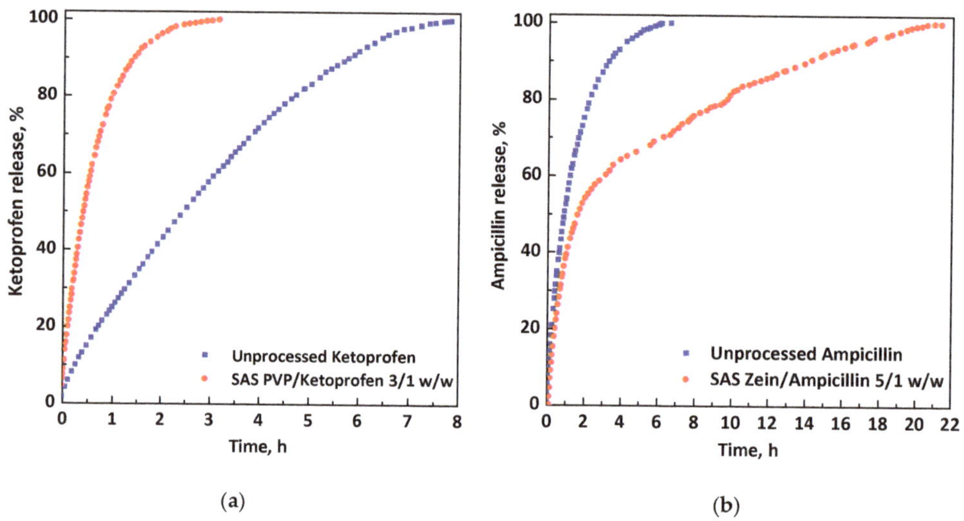

Figure 3. Dissolution kinetics of active principles from microparticles obtained through the SAS process: (**a**) Ketoprofen (pure and released from polyvinylpyrrolidone (PVP)); (**b**) ampicillin (pure and released from zein).

However, it is necessary to identify and to use carriers that can be processed by the SAS technique. Up to now, satisfactory results have been reached only with a reduced number of polymers, which are therefore identified as effective carriers for the SAS coprecipitation. PVP is currently the most used polymer and is considered one of the best carriers [8,9,11,31,33,35,65,70,80,81,93,94,97,103,115–117,123] since it is often allowed to obtain regular and spherical composite particles. PVP is followed by PLA [45,108,124] and PLLA [41,67,98,99,111–113,118–121,125–127] that are also suitable carriers often employed. Recently, zein [17,46] is also established as a good polymeric carrier for SAS coprecipitation; moreover, β-CD [22,128] seems to be a promising polymer, but its use has to be further investigated given the very few existing papers. Accurately selecting the solvent and the operating conditions, Eudragit L100-55 also led to the formation of regular composite particles. Therefore, it can be considered a novel carrier more suitable for SAS coprecipitation with respect to the previously tested Eudragits, including Eudragit RS100 and Eudragit RL100 [43,98,99,122]. Eudragit L100-55 is soluble at a pH higher than 5.5, corresponding to the first intestinal tract (duodenum), thus the active compound is protected against the acid gastric fluid; in the meanwhile, the side effects on the gastric tract are also avoided. Eudragit L100-55 allowed reaching a controlled release of NSAIDs, antibiotics, and bronchodilator drugs from SAS microspheres [44,104]. Until now, the SAS coprecipitation was attempted using other kinds of polymers, such as polyethylene glycol (PEG) [35,106,129,130], poly(lactide-co-glycolide) (PLGA) [36,37,109], ethyl cellulose (EC) [91,100–102,105], HPMC [34,39,42,70,107], and poly(hydroxybutyrate-co-hydroxyvalerate (PHBV) [95,96,131–135]. However, the use of these carriers must be further investigated, by changing the selected solvents or/and the operating conditions, because the morphology of the polymer/drug precipitated powder is not yet satisfactory.

Indeed, in addition to respecting the essential prerequisites of the SAS technique, it has to be considered that the SAS coprecipitation seems to be effective when composite microspheres are obtained, i.e., microparticles consisting of a polymeric matrix in which the drug is homogeneously dispersed [46]. In this case, in the meanwhile that the liquid solution is fed to the precipitator, the liquid jet break-up and the subsequent atomization quickly occur and prevail with respect to the surface tension vanishing. As a consequence, the polymer and the active compound are entrapped in the same droplet, which behaves as an isolated reactor. The droplet drying by $scCO_2$ leads to the attainment of the composite microsphere, as represented in Figure 4.

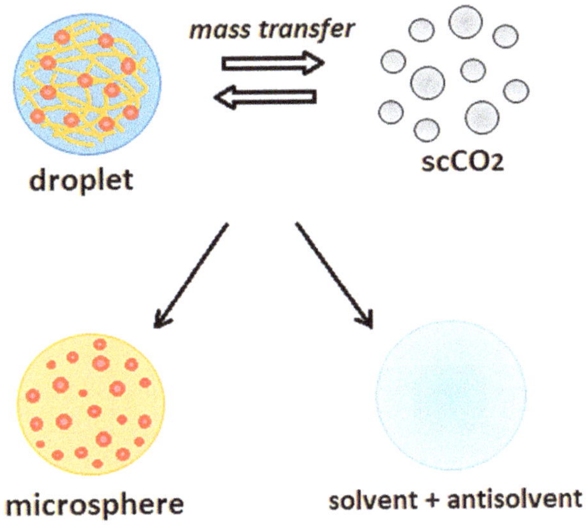

Figure 4. Mechanism of formation of a polymer/active compound microsphere by the SAS technique.

On the contrary, when nanoparticles precipitate, the coprecipitation fails at least in part. In particular, nanoparticles are formed when the surface tension vanishing is very fast, and the gas mixing prevails on the breakage of the liquid jet. Consequently, no droplets are formed and the precipitation of nanoparticles occurs by gas-to-particles nucleation and growth. In this case, a sort of physical mixture at the nanometric level is obtained, i.e., polymer nanoparticles mixed with nanoparticles of the active compound. In addition, in the case of the sub-microparticles, a specific portion of the recovered powder consists of the polymer and the drug precipitated together, but with an irregular distribution of the two materials. Hence, the achievement of an effective polymer/active compound coprecipitation seems to be strongly influenced by the size and the morphology of the particles produced, as well as by their shape. Indeed, the coprecipitation fails also in the case of the attainment of crystals, which occurs by operating in the miscibility gap between the solvent and the antisolvent, where the two compounds tend to precipitate separately, resulting in polymer crystals and drug crystals.

This correlation between a successful coprecipitation and the morphology/size of particles was proved in the study of Franco et al. [46] employing the dissolution tests. In particular, using zein as the carrier and diclofenac sodium as the model drug, it was observed that, increasing the polymer/drug ratio, the particles' diameter increased, whereas the release of the active compound was further slowed down. Moreover, as the particle size decreased, the burst-like effect, i.e., the dissolution of the drug near/on the particles' surface, which, therefore, dissolve as quickly as the unprocessed drug increased.

Several papers showed that the selection of a proper carrier could force even the morphology of active principles that cannot be processed alone by the SAS technique [5,8–13,44]. Moreover, the entrapment of the active compound into the amorphous polymeric matrix can also favor the inhibition of the drug recrystallization [136].

Nevertheless, it has to be taken into account that the presence of the active compound can modify the polymer processability even if, under the selected process conditions, the carrier can be micronized alone in well-defined particles [44]. This influence may be a consequence both of an interaction between the polymer and the drug, but usually, it is due to the presence of solutes (polymer and/or active principle) that can alter the high-pressure VLEs. Therefore, the critical point of the solvent/polymer/drug/antisolvent quaternary mixture can shift with respect to that of the solvent/scCO_2 binary mixture [82,83]. Each compound (i.e., carrier and drug) has a different influence on VLEs, which might be negligible or not; thus, the morphology of the precipitated composites also depends on this factor.

In the case of the coprecipitation, it is evident that the polymer/drug ratio plays a crucial role in the attainment of composite particles. Moreover, the polymer content strongly affects the dissolution rate; in general, the release of the active compound is mostly modified by increasing the polymer/drug ratio [23,41,46,99,110,117,119].

Aiming to obtain spherical microspheres that guarantee an effective coprecipitation, in some cases, it is not possible to reduce the amount of polymer in the composite particles beyond a specific value [97,123]. In particular, a high polymer content in the pharmaceutical formulations can be a limitation of the SAS coprecipitation when some carriers are employed. Among these, up to now, the use of Eudragit L100-55 allowed producing well-defined microparticles only at polymer/drug ratios equal to 20/1 and 10/1 *w/w* [44,104]. Similarly, in the studies of Chhouk et al. [123] and Ozkan et al. [97], polymer/active compound ratios equal to 30/1 and 20/1 *w/w* were indicated as optimal for the production of PVP/curcumin and PVP/quercetin particles, respectively. However, this drawback was found only in these few cases using PVP as the carrier. Alternatively, a high amount of polymer in the SAS coprecipitated powders can be avoided by selecting β-CD.

In general, cyclodextrins (CDs) are characterized by a hydrophilic external surface, which makes them very soluble in water, and a hydrophobic internal cavity, in which various molecules can be incorporated to form guest/host inclusion complexes by non-covalent interactions. The production of CDs-based inclusion complexes by the SAS technique has been exploited to mask the bitter taste and/or to increase the dissolution of various active compounds, such as antioxidants [114,128,137,138], antihistamine drugs [22], antibiotics [139], and others [140]. In these papers, hydroxypropyl-β-cyclodextrin (HP-β-CD) [114,137–140] is mainly employed, followed by β-CD [22,128]. In particular, from the few available studies, β-CD seems to be a more suitable carrier for the SAS process with respect to H-β-CD, since composite particles of regular morphology and size were successfully produced only with β-CD. In addition, inclusion complexes are generally formed by using a low polymer content, i.e., from molar ratio polymer/drug equal to 3/1 up to 1/2 mol/mol. Therefore, the application of β-CD may be a useful tool to reduce the amount of polymeric carrier in the SAS-prepared composite particles significantly, when a rapid drug release is desired.

In addition to the polymer/active compound ratio, other parameters influence the morphology and size of composite particles, mainly the pressure and the overall concentration of solutes in the liquid solution [8,9,44,65,80,81,93,94,97,110,117]. If the pressure is too high, the disappearance of the surface tension prevails over the breakage of the liquid jet, and the coprecipitation partly fails with the formation of nanoparticles, as previously mentioned. This failure also occurs if the pressure is too low and below the MCP, hence the operating point lies in the biphasic region, resulting in the precipitation of crystals or aggregates; even, the liquid can be recovered in the precipitator. Regarding the total concentration, undoubtedly, the upper limit is represented by the solubility of the solutes in the selected solvent. Nanoparticles or coalescent particles often precipitate when the concentration of solutes is too low, because of the fast vanishing of the surface tension and the low viscosity of the injected solution, not allowing the generation of well-defined droplets. Moreover, in the case of a low concentration in the solution, the content of the polymeric carrier, which acts as a microsphere-forming agent, is not enough to force the morphology of the composite system and to trap the active compound into the polymeric matrix. On the contrary, if the total concentration is too high, aggregates or even crystals can be obtained due to the influence that solutes may have on VLEs, as explained above.

In some cases, the variation of the temperature can favor the attainment of well-separated microparticles, as occurs with zein [17,46]. As demonstrated in the study of Franco et al. [17], by coprecipitating ampicillin (a model antibiotic) and zein at a fixed pressure (90 bar) and total concentration (50 mg/mL), polymer/drug 20/1 *w/w* sub-microparticles (mean size: 0.36 ± 0.30 μm) were produced working at 40 °C. Instead, spherical microparticles with a mean size in the range 19–9 μm were obtained at 50 °C, also managing to reduce the ampicillin/zein ratio up to 5/1 *w/w*. The morphology and shape of zein-based particles prepared by the SAS process are also definitely better than aggregates and irregular or collapsed particles obtained up to now with the techniques commonly used in the pharmaceutical field, namely spray freeze-drying and spray-drying [141–143].

A similar trend to that shown in the study of Franco et al. [17] was also reported in the paper of Montes et al. [99] since the dimensions of ibuprofen-loaded particles produced using PLLA or Eudragit L100 increased by increasing the temperature. Instead, Kalogiannis et al. [59] and Patomchaiviwat et al. [119] observed a rise in the degree of coalescence of the PLLA-based particles by increasing the temperature up to 50 °C, probably due to the PLLA plasticization. Indeed, it is well known that the glass transition temperature (T_g) of a polymer can be depressed in the presence of the $scCO_2$, because of its absorption into the polymeric matrix that leads to a weakening of the intramolecular and intermolecular attractions between the polymeric segments [144–146]. Therefore, the polymer T_g decreases in proportion to the amount of $scCO_2$ that the polymer is able to absorb. The CO_2 absorption and, consequently, its plasticizing effect in the presence of semicrystalline polymers also depends on the operating temperature, which influences the $scCO_2$ properties.

A glaring gap emerged from the SAS literature: Up to now, the use of polymeric blends has been employed to produce drug delivery systems in very few papers [23,37,40,41]. Polymer blends can be exploited to tune the drug release as desired, by modulating the ratio between the hydrophilic and hydrophobic polymers that constitute the microparticles. In addition, they are useful to improve other features of the final product, such as mechanical properties and moisture absorption. For example, the very hydrophilic PVP can induce a hygroscopic effect, which could lead to low stability over time and high water uptake in the produced composite powders. This critical issue can be overcome by introducing hydrophobic polymers in the formulation, thus increasing the particles' resistance to humidity. Lee et al. [37] selected PLGA and PLLA to produce microspheres loaded with bupivacaine HCl, a local anesthetic drug generally administered by parental route for the regional anaesthesia, i.e., the local pain control. Various PLGA/PLLA mass ratios were investigated, as well as the use of PLLA at two different molecular weights (MW 2000 or 50,000). It was proved that the release of bupivacaine HCl from particles obtained by SAS depends on the PLGA/PLLA ratio and on the molecular weight of PLLA. In particular, by increasing the PLGA content, the drug dissolution rate increased, until having a rapid release (in about 4 h) when only PLGA was used.

On the contrary, both molecular weight PLLA prolonged the bupivacaine dissolution, but a slower drug release with a lower burst-like effect was achieved when the lower molecular weight PLLA (MW 2000) was employed. These results are partly also since the particles' morphology improved as the PLGA content decreased. Amphiphilic block copolymers, obtained by combining hydrophilic and hydrophobic polymeric segments, can also be employed to deliver active compounds. For example, in the paper of Jung et al. [41], PLLA, methoxy poly-(ethylene glycol)-b-poly(L-lactide) (mPEG-PLLA) block copolymers or PLLA/PEG blends were chosen as carriers for controlled delivery of leuprolide acetate. The last-mentioned one is a non-steroidal antiandrogen drug used to treat prostate and breast cancers and endometriosis, which is usually administered via the parenteral route, being not orally bioavailable. However, leuprolide-based therapies require long-term daily injections and frequent administrations because this drug is very water-soluble and it has a short half-life when administrated via the parenteral route. The mPEG-PLLA diblock copolymers were synthesized by ring opening polymerization of L-lactide in the presence of PEG chain lengths. The effect of the mPEG block length, the PEG/PLLA blending, and the polymer/drug ratio on the leuprolide acetate release were investigated. A dichloromethane/methanol 1/1 *v/v* mixture was used for the SAS coprecipitation, and in all the investigated conditions, spherical microparticles were produced. It was observed that the mean diameter of PLLA-based microparticles was higher than the one of microparticles prepared using mPEG-PLLA.

Moreover, an increase in the mean particle size was noted as the PEG block length increased in the mPEG-PLLA diblock copolymer. The drug entrapment efficiency was not significantly influenced by the mPEG block length in mPEG-PLLA, the blending of PLLA and PEG, as well as the polymer/drug ratio. However, these factors influence the release of leuprolide acetate from SAS microparticles. Indeed, the drug dissolution rate increased by increasing the PEG block length in the mPEG-PLLA. Microparticles based on PEG/PLLA blends showed a faster drug release with respect to the mPEG-PLLA copolymer containing an equal mass ratio PEG/PLLA. Concerning the effect of the polymer/drug ratio, a higher amount of mPEG-PLLA allowed a slower drug release rate.

Recently, some changes to the SAS process have been suggested [13,47]. In particular, Matos et al. [13] proposed a single-step coprecipitation and coating by using the supercritical antisolvent. A SAS coprecipitation of PVP/curcumin particles was attempted to coat the surface of different polymers, namely microcrystalline cellulose (MCC) (size of crystals/irregular particles: 175 μm), corn starch (size: 15 μm), or lactose (size: <5 μm). Briefly, before starting the experiment, irregular particles/crystals of MCC, starch, or lactose were located into the precipitator; then, the solution containing PVP and curcumin is injected and, because of the antisolvent effect of the $scCO_2$, composite particles precipitated onto the different polymers used as support. During the single-step coprecipitation and coating, the stirring into the precipitator ensured a homogeneous coating onto the polymer surface. However, a proper morphology of the final product was not guaranteed, since it strongly depends on the shape of the polymeric supports initially charged into the precipitator. Moreover, from the release tests, it can be noted that only about 10% of the unprocessed curcumin dissolved in 60 min. In contrast, the drug released from PVP-based particles precipitated by a conventional SAS coprecipitation (i.e., without other polymeric supports) already reached more than 95% of dissolution in 5 min. The dissolution of curcumin did not show a further improvement when PVP/curcumin particles were precipitated onto the polymeric supports. Instead, Liu et al. [47] attempted to incorporate an anticancer compound, namely 10-hydroxycamptothecin (HCPT), into zein-based particles. By coprecipitating the polymer and the drug through the SAS technique, rods-like crystals and drug nanoparticles were obtained. These results suggest that coprecipitation did not occur because the polymer and the drug precipitated separately and not in the form of composite microspheres. An unsatisfactory morphology was also achieved using the built-in ultrasonic dialysis process (BUDP), which couples the ultrasonic emulsification with the dialysis technologies [147]. Then, the authors combined the supercritical process with the BUDP by dispersing the SAS coprecipitated zein/HCPT powders into an ethanol/water mixture, which was used as the dialysis solution for BUDP. In this way, well-defined zein/HCPT microspheres were produced, also prolonging the drug release. However, nowadays, aiming at the process intensification, the use of a multistage-procedure is not the best option, both in terms of costs and of process times.

A summary of the polymer/active compound composites produced by the SAS technique for medical purposes is reported in Table 2. The polymeric carriers and the active principles processed were indicated, as well as the main process parameters, including the liquid solvent used, the pressure (P), the temperature (T), and the overall concentration of solutes (polymer + drug) in the liquid solution (C_{tot}). The morphology and the dimensions of SAS-obtained samples are also specified. In particular, the literature results are listed in Table 2, firstly reporting the polymers that revealed to be the most used and the most suitable for the SAS technique.

Table 2. An overview of the polymer/active compound composites obtained by the SAS technique. C: Crystals; m.s.: Mean size; c: Coalescent; NP: Nanoparticles; SMP: Sub-microparticles; MP: Microparticles; EMP: Expanded microparticles (*balloons*); AGG: Aggregates; AC: Acetone; CHF: Chloroform; DCM: Dichloromethane; DMF: Dimethylformamide; DMSO: Dimethylsulfoxide; EtOH: Ethanol; EtAc: Ethyl acetate; MeOH: Methanol; NMP: N-methyl-2-pyrrolidone; THF: Tetrahydrofuran; PVP: Poly(vinylpyrrolidone; MCC: Microcrystalline cellulose; β-CD: β-Cyclodextrin; HP-β-CD: Hydroxypropyl-β-cyclodextrin; PLA: Polylactic acid; PCL: Poly(ε-caprolactone); PLGA: Poly(lactide-co-glycolide); PLLA: Poly(L-lactic acid); PEG: Poly-ethylene glycol; mPEG-PLLA: Methoxy poly(ethylene glycol)-b-poly(L-lactide); HPMC: Hydroxypropylmethyl cellulose; HPC: Hydroxypropyl cellulose; HCO-60: Polyoxyethylene (60) hydrogenated castor oil; CAP: Cellulose acetate phthalate; PHBV: Poly(hydroxybutirate-co-hydroxyvalerate); PVM/MA: Polymer poly(methyl vinyl ether-co-maleic anhydride); PMMA: Poly(methyl methacrylate; HCT: Hydrochlorothiazide; HCPT: 10-Hydroxycamptothecin; 17α-MT: 17α-Methyltestosterone; PZA: Pyrazinamide).

Polymeric Carrier	Active Compound	Solvent	P [MPa]	T [°C]	C_{tot} [mg/mL]	Morphology	Reference
PVP	Cefuroxime axetil	MeOH	7–20	35–50	50–150	cMP, MP (m.s. 1.88–3.97 μm)	[103]
	Ezetimibe	EtOH	15	40	25	NP (m.s. 0.21–0.23 μm)	[110]
	Dexamethasone, prednisolone, budesonide	EtOH	9–15	40	10–30	Dexamethasone MP (m.s. 1.82–2.51 μm), prednisolone MP (m.s. 1.96–3.03 μm), budesonide MP (m.s. 3.06–3.58 μm)	[93]
	Telmisartan	EtOH/DCM	12	45	25	SMP, MP (m.s. 0.38–0.60 μm)	[70]
	HCT	DMSO, DMSO/AC	8.6–19	30–40	10–30	NP (0.05–0.21 μm)	[33]
	Oxeglitazar	EtOH/CHF	8	35	30	C	[35]
	Nimesulide	DMSO	9–15	35–45	20–35	AGG or MP (m.s. 1.67–4.04 μm)	[8]
	Piroxicam	DCM	9.7	25	0.05	MP (0.1–5.0 μm)	[12]
	Ketoprofen	DMSO	9–15	40	10–50	MP (m.s. 2.41–3.81 μm)	[65]
	Diflunisal	AC/DCM	12–14	35	18–36	cNP, cMP (size: 0.4–8.1 μm)	[11]
	Folic Acid	DMSO	9–15	35–40	20–40	NP 0.05–0.20, SMP, MP (m.s. 0.30–3.80 μm)	[94]
	α-tocopherol, menadione	DMSO	9–15	35–50	20–60	α-tocopherol MP (m.s. 1.80–4.08 μm), menadione MP (m.s. 2.64–5.09 μm)	[9]
	Quercetin, rutin	DMSO	9	40–50	20–40	Quercetin MP (m.s. 0.47–9.52 μm), rutin MP (m.s. 0.84–8.17 μm)	[97]
	Fisetin	EtOH/DCM	10	45	Not reported	NP, SMP (0.08–0.72 μm)	[31]
	Chrysin	AC/EtOH	12	40	1–3	SMP (m.s. 0.27–0.96 μm)	[116]
	β-carotene	AC/EtOH	8.5–10	40	5–7	NPs (m.s. 0.25– μm), MP (m.s. 0.81–2.43 μm)	[81]
	Anthocyanins	EtOH	10	40	Not reported	C, AGG	[21]
	Curcuma	AC/EtOH	15–21	35–45	2–10	NP (m.s.0.11–0.21 μm)	[115]

Table 2. *Cont.*

Polymeric Carrier	Active Compound	Solvent	P [MPa]	T [°C]	C_{tot} [mg/mL]	Morphology	Reference
	Curcumin	AC/EtOH	8–12	40–60	1	cSMP, SMP (size <1 μm)	[117]
	Curcumin	AC/EtOH	10–20	30–50	1–10	NP, SMP (m.s. 0.03–0.34 μm)	[123]
	Curcumin	EtOH, AC/EtOH	2–12	35–50	5–20	AGG, NP or SMP (m.s. 0.05–0.33 μm)	[80]
PVP and MCC, starch or lactose	Curcumin	AC/EtOH	9–12	40	20	irregular particles/C of MCC (size: 175 μm), starch (size: 15 μm) or lactose (size <5 μm) coated with PVP/curcumin particles	[13]
PLA	Budesonide	DCM	8.6	40	14	MP (1.26 μm)	[32]
	Cholesterol	DCM	9	45	10, 46	MP (1.70 μm), C (8.0 μm)	[124]
	Insulin	DMSO/ DCM	8.5–13	20–38	1	MP (0.50–2.0 μm)	[108]
PLA/PCL	Lutein	EtAc	10	17	21.8–22.2	cMP (m.s. 1.0–10.0 μm)	[45]
	17α-MT	DCM	8	40	0.01	MP (23.0–54.0 μm)	[40]
PLLA	Ibuprofen	DCM	12–20	40–50	5–10	MP (0.93–1.97 μm)	[99]
	Naproxen	DCM	10–20	40–50	5–8	MP (0.56–1.43 μm)	[98]
	Amoxicillin	DMSO/ DCM	10–20	29–50	2–9	MP	[125]
	Rifampicin	DCM	14–21	33–50	10–30	MP (m.s. 3.26–30.53 μm)	[119]
	Azacytidine	DMSO/ DCM	11	40	19	C+MP (2.0 μm)	[111]
	Leuprolide acetate	DCM/ MeOH	13	35	11–12	MP	[41]
	Zidovudine	EtOH/ DCM	8.5–13.5	45	0.2	Filaments	[120]
	5-fluorouracil	EtOH/ DCM	12	33	4	MP (0.98 μm)	[112]
	5-fluorouracil	DMSO/DCM	12–25	35–50	0.1–0.2	AGG	[67]
	HCPT	EtOH/DCM	7.5–12	30–40	1–9	MP (0.57–1.37 μm)	[118]
	Paclitaxel	DCM, DCM/ DMSO, DCM/EtOH	8–14	30–45	7–14	MP (0.83–1.43 μm)	[126]
	Tamoxifen citrate	DCM	13	38	13.5	MP	[113]

Table 2. *Cont.*

Polymeric Carrier	Active Compound	Solvent	P [MPa]	T [°C]	C_{tot} [mg/mL]	Morphology	Reference
zein	Astaxanthin	DCM/AC	8–12	30–42	5–12	Irregular MP (size: 0.6–2.0 µm)	[127]
	Gefitinib	EtOH/DCM	9–12	33–48	6–13	MP (size: 1.1–3.8 µm)	[121]
	Lutein	AC/DMSO	10	32–45	10–20	SMP (m.s. 0.20–0.36 µm)	[49]
	riboflavin, δ-tocopherol, β-carotenein	EtOH	16	40	22–270	EMP, MP (m.s. 8–18 µm)	[148]
	HCPT	DMSO, DMSO/EtOH	8–14	35–45	6–21	Rods, C+NP	[47]
	Lysozyme	EtOH/H_2O	10	40	0.05	Collapsed MP/EMP with internal porosity (size up to 50 µm	[48]
	Diclofenac sodium	DMSO	9	40	30–50	SMP, MP (m.s. 0.42–1.3 µm)	[46]
	Amoxicillin, Ampicillin	DMSO	9	40–50	50	SMP, MP (m.s.: 0.65–12.0 µm for amoxicillin, 0.36–19 µm for ampicillin)	[17]
β-CD	Lycopene	DMF	10–14	40–50	1	NP (m.s. 0.04–0.12 µm)	[128]
	Cetirizine dihydrochloride	DMSO	15	35	10–24	SMP-MP (0.29–4.16 µm)	[22]
	Apigenin	DMF	10–25	35–65	67–402	needles + cSMP (m.s. 0.4 µm)	[114]
	Simvastatin	DCM/EtOH	12	40	Not reported	AGG	[140]
H-β-CD	Tosufloxacin tosylate	DCM/DMF	8–16	35–55	Not reported	C	[139]
	Resveratrol	EtOH	12	40	0.03	AGG/cNP	[137]
	Baicalein	AC/EtOH	8–14	35–50	Not reported	C, AGG (m.s. 0.3–1.0 µm)	[138]
Eudragit RS100, Eudragit RL100	Acetazolamide	AC	5.7–9	27–40	15–24	Needles + cMP, AGG	[43]
Eudragit L100	Ellagic acid	NMP	15	40	20	MP+C	[122]
	Ibuprofen	AC	12–20	40, 50	5–10	NP (0.08–0.21 µm), SMP (51 µm)	[99]
	Naproxen	EtOH	10–20	40–50	5–8	NP (0.08–0.15 µm) SMP (0.31 µm)	[98]
Eudragit L100–55	Ampicillin	DMSO	9–10	40	20–40	cMP, MP (m.s.: 1.5–2.5 µm	[104]

Table 2. *Cont.*

Polymeric Carrier	Active Compound	Solvent	P [MPa]	T [°C]	C_{tot} [mg/mL]	Morphology	Reference
PLGA	Diclofenac sodium, theophylline	DMSO	9–15	40	20–50	C, EMP, cMP, MP (m.s.: 1.5–2.9 µm for diclofenac, 1.6–6.8 µm for theophylline)	[44]
	5-Fluoracil	Acetone, DCM/methanol	11	36, 36–45	6, 6–30	Film, cMP	[109]
	Hydrocortisone	AC, MeOH/DCM	9	33	20	C	[36]
PLGA, PLGA/PLLA	Bupivacaine HCl	EtOH/DCM	8	40	Not reported	MP + Fibers (5.56–7.07 µm), MP (4.39–10.9 µm)	[37]
PEG	β-carotene, Lutein	DCM	8–10	15	13–17	MP and C for β-carotene, AGG and irregular cMP for lutein	[129]
	Carotene	DCM	16	35–50	6	C+cMP (m.s. 1–10 µm)	[149]
	Emodin	DCM/MeOH	8–20	35–50	1	C (m.s. 3–20µm)	[130]
mPEG/PLLA	Itraconazole	AC	19	40	2	C+ MP (m.s. 3 µm)	[106]
	Oxeglitazar	CHF	8	35	30	C	[35]
	Leuprolide acetate	DCM/MeOH	13	35, 15	11–12	MP (m.s. 2.86–5.63 µm)	[41]
Ethyl cellulose	Amoxicillin	DMSO/DCM	10–25	35–65	10	cMP (m.s. 0.23–2.66 µm)	[102]
	Ampicillin	DMSO/DCM	10–25	35–55	11	cMP (m.s. 1.0–3.0 µm)	[100]
	Amoxicillin	DMSO/DCM	10–15	35–50	23–24	AGG, MP	[101]
PMC	Quercetin	EtAc	10	35	1.4	cNP, cSMP (m.s. 0.18–0.34 µm)	[91]
	Rifampicin	EtAc, EtAc/DMSO	10	35–60	13–15	NP+C NP (m.s. 0.19–0.23 µm)	[105]
	Insulin	DMSO/AC	12	32	2, 4–7	NP (0.14 µm), SMP (m.s. 0.27–0.34 µm)	[107]
	Itraconazole	EtOH/DCM	8–15	45–60	Not reported	C+NP/cSMP (size: 0.1–0.5 µm)	[42]
	Lercanidipine	EtOH/DCM	15	40	50	NP, SMP (m.s. 0.22–0.44 µm)	[34]
	Megestrol acetate	EtOH/DCM	15	40	52	NP, SMP (m.s. 0.14–0.50 µm)	[39]

Table 2. *Cont.*

Polymeric Carrier	Active Compound	Solvent	P [MPa]	T [°C]	C_{tot} [mg/mL]	Morphology	Reference
HPMC/poloxamers/ HCO-60	Telmisartan	EtOH/ DCM	12	45	25	SMP (m.s. 0.45–0.50 µm)	[70]
	Felodipine	EtOH/ DCM	10	45	Not reported	cNP/cSMP (m.s. 0.20–0.25 µm)	[23]
	Oxeglitazar	DMC, EtOH/CHF	8	35	23	C	[35]
Poloxamers	Rosemary extracts	EtOH	10	50	Not reported	cSMP	[78]
PHBV	Astaxanthin	AC	8–12	35–40	6	cMP	[150]
	Quercetin	AC	10	40	0.03–0.2	cMP (1 µm)	[151]
	β-carotene	DCM	8	40	21–48	Irregular cMP, EMP, AGG	[135]
	β-carotene	DCM	8–20	40	32–61	Not reported	[96]
	Astaxanthin	DCM	8–10	35	5–10	SMP (m.s. 0.22–0.40 µm)	[131]
	Bixin	DCM	8–10	35–40	1.4–20.4	SMP (m.s. 0.20–0.55 µm)	[132]
	Grape seed extract	DCM	8–12	35–45	27–40	SMP (m.s. 0.62–0.72 µm)	[133]
	Pink pepper extract	DCM	8–12.5	35–55	30	cSMP, MP (m.s. 0.39–25.4 µm)	[134]
CAP	Quercetin	AC	9	40	20	NP+C	[152]
	Mangiferin	AC/DMSO	18	50	8–26	cSMP (m.s. 0.25–0.41 µm)	[153]
HPC	Ezetimibe	EtOH	12–18	40–50	10–25, 50–100	NP (m.s. 0.15–0.24 µm) SMP, MP (m.s. 0.33–0.91 µm)	[110]
	sulfamethoxazole	AC	10	35	88	AGG	[74]
Lactose	Rifampicin	MeOH	12.4	40	1–5	cMP < 8 µm	[154]
Lecithin/α-tocopherol	Lycopene	DMF	8–12	35	20–30	C	[155]
PVM/MA	Gentamicin	AC	10	25	140	NP, MP (m.s. 0.05–0.93 µm)	[156]
PMMA	ivermectin	AC	9–11	40–60	33	cNP, NP (m.s. 0.05–0.17 µm)	[157]
Urea, thiourea or PZA	5-fluorouracil	MeOH	7–15	40	4–8	C	[158]
L-arginine	glimepiride	DCM/EtOH	12	60	7	C (m.s. 5–79 µm)	[24]
Stearic acid	hesperidin	DMSO	8–20	35–45	75	NP-SMP (m.s. 0.15–0.39 µm)	[159]
Saccharin	Indomethacin	EtOH, MeOH, EtAc	8.4–9	50	Not reported	C	[160]

5. Conclusions

This review is focused on the application of the SAS technique in the biomedical field. In this context, the micronization and coprecipitation by the supercritical antisolvent have been proposed for several purposes. Regarding the production of composite particles, the opportunity of modulating the drug release by choosing a carrier based on the required therapy is particularly impressive. In this way, it is possible to release the active principle to a specific site of action and/or at the desired dissolution rate, therefore reducing the side effects caused by drug overdoses and improving the patient's compliance. However, to date, few polymers can be defined as valid for a successful SAS coprecipitation in the form of microspheres, including PVP, PLA, PLLA, zein, and Eudragit L100-55. From the very few conducted studies, β-CD also seems a promising SAS carrier, but its use has yet to be deeply investigated. The main challenge is, therefore, to identify new polymeric carriers suitable for the SAS coprecipitation, in order to develop novel polymer/active compound systems and overcome the main issues still existing in the biomedical field. A key role may be played by the use of polymeric blends, which is still practically unexplored with regard to the SAS technique. Maintaining the microparticles' morphology and varying the content of hydrophobic and hydrophilic polymers, the release can be tuned according to the specific application. Anyway, the use of the SAS technique in the biomedical field is strategic to produce composite systems without residues of solvents that are toxic for human health and to produce particles in a controlled manner with a narrow PSD.

Author Contributions: Conceptualization, P.F. and I.D.M.; methodology, P.F.; writing—original draft preparation, P.F.; writing—review and editing, I.D.M.; supervision, I.D.M. All authors have read and agreed to the submitted version of the manuscript.

Funding: This research received no external funding.

Conflicts of Interest: The authors declare no conflict of interest.

References

1. Montes, A.; Wehner, L.; Pereyra, C.; De La Ossa, E.M. Precipitation of submicron particles of rutin using supercritical antisolvent process. *J. Supercrit. Fluids* **2016**, *118*, 1–10. [CrossRef]
2. Chen, Y.-M.; Tang, M.; Chen, Y.-P. Recrystallization and micronization of sulfathiazole by applying the supercritical antisolvent technology. *Chem. Eng. J.* **2010**, *165*, 358–364. [CrossRef]
3. Zhao, X.; Zu, Y.; Jiang, R.; Wang, Y.; Li, Y.; Li, Q.; Zhao, D.; Zu, B.; Zhang, B.; Sun, Z. Preparation and physicochemical properties of 10-hydroxycamptothecin (HCPT) nanoparticles by supercritical antisolvent (SAS) process. *Int. J. Mol. Sci.* **2011**, *12*, 2678–2691. [CrossRef] [PubMed]
4. Zhao, X.; Zu, Y.; Li, Q.; Wang, M.; Zu, B.; Zhang, X.; Jiang, R.; Zu, C. Preparation and characterization of camptothecin powder micronized by a supercritical antisolvent (SAS) process. *J. Supercrit. Fluids* **2010**, *51*, 412–419. [CrossRef]
5. Zhao, C.; Wang, L.; Zu, Y.; Li, C.; Liu, S.; Yang, L.; Zhao, X.; Zu, B. Micronization of Ginkgo biloba extract using supercritical antisolvent process. *Powder Technol.* **2011**, *209*, 73–80. [CrossRef]
6. Zu, Y.; Zhang, Q.; Zhao, X.; Wang, D.; Li, W.; Sui, X.; Zhang, Y.; Jiang, S.; Wang, Q.; Gu, C. Preparation and characterization of vitexin powder micronized by a supercritical antisolvent (SAS) process. *Powder Technol.* **2012**, *228*, 47–55. [CrossRef]
7. Ha, E.-S.; Park, H.; Lee, S.-K.; Sim, W.-Y.; Jeong, J.-S.; Baek, I.-h.; Kim, M.-S. Pure Trans-Resveratrol Nanoparticles Prepared by A Supercritical Antisolvent Process Using Alcohol and Dichloromethane Mixtures: Effect of Particle Size on Dissolution and Bioavailability in Rats. *Antioxidants* **2020**, *9*, 342. [CrossRef]
8. Prosapio, V.; Reverchon, E.; De Marco, I. Formation of PVP/nimesulide microspheres by supercritical antisolvent coprecipitation. *J. Supercrit. Fluids* **2016**, *118*, 19–26. [CrossRef]
9. Prosapio, V.; Reverchon, E.; De Marco, I. Incorporation of liposoluble vitamins within PVP microparticles using supercritical antisolvent precipitation. *J. CO2 Util.* **2017**, *19*, 230–237. [CrossRef]
10. Zhang, H.-X.; Wang, J.-X.; Zhang, Z.-B.; Le, Y.; Shen, Z.-G.; Chen, J.-F. Micronization of atorvastatin calcium by antisolvent precipitation process. *Int. J. Pharm.* **2009**, *374*, 106–113. [CrossRef]

11. Zahran, F.; Cabañas, A.; Cheda, J.A.R.; Renuncio, J.A.R.; Pando, C. Dissolution rate enhancement of the anti-inflammatory drug diflunisal by coprecipitation with a biocompatible polymer using carbon dioxide as a supercritical fluid antisolvent. *J. Supercrit. Fluids* **2014**, *88*, 56–65. [CrossRef]
12. Wu, K.; Li, J.; Wang, W.; Winstead, D.A. Formation and characterization of solid dispersions of piroxicam and polyvinylpyrrolidone using spray drying and precipitation with compressed antisolvent. *J. Pharm. Sci.* **2009**, *98*, 2422–2431. [CrossRef] [PubMed]
13. Matos, R.L.; Lu, T.; Leeke, G.; Prosapio, V.; McConville, C.; Ingram, A. Single-step coprecipitation and coating to prepare curcumin formulations by supercritical fluid technology. *J. Supercrit. Fluids* **2020**, *159*, 104758. [CrossRef]
14. Stebbins, N.D.; Ouimet, M.A.; Uhrich, K.E. Antibiotic-containing polymers for localized, sustained drug delivery. *Adv. Drug Del. Rev.* **2014**, *78*, 77–87. [CrossRef] [PubMed]
15. Gürsel, İ.; Korkusuz, F.; Türesin, F.; Alaeddinoğlu, N.G.; Hasırcı, V. In vivo application of biodegradable controlled antibiotic release systems for the treatment of implant-related osteomyelitis. *Biomaterials* **2000**, *22*, 73–80. [CrossRef]
16. Liu, Z.; Lu, W.; Qian, L.; Zhang, X.; Zeng, P.; Pan, J. In vitro and in vivo studies on mucoadhesive microspheres of amoxicillin. *J. Control. Release* **2005**, *102*, 135–144. [CrossRef]
17. Franco, P.; Reverchon, E.; De Marco, I. Production of zein/antibiotic microparticles by supercritical antisolvent coprecipitation. *J. Supercrit. Fluids* **2019**, *145*, 31–38. [CrossRef]
18. Altman, R.; Bosch, B.; Brune, K.; Patrignani, P.; Young, C. Advances in NSAID development: Evolution of diclofenac products using pharmaceutical technology. *Drugs* **2015**, *75*, 859–877. [CrossRef]
19. Gupta, A.; Bah, M. NSAIDs in the treatment of postoperative pain. *Curr. Pain Headache Rep.* **2016**, *20*, 62. [CrossRef]
20. Maniar, K.H.; Jones, I.A.; Gopalakrishna, R.; Vangsness, C.T., Jr. Lowering side effects of NSAID usage in osteoarthritis: Recent attempts at minimizing dosage. *Expert Opin. Pharmacother.* **2018**, *19*, 93–102. [CrossRef]
21. Da Fonseca Machado, A.P.; Rezende, C.A.; Rodrigues, R.A.; Barbero, G.F.; e Rosa, P.d.T.V.; Martínez, J. Encapsulation of anthocyanin-rich extract from blackberry residues by spray-drying, freeze-drying and supercritical antisolvent. *Powder Technol.* **2018**, *340*, 553–562. [CrossRef]
22. Lee, C.-W.; Kim, S.-J.; Youn, Y.-S.; Widjojokusumo, E.; Lee, Y.-H.; Kim, J.; Lee, Y.-W.; Tjandrawinata, R.R. Preparation of bitter taste masked cetirizine dihydrochloride/β-cyclodextrin inclusion complex by supercritical antisolvent (SAS) process. *J. Supercrit. Fluids* **2010**, *55*, 348–357. [CrossRef]
23. Won, D.-H.; Kim, M.-S.; Lee, S.; Park, J.-S.; Hwang, S.-J. Improved physicochemical characteristics of felodipine solid dispersion particles by supercritical anti-solvent precipitation process. *Int. J. Pharm.* **2005**, *301*, 199–208. [CrossRef] [PubMed]
24. Park, H.; Seo, H.J.; Hong, S.-h.; Ha, E.-S.; Lee, S.; Kim, J.-S.; Baek, I.-h.; Kim, M.-S.; Hwang, S.-J. Characterization and therapeutic efficacy evaluation of glimepiride and L-arginine co-amorphous formulation prepared by supercritical antisolvent process: Influence of molar ratio and preparation methods. *Int. J. Pharm.* **2020**, *581*, 119232. [CrossRef] [PubMed]
25. Park, J.; Park, H.J.; Cho, W.; Cha, K.-H.; Kang, Y.-S.; Hwang, S.-J. Preparation and pharmaceutical characterization of amorphous cefdinir using spray-drying and SAS-process. *Int. J. Pharm.* **2010**, *396*, 239–245. [CrossRef] [PubMed]
26. Yeo, S.D.; Debenedetti, P.G.; Radosz, M.; Schmidt, H.W. Supercritical Antisolvent Process for Substituted Para-Linked Aromatic Polyamides: Phase Equilibrium and Morphology Study. *Macromolecules* **1993**, *26*, 6207–6210. [CrossRef]
27. Yeo, S.D.; Debenedetti, P.G.; Radosz, M.; Giesa, R.; Schmidt, H.W. Supercritical Antisolvent Process for a Series of Substituted Para-Linked Aromatic Polyamides. *Macromolecules* **1995**, *28*, 1316–1317. [CrossRef]
28. Bertucco, A.; Pallado, P.; Benedetti, L. Formation of biocompatible polymer microspheres for controlled drug delivery by a supercritical antisolvent technique. *Process Technol. Proc.* **1996**, *12*, 217–222.
29. Reverchon, E.; Della Porta, G.; Di Trolio, A.; Pace, S. Supercritical Antisolvent Precipitation of Nanoparticles of Superconductor Precursors. *Ind. Eng. Chem. Res.* **1998**, *37*, 952–958. [CrossRef]
30. Reverchon, E.; Celano, C.; Della Porta, G.; Di Trolio, A.; Pace, S. Supercritical antisolvent precipitation: A new technique for preparing submicronic yttrium powders to improve YBCO superconductor. *J. Mater. Res.* **1998**, *13*, 284–289. [CrossRef]

31. Chen, L.-F.; Xu, P.-Y.; Fu, C.-P.; Kankala, R.K.; Chen, A.-Z.; Wang, S.-B. Fabrication of Supercritical Antisolvent (SAS) Process-Assisted Fisetin-Encapsulated Poly (Vinyl Pyrrolidone)(PVP) Nanocomposites for Improved Anticancer Therapy. *Nanomaterials* **2020**, *10*, 322. [CrossRef] [PubMed]
32. Martin, T.M.; Bandi, N.; Shulz, R.; Roberts, C.B.; Kompella, U.B. Preparation of budesonide and budesonide-PLA microparticles using supercritical fluid precipitation technology. *AAPS PharmSciTech* **2002**, *3*, 16–26. [CrossRef] [PubMed]
33. Park, H.J.; Yoon, T.J.; Kwon, D.E.; Yu, K.; Lee, Y.-W. Coprecipitation of hydrochlorothiazide/PVP for the dissolution rate improvement by precipitation with compressed fluid antisolvent process. *J. Supercrit. Fluids* **2017**, *126*, 37–46. [CrossRef]
34. Ha, E.-S.; Choo, G.-H.; Baek, I.-h.; Kim, J.-S.; Cho, W.; Jung, Y.S.; Jin, S.-E.; Hwang, S.-J.; Kim, M.-S. Dissolution and bioavailability of lercanidipine–hydroxypropylmethyl cellulose nanoparticles with surfactant. *Int. J. Biol. Macromol.* **2015**, *72*, 218–222. [CrossRef]
35. Majerik, V.; Charbit, G.; Badens, E.; Horváth, G.; Szokonya, L.; Bosc, N.; Teillaud, E. Bioavailability enhancement of an active substance by supercritical antisolvent precipitation. *J. Supercrit. Fluids* **2007**, *40*, 101–110. [CrossRef]
36. Wang, Y.; Wang, Y.; Yang, J.; Pfeffer, R.; Dave, R.; Michniak, B. The application of a supercritical antisolvent process for sustained drug delivery. *Powder Technol.* **2006**, *164*, 94–102. [CrossRef]
37. Lee, S.; Kim, M.; Kim, J.; Park, H.; Woo, J.; Lee, B.; Hwang, S.J. Controlled delivery of a hydrophilic drug from a biodegradable microsphere system by supercritical anti-solvent precipitation technique. *J. Microencapsul.* **2006**, *23*, 741–749. [CrossRef]
38. Liu, M.; Liu, Y.; Ge, Y.; Zhong, Z.; Wang, Z.; Wu, T.; Zhao, X.; Zu, Y. Solubility, Antioxidation, and Oral Bioavailability Improvement of Mangiferin Microparticles Prepared Using the Supercritical Antisolvent Method. *Pharmaceutics* **2020**, *12*, 90. [CrossRef]
39. Ha, E.S.; Kim, J.S.; Baek, I.H.; Yoo, J.W.; Jung, Y.; Moon, H.R.; Kim, M.S. Development of megestrol acetate solid dispersion nanoparticles for enhanced oral delivery by using a supercritical antisolvent process. *Drug Des. Dev. Ther.* **2015**, *9*, 4269–4277.
40. Sacchetin, P.S.C.; Setti, R.F.; e Rosa, P.D.T.V.; Moraes, Â.M. Properties of PLA/PCL particles as vehicles for oral delivery of the androgen hormone 17α-methyltestosterone. *Mater. Sci. Eng. C* **2016**, *58*, 870–881. [CrossRef]
41. Jung, I.-I.; Haam, S.; Lim, G.; Ryu, J.-H. Preparation of peptide-loaded polymer microparticles using supercritical carbon dioxide. *Biotechnol. Bioprocess Eng.* **2012**, *17*, 185–194. [CrossRef]
42. Lee, S.; Nam, K.; Kim, M.S.; Jun, S.W.; Park, J.-S.; Woo, J.S.; Hwang, S.-J. Preparation and characterization of solid dispersions of itraconazole by using aerosol solvent extraction system for improvement in drug solubility and bioavailability. *Arch. Pharm. Res.* **2005**, *28*, 866–874. [CrossRef] [PubMed]
43. Duarte, A.R.C.; Roy, C.; Vega-González, A.; Duarte, C.M.; Subra-Paternault, P. Preparation of acetazolamide composite microparticles by supercritical anti-solvent techniques. *Int. J. Pharm.* **2007**, *332*, 132–139. [CrossRef] [PubMed]
44. Franco, P.; de Marco, I. Eudragit: A Novel Carrier for Controlled Drug Delivery in Supercritical Antisolvent Coprecipitation. *Polymers* **2020**, *12*, 234. [CrossRef] [PubMed]
45. Miguel, F.; Martín, A.; Mattea, F.; Cocero, M.J. Precipitation of lutein and co-precipitation of lutein and poly-lactic acid with the supercritical anti-solvent process. *Chem. Eng. Process.* **2008**, *47*, 1594–1602. [CrossRef]
46. Franco, P.; Reverchon, E.; De Marco, I. Zein/diclofenac sodium coprecipitation at micrometric and nanometric range by supercritical antisolvent processing. *J. CO2 Util.* **2018**, *27*, 366–373. [CrossRef]
47. Liu, G.; Li, S.; Huang, Y.; Wang, H.; Jiang, Y. Incorporation of 10-hydroxycamptothecin nanocrystals into zein microspheres. *Chem. Eng. Sci.* **2016**, *155*, 405–414. [CrossRef]
48. Zhong, Q.; Jin, M.; Davidson, P.M.; Zivanovic, S. Sustained release of lysozyme from zein microcapsules produced by a supercritical anti-solvent process. *Food Chem.* **2009**, *115*, 697–700. [CrossRef]
49. Hu, D.; Lin, C.; Liu, L.; Li, S.; Zhao, Y. Preparation, characterization, and in vitro release investigation of lutein/zein nanoparticles via solution enhanced dispersion by supercritical fluids. *J. Food Eng.* **2012**, *109*, 545–552. [CrossRef]
50. Abuzar, S.M.; Hyun, S.-M.; Kim, J.-H.; Park, H.J.; Kim, M.-S.; Park, J.-S.; Hwang, S.-J. Enhancing the solubility and bioavailability of poorly water-soluble drugs using supercritical antisolvent (SAS) process. *Int. J. Pharm.* **2018**, *538*, 1–13. [CrossRef]

51. Kalani, M.; Yunus, R. Application of supercritical antisolvent method in drug encapsulation: A review. *Int. J. Nanomed.* **2011**, *6*, 1429. [CrossRef] [PubMed]
52. Gonzalez-Coloma, A.; Martín, L.; Mainar, A.; Urieta, J.; Fraga, B.M.; Rodríguez-Vallejo, V.; Díaz, C.E. Supercritical extraction and supercritical antisolvent fractionation of natural products from plant material: Comparative results on Persea indica. *Phytochem. Rev.* **2012**, *11*, 433–446. [CrossRef]
53. Reverchon, E. Supercritical antisolvent precipitation of micro-and nano-particles. *J. Supercrit. Fluids* **1999**, *15*, 1–21. [CrossRef]
54. Savjani, K.T.; Gajjar, A.K.; Savjani, J.K. Drug solubility: Importance and enhancement techniques. *ISRN Pharm.* **2012**, *2012*, 195727. [CrossRef] [PubMed]
55. Vieth, M.; Siegel, M.G.; Higgs, R.E.; Watson, I.A.; Robertson, D.H.; Savin, K.A.; Durst, G.L.; Hipskind, P.A. Characteristic physical properties and structural fragments of marketed oral drugs. *J. Med. Chem.* **2004**, *47*, 224–232. [CrossRef]
56. Rodriguez-Aller, M.; Guillarme, D.; Veuthey, J.-L.; Gurny, R. Strategies for formulating and delivering poorly water-soluble drugs. *J. Drug Deliv. Sci. Technol.* **2015**, *30*, 342–351. [CrossRef]
57. Takagi, T.; Ramachandran, C.; Bermejo, M.; Yamashita, S.; Yu, L.X.; Amidon, G.L. A provisional biopharmaceutical classification of the top 200 oral drug products in the United States, Great Britain, Spain, and Japan. *J. CO2 Util. Pharm.* **2006**, 3, 631–643. [CrossRef]
58. Tenorio, A.; Gordillo, M.D.; Pereyra, C.; Martinez de la Ossa, E.J. Controlled submicro particle formation of ampicillin by supercritical antisolvent precipitation. *J. Supercrit. Fluids* **2007**, *40*, 308–316. [CrossRef]
59. Kalogiannis, C.G.; Pavlidou, E.; Panayiotou, C.G. Production of amoxicillin microparticles by supercritical antisolvent precipitation. *Ind. Eng. Chem. Res.* **2005**, *44*, 9339–9346. [CrossRef]
60. Reverchon, E.; Della Porta, G.; Falivene, M. Process parameters and morphology in amoxicillin micro and submicro particles generation by supercritical antisolvent precipitation. *J. Supercrit. Fluids* **2000**, *17*, 239–248. [CrossRef]
61. Reverchon, E.; De Marco, I. Supercritical antisolvent precipitation of cephalosporins. *Powder Technol.* **2006**, *164*, 139–146. [CrossRef]
62. Reverchon, E.; De Marco, I.; Della Porta, G. Rifampicin microparticles production by supercritical antisolvent precipitation. *Int. J. Pharm.* **2002**, *243*, 83–91. [CrossRef]
63. Cardoso, M.T.; Monteiro, G.; Cardoso, J.; Prazeres, T.; Figueiredo, J.; Martinho, J.; Cabral, J.; Palavra, A. Supercritical antisolvent micronization of minocycline hydrochloride. *J. Supercrit. Fluids* **2008**, *44*, 238–244. [CrossRef]
64. Cardoso, M.T.; Geraldes, V.; Cabral, J.; Palavra, A. Characterization of minocycline powder micronized by a supercritical antisolvent (SAS) process. *J. Supercrit. Fluids* **2008**, *46*, 71–76. [CrossRef]
65. Franco, P.; Reverchon, E.; De Marco, I. PVP/ketoprofen coprecipitation using supercritical antisolvent process. *Powder Technol.* **2018**, *340*, 1–7. [CrossRef]
66. Montes, A.; Wehner, L.; Pereyra, C.; Martínez de la Ossa, E.J. Mangiferin nanoparticles precipitation by supercritical antisolvent process. *J. Supercrit. Fluids* **2016**, *112*, 44–50. [CrossRef]
67. Cuadra, I.A.; Zahran, F.; Martín, D.; Cabañas, A.; Pando, C. Preparation of 5-fluorouracil microparticles and 5-fluorouracil/poly(l-lactide) composites by a supercritical CO_2 antisolvent process. *J. Supercrit. Fluids* **2019**, *143*, 64–71. [CrossRef]
68. Widjojokusumo, E.; Veriansyah, B.; Tjandrawinata, R.R. Supercritical anti-solvent (SAS) micronization of Manilkara kauki bioactive fraction (DLBS2347). *J. CO2 Util.* **2013**, *3-4*, 30–36. [CrossRef]
69. Zhao, X.; Chen, X.; Zu, Y.; Jiang, R.; Zhao, D. Recrystallization and micronization of taxol using the supercritical antisolvent (SAS) process. *Ind. Eng. Chem. Res.* **2012**, *51*, 9591–9597. [CrossRef]
70. Park, J.; Cho, W.; Cha, K.-H.; Ahn, J.; Han, K.; Hwang, S.-J. Solubilization of the poorly water soluble drug, telmisartan, using supercritical anti-solvent (SAS) process. *Int. J. Pharm.* **2013**, *441*, 50–55. [CrossRef]
71. Reverchon, E.; Della Porta, G.; Pallado, P. Supercritical antisolvent precipitation of salbutamol microparticles. *Powder Technol.* **2001**, *114*, 17–22. [CrossRef]
72. Kim, M.-S.; Jin, S.-J.; Kim, J.-S.; Park, H.J.; Song, H.-S.; Neubert, R.H.; Hwang, S.-J. Preparation, characterization and in vivo evaluation of amorphous atorvastatin calcium nanoparticles using supercritical antisolvent (SAS) process. *Eur. J. Pharm. Biopharm.* **2008**, *69*, 454–465. [CrossRef] [PubMed]

73. Rogers, T.L.; Johnston, K.P.; Williams, R.O., III. Solution-based particle formation of pharmaceutical powders by supercritical or compressed fluid CO_2 and cryogenic spray-freezing technologies. *Drug Dev. Ind. Pharm.* **2001**, *27*, 1003–1015. [CrossRef] [PubMed]
74. Chang, Y.-P.; Tang, M.; Chen, Y.-P. Micronization of sulfamethoxazole using the supercritical anti-solvent process. *J. Mater. Sci.* **2008**, *43*, 2328–2335. [CrossRef]
75. Hiendrawan, S.; Veriansyah, B.; Widjojokusumo, E.; Soewandhi, S.; Wikarsa, S.; Tjandrawinata, R.R. Simultaneous cocrystallization and micronization of paracetamol-dipicolinic acid cocrystal by supercritical antisolvent (SAS). *Int. J. Pharm. Pharm. Sci.* **2016**, *8*, 89–98.
76. Kim, M.-S.; Lee, S.; Park, J.-S.; Woo, J.-S.; Hwang, S.-J. Micronization of cilostazol using supercritical antisolvent (SAS) process: Effect of process parameters. *Powder Technol.* **2007**, *177*, 64–70. [CrossRef]
77. Kudryashova, E.; Sukhoverkov, K.; Deygen, I.; Vorobei, A.; Pokrovskiy, O.; Parenago, O.; Presnov, D.; Egorov, A. Moxifloxacin Micronization via Supercritical Antisolvent Precipitation. *Russ. J. Phys. Chem. B* **2017**, *11*, 1153–1162. [CrossRef]
78. Visentin, A.; Rodríguez-Rojo, S.; Navarrete, A.; Maestri, D.; Cocero, M.J. Precipitation and encapsulation of rosemary antioxidants by supercritical antisolvent process. *J. Food Eng.* **2012**, *109*, 9–15. [CrossRef]
79. De Marco, I.; Prosapio, V.; Cice, F.; Reverchon, E. Use of solvent mixtures in supercritical antisolvent process to modify precipitates morphology: Cellulose acetate microparticles. *J. Supercrit. Fluids* **2013**, *83*, 153–160. [CrossRef]
80. Matos, R.L.; Lu, T.; Prosapio, V.; McConville, C.; Leeke, G.; Ingram, A. Coprecipitation of curcumin/PVP with enhanced dissolution properties by the supercritical antisolvent process. *J. CO2 Util.* **2019**, *30*, 48–62. [CrossRef]
81. Prosapio, V.; Reverchon, E.; De Marco, I. Coprecipitation of polyvinylpyrrolidone/β-carotene by supercritical antisolvent processing. *Ind. Eng. Chem. Res.* **2015**, *54*, 11568–11575. [CrossRef]
82. Campardelli, R.; Reverchon, E.; De Marco, I. Dependence of SAS particle morphologies on the ternary phase equilibria. *J. Supercrit. Fluids* **2017**, *130*, 273–281. [CrossRef]
83. Campardelli, R.; Reverchon, E.; De Marco, I. PVP microparticles precipitation from acetone-ethanol mixtures using SAS process: Effect of phase behavior. *J. Supercrit. Fluids* **2019**, *143*, 321–329. [CrossRef]
84. Martín, A.n.; Scholle, K.; Mattea, F.; Meterc, D.; Cocero, M.a.J. Production of polymorphs of ibuprofen sodium by supercritical antisolvent (SAS) precipitation. *Cryst. Growth Des.* **2009**, *9*, 2504–2511. [CrossRef]
85. Reverchon, E.; Della Porta, G. Production of antibiotic micro-and nano-particles by supercritical antisolvent precipitation. *Powder Technol.* **1999**, *106*, 23–29. [CrossRef]
86. Chattopadhyay, P.; Gupta, R.B. Production of griseofulvin nanoparticles using supercritical CO2 antisolvent with enhanced mass transfer. *Int. J. Pharm.* **2001**, *228*, 19–31. [CrossRef]
87. Bagratashvili, V.; Egorov, A.; Krotova, L.; Mironov, A.; Panchenko, V.Y.; Parenago, O.; Popov, V.; Revelsky, I.; Timashev, P.; Tsypina, S. Supercritical fluid micronization of risperidone pharmaceutical substance. *Russ. J. Phys. Chem. B* **2012**, *6*, 804–812. [CrossRef]
88. Su, C.S.; Lo, W.S.; Lien, L.H. Micronization of fluticasone propionate using supercritical antisolvent (SAS) process. *Chem. Eng. Technol.* **2011**, *34*, 535–541. [CrossRef]
89. Quintana, S.E.; Hernández, D.M.; Villanueva-Bermejo, D.; García-Risco, M.R.; Fornari, T. Fractionation and precipitation of licorice (Glycyrrhiza glabra L.) phytochemicals by supercritical antisolvent (SAS) technique. *LWT* **2020**, 109315. [CrossRef]
90. Miguel, F.; Martin, A.; Gamse, T.; Cocero, M.J. Supercritical anti solvent precipitation of lycopene: Effect of the operating parameters. *J. Supercrit. Fluids* **2006**, *36*, 225–235. [CrossRef]
91. Fernández-Ponce, M.T.; Masmoudi, Y.; Djerafi, R.; Casas, L.; Mantell, C.; de La Ossa, E.M.; Badens, E. Particle design applied to quercetin using supercritical anti-solvent techniques. *J. Supercrit. Fluids* **2015**, *105*, 119–127. [CrossRef]
92. Montes, A.; Pereyra, C.; de la Ossa, E.M. Screening design of experiment applied to the supercritical antisolvent precipitation of quercetin. *J. Supercrit. Fluids* **2015**, *104*, 10–18. [CrossRef]
93. Prosapio, V.; De Marco, I.; Reverchon, E. PVP/corticosteroid microspheres produced by supercritical antisolvent coprecipitation. *Chem. Eng. J.* **2016**, *292*, 264–275. [CrossRef]
94. Prosapio, V.; De Marco, I.; Scognamiglio, M.; Reverchon, E. Folic acid–PVP nanostructured composite microparticles by supercritical antisolvent precipitation. *Chem. Eng. J.* **2015**, *277*, 286–294. [CrossRef]

95. Franceschi, E.; De Cesaro, A.M.; Feiten, M.; Ferreira, S.R.S.; Dariva, C.; Kunita, M.H.; Rubira, A.F.; Muniz, E.C.; Corazza, M.L.; Oliveira, J.V. Precipitation of β-carotene and PHBV and co-precipitation from SEDS technique using supercritical CO_2. *J. Supercrit. Fluids* **2008**, *47*, 259–269. [CrossRef]
96. Priamo, W.L.; De Cezaro, A.M.; Ferreira, S.R.S.; Oliveira, J.V. Precipitation and encapsulation of β-carotene in PHBV using carbon dioxide as anti-solvent. *J. Supercrit. Fluids* **2010**, *54*, 103–109. [CrossRef]
97. Ozkan, G.; Franco, P.; Capanoglu, E.; De Marco, I. PVP/flavonoid coprecipitation by supercritical antisolvent process. *Chem. Eng. Process.* **2019**, *146*, 107689. [CrossRef]
98. Montes, A.; Kin, N.; Gordillo, M.D.; Pereyra, C.; Martínez de la Ossa, E.J. Polymer–naproxen precipitation by supercritical antisolvent (SAS) process. *J. Supercrit. Fluids* **2014**, *89*, 58–67. [CrossRef]
99. Montes, A.; Gordillo, M.D.; Pereyra, C.; De los Santos, D.M.; Martínez de la Ossa, E.J. Ibuprofen–polymer precipitation using supercritical CO_2 at low temperature. *J. Supercrit. Fluids* **2014**, *94*, 91–101. [CrossRef]
100. Montes, A.; Gordillo, M.D.; Pereyra, C.; Martínez de la Ossa, E.J. Polymer and ampicillin co-precipitation by supercritical antisolvent process. *J. Supercrit. Fluids* **2012**, *63*, 92–98. [CrossRef]
101. Montes, A.; Nunes, A.; Gordillo, M.; Pereyra, C.; Duarte, C.M.; Martinez de la Ossa, E.J. Amoxicillin and ethyl cellulose precipitation by two supercritical antisolvent processes. *Chem. Eng. Technol.* **2013**, *36*, 665–672. [CrossRef]
102. Montes, A.; Gordillo, M.D.; Pereyra, C.; Martínez de la Ossa, E.J. Co-precipitation of amoxicillin and ethyl cellulose microparticles by supercritical antisolvent process. *J. Supercrit. Fluids* **2011**, *60*, 75–80. [CrossRef]
103. Uzun, İ.N.; Sipahigil, O.; Dinçer, S. Coprecipitation of Cefuroxime Axetil–PVP composite microparticles by batch supercritical antisolvent process. *J. Supercrit. Fluids* **2011**, *55*, 1059–1069. [CrossRef]
104. De Marco, I.; Franco, P. Production of Eudragit/ampicillin Microparticles by Supercritical Antisolvent Coprecipitation. *Chem. Eng. Trans.* **2020**, *79*, 229–234.
105. Djerafi, R.; Swanepoel, A.; Crampon, C.; Kalombo, L.; Labuschagne, P.; Badens, E.; Masmoudi, Y. Supercritical antisolvent co-precipitation of rifampicin and ethyl cellulose. *Eur. J. Pharm. Sci.* **2017**, *102*, 161–171. [CrossRef]
106. Barrett, A.M.; Dehghani, F.; Foster, N.R. Increasing the Dissolution Rate of Itraconazole Processed by Gas Antisolvent Techniques using Polyethylene Glycol as a Carrier. *Pharm. Res.* **2007**, *25*, 1274–1289. [CrossRef]
107. Jin, H.Y.; Xia, F.; Zhao, Y.P. Preparation of hydroxypropyl methyl cellulose phthalate nanoparticles with mixed solvent using supercritical antisolvent process and its application in co-precipitation of insulin. *Adv. Powder Technol.* **2012**, *23*, 157–163. [CrossRef]
108. Elvassore, N.; Bertucco, A.; Caliceti, P. Production of protein-loaded polymeric microcapsules by compressed CO_2 in a mixed solvent. *Ind. Eng. Chem. Res.* **2001**, *40*, 795–800. [CrossRef]
109. Kalantarian, P.; Haririan, I.; Najafabadi, A.R.; Shokrgozar, M.A.; Vatanara, A. Entrapment of 5-fluorouracil into PLGA matrices using supercritical antisolvent processes. *J. Pharm. Pharmacol.* **2011**, *63*, 500–506. [CrossRef]
110. Ha, E.-S.; Kim, J.-S.; Baek, I.-h.; Hwang, S.-J.; Kim, M.-S. Enhancement of dissolution and bioavailability of ezetimibe by amorphous solid dispersion nanoparticles fabricated using supercritical antisolvent process. *J. Pharm. Investig.* **2015**, *45*, 641–649. [CrossRef]
111. Argemí, A.; Vega, A.; Subra-Paternault, P.; Saurina, J. Characterization of azacytidine/poly(l-lactic) acid particles prepared by supercritical antisolvent precipitation. *J. Pharm. Biomed. Anal.* **2009**, *50*, 847–852. [CrossRef] [PubMed]
112. Chen, A.Z.; Pu, X.M.; Kang, Y.Q.; Liao, L.; Yao, Y.D.; Yin, G.F. Preparation of 5–Fluorouracil–Poly (L–lactide) Microparticles Using Solution–Enhanced Dispersion by Supercritical CO_2. *Macromol. Rapid Commun.* **2006**, *27*, 1254–1259. [CrossRef]
113. Alias, D.; Yunus, R.; Chong, G.H.; Che Abdullah, C.A. Single step encapsulation process of tamoxifen in biodegradable polymer using supercritical anti-solvent (SAS) process. *Powder Technol.* **2017**, *309*, 89–94. [CrossRef]
114. Huang, Y.; Zu, Y.; Zhao, X.; Wu, M.; Feng, Z.; Deng, Y.; Zu, C.; Wang, L. Preparation of inclusion complex of apigenin-hydroxypropyl-β-cyclodextrin by using supercritical antisolvent process for dissolution and bioavailability enhancement. *Int. J. Pharm.* **2016**, *511*, 921–930. [CrossRef]
115. Lestari, S.D.; Machmudah, S.; Winardi, S.; Kanda, H.; Goto, M. Particle micronization of Curcuma mangga rhizomes ethanolic extract/biopolymer PVP using supercritical antisolvent process. *J. Supercrit. Fluids* **2019**, *146*, 226–239. [CrossRef]

116. Pan, Y.-J.; Xu, P.-Y.; Chen, B.-Q.; Fu, C.-P.; Kankala, R.K.; Chen, A.-Z.; Wang, S.-B. Supercritical Antisolvent Process-assisted Fabrication of Chrysin-polyvinylpyrrolidone Sub-microparticles for Improved Anticancer Efficiency. *J. Supercrit. Fluids* **2020**, *162*, 104847. [CrossRef]
117. Machmudah, S.; Winardi, S.; Wahyudiono; Kanda, H.; Goto, M. Formation of Fine Particles from Curcumin/PVP by the Supercritical Antisolvent Process with a Coaxial Nozzle. *ACS Omega* **2020**, *5*, 6705–6714. [CrossRef]
118. Wang, W.; Liu, G.; Wu, J.; Jiang, Y. Co-precipitation of 10-hydroxycamptothecin and poly (L-lactic acid) by supercritical CO_2 anti-solvent process using dichloromethane/ethanol co-solvent. *J. Supercrit. Fluids* **2013**, *74*, 137–144. [CrossRef]
119. Patomchaiviwat, V.; Paeratakul, O.; Kulvanich, P. Formation of inhalable rifampicin–poly (L-lactide) microparticles by supercritical anti-solvent process. *AAPS PharmSciTech* **2008**, *9*, 1119–1129. [CrossRef]
120. Yoshida, V.M.H.; Balcão, V.M.; Vila, M.M.D.C.; Oliveira Júnior, J.M.; Aranha, N.; Chaud, M.V.; Gremião, M.P.D. Zidovudine-Poly(L-Lactic Acid) Solid Dispersions with Improved Intestinal Permeability Prepared by Supercritical Antisolvent Process. *J. Pharm. Sci.* **2015**, *104*, 1691–1700. [CrossRef]
121. Lin, Q.; Liu, G.; Zhao, Z.; Wei, D.; Pang, J.; Jiang, Y. Design of gefitinib-loaded poly (L-lactic acid) microspheres via a supercritical anti-solvent process for dry powder inhalation. *Int. J. Pharm.* **2017**, *532*, 573–580. [CrossRef] [PubMed]
122. Montes, A.; Wehner, L.; Pereyra, C.; Martínez de la Ossa, E.J. Generation of microparticles of ellagic acid by supercritical antisolvent process. *J. Supercrit. Fluids* **2016**, *116*, 101–110. [CrossRef]
123. Chhouk, K.; Kanda, H.; Kawasaki, S.-I.; Goto, M. Micronization of curcumin with biodegradable polymer by supercritical anti-solvent using micro swirl mixer. *Front. Chem. Sci. Eng.* **2018**, *12*, 184–193. [CrossRef]
124. Guha, R.; Vinjamur, M.; Mukhopadhyay, M. Demonstration of mechanisms for coprecipitation and encapsulation by supercritical antisolvent process. *Ind. Eng. Chem. Res.* **2011**, *50*, 1079–1088. [CrossRef]
125. Kalogiannis, C.G.; Michailof, C.M.; Panayiotou, C.G. Microencapsulation of amoxicillin in poly (l-lactic acid) by supercritical antisolvent precipitation. *Ind. Eng. Chem. Res.* **2006**, *45*, 8738–8743. [CrossRef]
126. Li, W.; Liu, G.; Li, L.; Wu, J.; LÜ, Y.; Jiang, Y. Effect of Process Parameters on Co-precipitation of Paclitaxel and Poly(L-lactic Acid) by Supercritical Antisolvent Process. *Chin. J. Chem. Eng.* **2012**, *20*, 803–813. [CrossRef]
127. Liu, G.; Hu, M.; Zhao, Z.; Lin, Q.; Wei, D.; Jiang, Y. Enhancing the stability of astaxanthin by encapsulation in poly (L-lactic acid) microspheres using a supercritical anti-solvent process. *Particuology* **2019**, *44*, 54–62. [CrossRef]
128. Nerome, H.; Machmudah, S.; Wahyudiono; Fukuzato, R.; Higashiura, T.; Youn, Y.S.; Lee, Y.W.; Goto, M. Nanoparticle formation of lycopene/β-cyclodextrin inclusion complex using supercritical antisolvent precipitation. *J. Supercrit. Fluids* **2013**, *83*, 97–103. [CrossRef]
129. Martín, A.; Mattea, F.; Gutiérrez, L.; Miguel, F.; Cocero, M.J. Co-precipitation of carotenoids and bio-polymers with the supercritical anti-solvent process. *J. Supercrit. Fluids* **2007**, *41*, 138–147. [CrossRef]
130. Lang, Z.M.; Hong, H.L.; Han, L.M.; Zhu, N.; Suo, Q.L. Preparation of emodin-polyethylene glycol composite microparticles using a supercritical antisolvent process. *Chem. Eng. Technol.* **2012**, *35*, 362–368. [CrossRef]
131. Machado, F.R.S.; Trevisol, T.C.; Boschetto, D.L.; Burkert, J.F.M.; Ferreira, S.R.S.; Oliveira, J.V.; Burkert, C.A.V. Technological process for cell disruption, extraction and encapsulation of astaxanthin from Haematococcus pluvialis. *J. Biotechnol.* **2016**, *218*, 108–114. [CrossRef] [PubMed]
132. Boschetto, D.L.; Aranha, E.M.; de Souza, A.A.U.; Souza, S.M.A.G.U.; Ferreira, S.R.S.; Priamo, W.L.; Oliveira, J.V. Encapsulation of bixin in PHBV using SEDS technique and in vitro release evaluation. *Ind Crops Prod.* **2014**, *60*, 22–29. [CrossRef]
133. Boschetto, D.L.; Dalmolin, I.; de Cesaro, A.M.; Rigo, A.A.; Ferreira, S.R.S.; Meireles, M.A.A.; Batista, E.A.C.; Vladimir Oliveira, J. Phase behavior and process parameters effect on grape seed extract encapsulation by SEDS technique. *Ind. Crops Prod.* **2013**, *50*, 352–360. [CrossRef]
134. Andrade, K.S.; Aguiar, G.P.S.; Rebelatto, E.A.; Lanza, M.; Oliveira, J.V.; Ferreira, S.R. Encapsulation of pink pepper extract by SEDS technique: Phase behavior data and process parameters. *J. Supercrit. Fluids* **2020**, *161*, 104822. [CrossRef]
135. Franceschi, E.; De Cezaro, A.; Ferreira, S.R.S.; Kunita, M.H.; Muniz, E.C.; Rubira, A.F.; Oliveira, J.V. Co-precipitation of beta-carotene and bio-polymer using supercritical carbon dioxide as antisolvent. *Open Chem. Eng. J.* **2010**, *4*, 11–20. [CrossRef]

136. Lee, S.Y.; Abdullah, L.C.; Rahman, R.A.; Abas, F.; Chong, G.H. Role of polymers as crystal growth inhibitors in coprecipitation via solution-enhanced dispersion by supercritical fluids (SEDS) to improve andrographolide dissolution from standardized Andrographis paniculata extract. *J. Drug Deliv. Sci. Technol.* **2019**, *50*, 145–154. [CrossRef]
137. Zhou, R.; Wang, F.; Guo, Z.; Zhao, Y. Preparation and characterization of resveratrol/hydroxypropyl–β–cyclodextrin inclusion complex using supercritical antisolvent technology. *J. Food Process Eng.* **2012**, *35*, 677–686. [CrossRef]
138. Yan, T.; Ji, M.; Sun, Y.; Yan, T.; Zhao, J.; Zhang, H.; Wang, Z. Preparation and characterization of baicalein/hydroxypropyl-β-cyclodextrin inclusion complex for enhancement of solubility, antioxidant activity and antibacterial activity using supercritical antisolvent technology. *J. Incl. Phenom. Macrocycl. Chem.* **2020**, *96*, 285–295. [CrossRef]
139. Sun, J.; Hong, H.; Zhu, N.; Han, L.; Suo, Q. Response surface methodology to optimize the preparation of tosufloxacin tosylate/hydroxypropyl-β-cyclodextrin inclusion complex by supercritical antisolvent process. *J. Mol. Struct.* **2019**, *1198*, 126939. [CrossRef]
140. Jun, S.W.; Kim, M.-S.; Kim, J.-S.; Park, H.J.; Lee, S.; Woo, J.-S.; Hwang, S.-J. Preparation and characterization of simvastatin/hydroxypropyl-β-cyclodextrin inclusion complex using supercritical antisolvent (SAS) process. *Eur. J. Pharm. Biopharm.* **2007**, *66*, 413–421. [CrossRef]
141. Zhong, Q.; Jin, M. Nanoscalar structures of spray-dried zein microcapsules and in vitro release kinetics of the encapsulated lysozyme as affected by formulations. *J. Agric. Food Chem.* **2009**, *57*, 3886–3894. [CrossRef] [PubMed]
142. Quispe-Condori, S.; Saldaña, M.D.; Temelli, F. Microencapsulation of flax oil with zein using spray and freeze drying. *LWT-Food Sci. Technol.* **2011**, *44*, 1880–1887. [CrossRef]
143. Wang, L.; Zhang, Y. Eugenol nanoemulsion stabilized with zein and sodium caseinate by self-assembly. *J. Agric. Food Chem.* **2017**, *65*, 2990–2998. [CrossRef] [PubMed]
144. Tu, L.S.; Dehghani, F.; Foster, N. Micronisation and microencapsulation of pharmaceuticals using a carbon dioxide antisolvent. *Powder Technol.* **2002**, *126*, 134–149. [CrossRef]
145. Kikic, I.; Vecchione, F. Supercritical impregnation of polymers. *Curr. Opin. Solid State Mater. Sci.* **2003**, *7*, 399–405. [CrossRef]
146. Lian, Z.; Epstein, S.A.; Blenk, C.W.; Shine, A.D. Carbon dioxide-induced melting point depression of biodegradable semicrystalline polymers. *J. Supercrit. Fluids* **2006**, *39*, 107–117. [CrossRef]
147. Liu, G.; Wei, D.; Wang, H.; Hu, Y.; Jiang, Y. Self-assembly of zein microspheres with controllable particle size and narrow distribution using a novel built-in ultrasonic dialysis process. *Chem. Eng. J.* **2016**, *284*, 1094–1105. [CrossRef]
148. Rosa, M.T.M.; Alvarez, V.H.; Albarelli, J.Q.; Santos, D.T.; Meireles, M.A.A.; Saldaña, M.D. Supercritical Anti-solvent Process as an Alternative Technology for Vitamin Complex Encapsulation Using Zein as Wall Material: Technical-economic Evaluation. *J. Supercrit. Fluids* **2019**. [CrossRef]
149. He, W.; Suo, Q.; Hong, H.; Shan, A.; Li, C.; Huang, Y.; Li, Y.; Zhu, M. Production of natural carotene-dispersed polymer microparticles by SEDS-PA co-precipitation. *J. Mater. Sci.* **2007**, *42*, 3495–3501. [CrossRef]
150. Mezzomo, N.; Paz, E.d.; Maraschin, M.; Martín, Á.; Cocero, M.J.; Ferreira, S.R.S. Supercritical anti-solvent precipitation of carotenoid fraction from pink shrimp residue: Effect of operational conditions on encapsulation efficiency. *J. Supercrit. Fluids* **2012**, *66*, 342–349. [CrossRef]
151. Fraile, M.; Buratto, R.; Gómez, B.; Martín, Á.; Cocero, M.J. Enhanced Delivery of Quercetin by Encapsulation in Poloxamers by Supercritical Antisolvent Process. *Ind. Eng. Chem. Res.* **2014**, *53*, 4318–4327. [CrossRef]
152. García-Casas, I.; Montes, A.; Pereyra, C.; Martínez de la Ossa, E.J. Generation of quercetin/cellulose acetate phthalate systems for delivery by supercritical antisolvent process. *Eur. J. Pharm. Sci.* **2017**, *100*, 79–86. [CrossRef] [PubMed]
153. García-Casas, I.; Montes, A.; Pereyra, C.; Martínez De La Ossa, E.J. Co-precipitation of mangiferin with cellulose acetate phthalate by supercritical antisolvent process. *J. CO2 Util.* **2017**, *22*, 197–207. [CrossRef]
154. Ober, C.A.; Kalombo, L.; Swai, H.; Gupta, R.B. Preparation of rifampicin/lactose microparticle composites by a supercritical antisolvent-drug excipient mixing technique for inhalation delivery. *Powder Technol.* **2013**, *236*, 132–138. [CrossRef]

155. Cheng, Y.-S.; Lu, P.-M.; Huang, C.-Y.; Wu, J.-J. Encapsulation of lycopene with lecithin and α-tocopherol by supercritical antisolvent process for stability enhancement. *J. Supercrit. Fluids* **2017**, *130*, 246–252. [CrossRef]
156. Elizondo, E.; Sala, S.; Imbuluzqueta, E.; González, D.; Blanco-Prieto, M.J.; Gamazo, C.; Ventosa, N.; Veciana, J. High loading of gentamicin in bioadhesive PVM/MA nanostructured microparticles using compressed carbon-dioxide. *Pharm. Res.* **2011**, *28*, 309–321. [CrossRef]
157. Junior, O.V.; Cardoso, F.A.R.; Giufrida, W.M.; de Souza, M.F.; Cardozo-Filho, L. Production and computational fluid dynamics-based modeling of PMMA nanoparticles impregnated with ivermectin by a supercritical antisolvent process. *J. CO2 Util.* **2020**, *35*, 47–58. [CrossRef]
158. Cuadra, I.A.; Cabañas, A.; Cheda, J.A.; Türk, M.; Pando, C. Cocrystallization of the anticancer drug 5-fluorouracil and coformers urea, thiourea or pyrazinamide using supercritical CO2 as an antisolvent (SAS) and as a solvent (CSS). *J. Supercrit. Fluids* **2020**, 104813. [CrossRef]
159. Saad, S.; Ahmad, I.; Kawish, S.M.; Khan, U.A.; Ahmad, F.J.; Ali, A.; Jain, G.K. Improved cardioprotective effects of hesperidin solid lipid nanoparticles prepared by supercritical antisolvent technology. *Colloids Surf. B Biointerfaces* **2020**, *187*, 110628. [CrossRef]
160. Padrela, L.; Rodrigues, M.A.; Velaga, S.P.; Matos, H.A.; de Azevedo, E.G. Formation of indomethacin–saccharin cocrystals using supercritical fluid technology. *Eur. J. Pharm. Sci.* **2009**, *38*, 9–17. [CrossRef]

MDPI
St. Alban-Anlage 66
4052 Basel
Switzerland
www.mdpi.com

Processes Editorial Office
E-mail: processes@mdpi.com
www.mdpi.com/journal/processes

www.ingramcontent.com/pod-product-compliance
Lightning Source LLC
Chambersburg PA
CBHW061451210926
51850CB00008B/1678
* 9 7 8 3 0 3 6 5 9 7 2 2 5 *